LONDON MATHEMATICAL SOCIETY LECTURE NOTE SERIES

Managing Editor: Professor M. Reid, Mathematics Institute,
University of Warwick, Coventry CV4 7AL, United Kingdom

The titles below are available from booksellers, or from Cambridge University Press at
http://www.cambridge.org/mathematics

312 Foundations of computational mathematics, Minneapolis 2002, F. CUCKER *et al* (eds)
313 Transcendental aspects of algebraic cycles, S. MÜLLER-STACH & C. PETERS (eds)
314 Spectral generalizations of line graphs, D. CVETKOVIC, P. ROWLINSON & S. SIMIC
315 Structured ring spectra, A. BAKER & B. RICHTER (eds)
316 Linear logic in computer science, T. EHRHARD, P. RUET, J.-Y. GIRARD & P. SCOTT (eds)
317 Advances in elliptic curve cryptography, I.F. BLAKE, G. SEROUSSI & N.P. SMART (eds)
318 Perturbation of the boundary in boundary-value problems of partial differential equations, D. HENRY
319 Double affine Hecke algebras, I. CHEREDNIK
320 L-functions and Galois representations, D. BURNS, K. BUZZARD & J. NEKOVÁŘ (eds)
321 Surveys in modern mathematics, V. PRASOLOV & Y. ILYASHENKO (eds)
322 Recent perspectives in random matrix theory and number theory, F. MEZZADRI & N.C. SNAITH (eds)
323 Poisson geometry, deformation quantisation and group representations, S. GUTT *et al* (eds)
324 Singularities and computer algebra, C. LOSSEN & G. PFISTER (eds)
325 Lectures on the Ricci flow, P. TOPPING
326 Modular representations of finite groups of Lie type, J.E. HUMPHREYS
327 Surveys in combinatorics 2005, B.S. WEBB (ed)
328 Fundamentals of hyperbolic manifolds, R. CANARY, D. EPSTEIN & A. MARDEN (eds)
329 Spaces of Kleinian groups, Y. MINSKY, M. SAKUMA & C. SERIES (eds)
330 Noncommutative localization in algebra and topology, A. RANICKI (ed)
331 Foundations of computational mathematics, Santander 2005, L.M PARDO, A. PINKUS, E. SÜLI & M.J. TODD (eds)
332 Handbook of tilting theory, L. ANGELERI HÜGEL, D. HAPPEL & H. KRAUSE (eds)
333 Synthetic differential geometry (2nd Edition), A. KOCK
334 The Navier–Stokes equations, N. RILEY & P. DRAZIN
335 Lectures on the combinatorics of free probability, A. NICA & R. SPEICHER
336 Integral closure of ideals, rings, and modules, I. SWANSON & C. HUNEKE
337 Methods in Banach space theory, J.M.F. CASTILLO & W.B. JOHNSON (eds)
338 Surveys in geometry and number theory, N. YOUNG (ed)
339 Groups St Andrews 2005 I, C.M. CAMPBELL, M.R. QUICK, E.F. ROBERTSON & G.C. SMITH (eds)
340 Groups St Andrews 2005 II, C.M. CAMPBELL, M.R. QUICK, E.F. ROBERTSON & G.C. SMITH (eds)
341 Ranks of elliptic curves and random matrix theory, J.B. CONREY, D.W. FARMER, F. MEZZADRI & N.C. SNAITH (eds)
342 Elliptic cohomology, H.R. MILLER & D.C. RAVENEL (eds)
343 Algebraic cycles and motives I, J. NAGEL & C. PETERS (eds)
344 Algebraic cycles and motives II, J. NAGEL & C. PETERS (eds)
345 Algebraic and analytic geometry, A. NEEMAN
346 Surveys in combinatorics 2007, A. HILTON & J. TALBOT (eds)
347 Surveys in contemporary mathematics, N. YOUNG & Y. CHOI (eds)
348 Transcendental dynamics and complex analysis, P.J. RIPPON & G.M. STALLARD (eds)
349 Model theory with applications to algebra and analysis I, Z. CHATZIDAKIS, D. MACPHERSON, A. PILLAY & A. WILKIE (eds)
350 Model theory with applications to algebra and analysis II, Z. CHATZIDAKIS, D. MACPHERSON, A. PILLAY & A. WILKIE (eds)
351 Finite von Neumann algebras and masas, A.M. SINCLAIR & R.R. SMITH
352 Number theory and polynomials, J. MCKEE & C. SMYTH (eds)
353 Trends in stochastic analysis, J. BLATH, P. MÖRTERS & M. SCHEUTZOW (eds)
354 Groups and analysis, K. TENT (ed)
355 Non-equilibrium statistical mechanics and turbulence, J. CARDY, G. FALKOVICH & K. GAWEDZKI
356 Elliptic curves and big Galois representations, D. DELBOURGO
357 Algebraic theory of differential equations, M.A.H. MACCALLUM & A.V. MIKHAILOV (eds)
358 Geometric and cohomological methods in group theory, M.R. BRIDSON, P.H. KROPHOLLER & I.J. LEARY (eds)
359 Moduli spaces and vector bundles, L. BRAMBILA-PAZ, S.B. BRADLOW, O. GARCÍA-PRADA & S. RAMANAN (eds)
360 Zariski geometries, B. ZILBER
361 Words: Notes on verbal width in groups, D. SEGAL
362 Differential tensor algebras and their module categories, R. BAUTISTA, L. SALMERÓN & R. ZUAZUA
363 Foundations of computational mathematics, Hong Kong 2008, F. CUCKER, A. PINKUS & M.J. TODD (eds)
364 Partial differential equations and fluid mechanics, J.C. ROBINSON & J.L. RODRIGO (eds)
365 Surveys in combinatorics 2009, S. HUCZYNSKA, J.D. MITCHELL & C.M. RONEY-DOUGAL (eds)
366 Highly oscillatory problems, B. ENGQUIST, A. FOKAS, E. HAIRER & A. ISERLES (eds)
367 Random matrices: High dimensional phenomena, G. BLOWER
368 Geometry of Riemann surfaces, F.P. GARDINER, G. GONZÁLEZ-DIEZ & C. KOUROUNIOTIS (eds)
369 Epidemics and rumours in complex networks, M. DRAIEF & L. MASSOULIÉ
370 Theory of p-adic distributions, S. ALBEVERIO, A.YU. KHRENNIKOV & V.M. SHELKOVICH
371 Conformal fractals, F. PRZYTYCKI & M. URBANSKI
372 Moonshine: The first quarter century and beyond, J. LEPOWSKY, J. MCKAY & M.P. TUITE (eds)

373 Smoothness, regularity and complete intersection, J. MAJADAS & A. G. RODICIO
374 Geometric analysis of hyperbolic differential equations: An introduction, S. ALINHAC
375 Triangulated categories, T. HOLM, P. JØRGENSEN & R. ROUQUIER (eds)
376 Permutation patterns, S. LINTON, N. RUŠKUC & V. VATTER (eds)
377 An introduction to Galois cohomology and its applications, G. BERHUY
378 Probability and mathematical genetics, N. H. BINGHAM & C. M. GOLDIE (eds)
379 Finite and algorithmic model theory, J. ESPARZA, C. MICHAUX & C. STEINHORN (eds)
380 Real and complex singularities, M. MANOEL, M.C. ROMERO FUSTER & C.T.C WALL (eds)
381 Symmetries and integrability of difference equations, D. LEVI, P. OLVER, Z. THOMOVA & P. WINTERNITZ (eds)
382 Forcing with random variables and proof complexity, J. KRAJÍČEK
383 Motivic integration and its interactions with model theory and non-Archimedean geometry I, R. CLUCKERS, J. NICAISE & J. SEBAG (eds)
384 Motivic integration and its interactions with model theory and non-Archimedean geometry II, R. CLUCKERS, J. NICAISE & J. SEBAG (eds)
385 Entropy of hidden Markov processes and connections to dynamical systems, B. MARCUS, K. PETERSEN & T. WEISSMAN (eds)
386 Independence-friendly logic, A.L. MANN, G. SANDU & M. SEVENSTER
387 Groups St Andrews 2009 in Bath I, C.M. CAMPBELL *et al* (eds)
388 Groups St Andrews 2009 in Bath II, C.M. CAMPBELL *et al* (eds)
389 Random fields on the sphere, D. MARINUCCI & G. PECCATI
390 Localization in periodic potentials, D.E. PELINOVSKY
391 Fusion systems in algebra and topology, M. ASCHBACHER, R. KESSAR & B. OLIVER
392 Surveys in combinatorics 2011, R. CHAPMAN (ed)
393 Non-abelian fundamental groups and Iwasawa theory, J. COATES *et al* (eds)
394 Variational problems in differential geometry, R. BIELAWSKI, K. HOUSTON & M. SPEIGHT (eds)
395 How groups grow, A. MANN
396 Arithmetic differential operators over the p-adic integers, C.C. RALPH & S.R. SIMANCA
397 Hyperbolic geometry and applications in quantum chaos and cosmology, J. BOLTE & F. STEINER (eds)
398 Mathematical models in contact mechanics, M. SOFONEA & A. MATEI
399 Circuit double cover of graphs, C.-Q. ZHANG
400 Dense sphere packings: a blueprint for formal proofs, T. HALES
401 A double Hall algebra approach to affine quantum Schur–Weyl theory, B. DENG, J. DU & Q. FU
402 Mathematical aspects of fluid mechanics, J.C. ROBINSON, J.L. RODRIGO & W. SADOWSKI (eds)
403 Foundations of computational mathematics, Budapest 2011, F. CUCKER, T. KRICK, A. PINKUS & A. SZANTO (eds)
404 Operator methods for boundary value problems, S. HASSI, H.S.V. DE SNOO & F.H. SZAFRANIEC (eds)
405 Torsors, étale homotopy and applications to rational points, A.N. SKOROBOGATOV (ed)
406 Appalachian set theory, J. CUMMINGS & E. SCHIMMERLING (eds)
407 The maximal subgroups of the low-dimensional finite classical groups, J.N. BRAY, D.F. HOLT & C.M. RONEY-DOUGAL
408 Complexity science: the Warwick master's course, R. BALL, V. KOLOKOLTSOV & R.S. MACKAY (eds)
409 Surveys in combinatorics 2013, S.R. BLACKBURN, S. GERKE & M. WILDON (eds)
410 Representation theory and harmonic analysis of wreath products of finite groups, T. CECCHERINI-SILBERSTEIN, F. SCARABOTTI & F. TOLLI
411 Moduli spaces, L. BRAMBILA-PAZ, O. GARCÍA-PRADA, P. NEWSTEAD & R.P. THOMAS (eds)
412 Automorphisms and equivalence relations in topological dynamics, D.B. ELLIS & R. ELLIS
413 Optimal transportation, Y. OLLIVIER, H. PAJOT & C. VILLANI (eds)
414 Automorphic forms and Galois representations I, F. DIAMOND, P.L. KASSAEI & M. KIM (eds)
415 Automorphic forms and Galois representations II, F. DIAMOND, P.L. KASSAEI & M. KIM (eds)
416 Reversibility in dynamics and group theory, A.G. O'FARRELL & I. SHORT
417 Recent advances in algebraic geometry, C.D. HACON, M. MUSTAŢĂ & M. POPA (eds)
418 The Bloch–Kato conjecture for the Riemann zeta function, J. COATES, A. RAGHURAM, A. SAIKIA & R. SUJATHA (eds)
419 The Cauchy problem for non-Lipschitz semi-linear parabolic partial differential equations, J.C. MEYER & D.J. NEEDHAM
420 Arithmetic and geometry, L. DIEULEFAIT *et al* (eds)
421 O-minimality and Diophantine geometry, G.O. JONES & A.J. WILKIE (eds)
422 Groups St Andrews 2013, C.M. CAMPBELL *et al* (eds)
423 Inequalities for graph eigenvalues, Z. STANIĆ
424 Surveys in combinatorics 2015, A. CZUMAJ *et al* (eds)
425 Geometry, topology and dynamics in negative curvature, C.S. ARAVINDA, F.T. FARRELL & J.-F. LAFONT (eds)
426 Lectures on the theory of water waves, T. BRIDGES, M. GROVES & D. NICHOLLS (eds)
427 Recent advances in Hodge theory, M. KERR & G. PEARLSTEIN (eds)
428 Geometry in a Fréchet context, C. T. J. DODSON, G. GALANIS & E. VASSILIOU
429 Sheaves and functions modulo p, L. TAELMAN
430 Recent progress in the theory of the Euler and Navier-Stokes equations, J.C. ROBINSON, J.L. RODRIGO, W. SADOWSKI & A. VIDAL-LÓPEZ (eds)
431 Harmonic and subharmonic function theory on the real hyperbolic ball, M. STOLL
432 Topics in graph automorphisms and reconstruction (2nd Edition), J. LAURI & R SCAPELLATO
433 Regular and irregular holonomic D-modules, M. KASHIWARA & P. SCHAPIRA
434 Analytic semigroups and semilinear initial boundary value problems (2nd Edition), K. TAIRA
435 Graded rings and graded Grothendieck groups, R. HAZRAT

London Mathematical Society Lecture Note Series: 435

Graded Rings and Graded Grothendieck Groups

ROOZBEH HAZRAT
Western Sydney University

Shaftesbury Road, Cambridge CB2 8EA, United Kingdom

One Liberty Plaza, 20th Floor, New York, NY 10006, USA

477 Williamstown Road, Port Melbourne, VIC 3207, Australia

314–321, 3rd Floor, Plot 3, Splendor Forum, Jasola District Centre, New Delhi – 110025, India

103 Penang Road, #05–06/07, Visioncrest Commercial, Singapore 238467

Cambridge University Press is part of Cambridge University Press & Assessment, a department of the University of Cambridge.

We share the University's mission to contribute to society through the pursuit of education, learning and research at the highest international levels of excellence.

www.cambridge.org
Information on this title: www.cambridge.org/9781316619582

First published 2016

A catalogue record for this publication is available from the British Library

Library of Congress Cataloging-in-Publication data
Names: Hazrat, Roozbeh, 1971–
Title: Graded rings and graded Grothendieck groups / Roozbeh Hazrat, University of Western Sydney.
Description: Cambridge : Cambridge University Press, 2016. | Series: London Mathematical Society lecture note series ; 435 | Includes bibliographical references and index.
Identifiers: LCCN 2016010216 | ISBN 9781316619582 (pbk. : alk. paper)
Subjects: LCSH: Grothendieck groups. | Graded rings. | Rings (Algebra)
Classification: LCC QA251.5 .H39 2016 | DDC 512/.4–dc23 LC record available at http://lccn.loc.gov/2016010216

ISBN 978-1-316-61958-2 Paperback

Contents

Introduction

A bird's eye view of the theory of graded modules over a graded ring might give the impression that it is nothing but ordinary module theory with all its statements decorated with the adjective "graded". Once the grading is considered to be trivial, the graded theory reduces to the usual module theory. From this perspective, the theory of graded modules can be considered as an extension of module theory. However, this simplistic overview might conceal the point that graded modules come equipped with a *shift*, thanks to the possibility of partitioning the structures and then rearranging the partitions. This adds an extra layer of structure (and complexity) to the theory. This monograph focuses on the theory of the graded Grothendieck group K_0^{gr}, that provides a sparkling illustration of this idea. Whereas the usual K_0 is an abelian group, the shift provides K_0^{gr} with a natural structure of a $\mathbb{Z}[\Gamma]$-module, where Γ is the group used for the grading and $\mathbb{Z}[\Gamma]$ its group ring. As we will see throughout this book, this extra structure carries substantial information about the graded ring.

Let Γ and Δ be abelian groups and $f : \Gamma \to \Delta$ a group homomorphism. Then for any Γ-graded ring A, one can consider a natural Δ-grading on A (see §1.1.2); in the same way, any Γ-graded A-module can be viewed as a Δ-graded A-module. These operations induce functors

$$U_f : \mathrm{Gr}^{\Gamma}\text{-}A \longrightarrow \mathrm{Gr}^{\Delta}\text{-}A,$$
$$(-)_{\Omega} : \mathrm{Gr}^{\Gamma}\text{-}A \longrightarrow \mathrm{Gr}^{\Omega}\text{-}A_{\Omega},$$

(see §1.2.8), where $\mathrm{Gr}^{\Gamma}\text{-}A$ is the category of Γ-graded right A-modules, $\mathrm{Gr}^{\Delta}\text{-}A$ that of Δ-graded right A-modules, and $\mathrm{Gr}^{\Omega}\text{-}A$ the category of Ω-graded right A_{Ω}-module with $\Omega = \ker(f)$.

One aim of the theory of graded rings is to investigate the ways in which these categories relate to one another, and which properties of one category can be lifted to another. In particular, in the two extreme cases when the group

$\Delta = 0$ or $f : \Gamma \to \Delta$ is the identity, we obtain the forgetful functors

$$U : \mathrm{Gr}^{\Gamma}\text{-}A \longrightarrow \text{Mod-}A,$$
$$(-)_0 : \mathrm{Gr}^{\Gamma}\text{-}A \longrightarrow \text{Mod-}A_0.$$

The category $\mathrm{Pgr}^{\Gamma}\text{-}A$ of graded finitely generated projective A-modules is an exact category. Thus Quillen's K-theory machinery [81] defines graded K-groups

$$K_i^{\mathrm{gr}}(A) := K_i(\mathrm{Pgr}^{\Gamma}\text{-}A),$$

for $i \in \mathbb{N}$. On the other hand, the shift operation on modules induces a functor on $\mathrm{Gr}^{\Gamma}\text{-}A$ that is an auto-equivalence (§1.2.2), so that these K-groups also carry a Γ-module structure. One can treat the groups $K_i(A)$ and $K_i(A_0)$ in a similar way. Quillen's K-theory machinery allows us to establish relations between these K-groups. In particular:

Relating $K_*^{\mathrm{gr}}(A)$ to $K_*(A)$ for a positively graded rings §6.1. For a $\mathbb{Z}$-graded ring with the positive support, there is a $\mathbb{Z}[x, x^{-1}]$-module isomorphism,

$$K_i^{\mathrm{gr}}(A) \cong K_i(A_0) \otimes_{\mathbb{Z}} \mathbb{Z}[x, x^{-1}].$$

Relating $K_*^{\mathrm{gr}}(A)$ to $K_*(A_0)$ for graded Noetherian regular rings §6.3. Consider the full subcategory $\mathrm{Gr}_0\text{-}A$ of Gr-A, of all graded modules M as objects such that $M_0 = 0$. This is a Serre subcategory of Gr-A. One can show that $\text{Gr-}A/\mathrm{Gr}_0\text{-}A \cong \text{Mod-}A_0$. If A is a (right) regular Noetherian ring, the quotient category identity above holds for the corresponding graded finitely generated modules, *i.e.*, $\text{gr-}A/\mathrm{gr}_0\text{-}A \cong \text{mod-}A_0$ and the localisation theorem gives a long exact sequence of abelian groups,

$$\cdots \longrightarrow K_{n+1}(A_0) \xrightarrow{\delta} K_n(\mathrm{gr}_0\text{-}A) \longrightarrow K_n^{\mathrm{gr}}(A) \longrightarrow K_n(A_0) \longrightarrow \cdots.$$

Relating $K_*^{\mathrm{gr}}(A)$ to $K_*(A)$ for graded Noetherian regular rings §6.4. For a $\mathbb{Z}$-graded ring A which is right regular Noetherian, there is a long exact sequence of abelian groups

$$\cdots \longrightarrow K_{n+1}(A) \longrightarrow K_n^{\mathrm{gr}}(A) \xrightarrow{\bar{i}} K_n^{\mathrm{gr}}(A) \xrightarrow{U} K_n(A) \longrightarrow \cdots.$$

The main emphasis of this book is on the group K_0^{gr} as a powerful invariant in the classification problems. This group is equipped with the extra structure of the action of the grade group induced by the shift. In many important examples, in fact this shift is all the difference between the graded Grothendieck group and the usual Grothendieck group, *i.e.*,

$$K_0^{\mathrm{gr}}(A)/\langle [P] - [P(1)] \rangle \cong K_0(A),$$

where P is a graded projective A-module and $P(1)$ is the shifted module (Chapter 6, see Corollary 6.4.2).

The motivation to write this book came from recent activities that adopt the graded Grothendieck group as an invariant to classify the Leavitt path algebras [47, 48, 79]. Surprisingly, not much is recorded about the graded version of the Grothendieck group in the literature, despite the fact that K_0 has been used on many occasions as a crucial invariant, and there is a substantial amount of information about the graded version of other invariants such as (co)homology groups, Brauer groups, etc. The other surge of interest in this group stems from the recent activities on the (graded) representation theory of Hecke algebras. In particular for a quiver Hecke algebra, its graded Grothendieck group is closely related to its corresponding quantised enveloping algebra. For this line of research see the survey [54].

This book tries to fill this gap, by systematically developing the theory of graded Grothendieck groups. In order to do this, we have to carry over and work out the details of known results in the nongraded case to the graded setting, and gather together important results on the graded theory scattered in research papers.

The group K_0 has been successfully used in operator theory to classify certain classes of C^*-algebras. Building on work of Bratteli, Elliott in [36] used the pointed ordered K_0-groups (called dimension groups) as a complete invariant for AF C^*-algebras. Another cornerstone of using K-groups for the classifications of a wider range of C^*-algebras was the work of Kirchberg and Phillips [80], who showed that K_0 and K_1-groups together are a complete invariant for a certain type of C^*-algebras. The Grothendieck group considered as a module induced by a group action was used by Handelman and Rossmann [45] to give a complete invariant for the class of direct limits of finite dimensional, representable dynamical systems. Krieger [56] introduced (past) dimension groups as a complete invariant for the shift equivalence of topological Markov chains (shift of finite types) in symbolic dynamics. Surprisingly, we will see that Krieger's groups are naturally expressed by graded Grothendieck groups (§3.11).

We develop the theory for rings graded by abelian groups rather than arbitrary groups for two reasons, although most of the results could be carried over to nonabelian grade groups. One reason is that using the abelian grading makes the presentation and proofs much more transparent. In addition, in most applications of graded K-theory, the ring has an abelian grading (often a $\mathbb{Z}$-grading).

In Chapter 1 we study the basic theory of graded rings. Chapter 2 concentrates on graded Morita theory. In Chapter 3 we compute K_0^{gr} for certain graded rings, such as graded local rings and (Leavitt) path algebras. We study the pre-ordering available on K_0^{gr} and determine the action of Γ on this group. Chapter 4 studies graded Picard groups and in Chapter 5 we prove that for the so-called graded ultramatricial algebras, the graded Grothendieck group is a complete invariant. Finally, in Chapter 6, we explore the relations between (higher) K_n^{gr} and K_n, for the class of $\mathbb{Z}$-graded rings. We describe a generalisation of the Quillen and van den Bergh theorems. The latter theorem uses the techniques employed in the proof of the fundamental theorem of K-theory, where the graded K-theory appears. For this reason we present a proof of the fundamental theorem in this chapter.

Conventions Throughout this book, unless it is explicitly stated, all rings have identities, homomorphisms preserve the identity and all modules are unitary. Moreover, all modules are considered right modules. For a ring A, the category of right A-modules is denoted by Mod-A. A full subcategory of Mod-A consisted of all finitely generated A-modules is denoted by mod-A. By Pr-A we denote the category of finitely generated projective A-modules.

For a set Γ, we write $\bigoplus_\Gamma \mathbb{Z}$ or $\mathbb{Z}^\Gamma$ to mean $\bigoplus_{\gamma\in\Gamma} \mathbb{Z}_\gamma$, where $\mathbb{Z}_\gamma = \mathbb{Z}$ for each $\gamma \in \Gamma$. We denote the cyclic group $\mathbb{Z}/n\mathbb{Z}$ with n elements by $\mathbb{Z}_n$.

Acknowledgements I would like to acknowledge Australian Research Council grants DP150101598 and DP160101481. Part of this work was done at the University of Bielefeld, where the author was a Humboldt Fellow.

I learned about the graded techniques in algebra from Adrian Wadsworth. Judith Millar worked with me to study the graded K-theory of Azumaya algebras. Gene Abrams was a source of encouragement that the graded techniques would be fruitful in the study of Leavitt path algebras. Andrew Mathas told me how graded Grothendieck groups are relevant in representation theory and pointed me to the relevant literature. The discussions with Zuhong Zhang, who kindly invited me to the Beijing Institute of Technology in July 2013 and 2014, helped to improve the presentation. To all of them, I am grateful.

1
Graded rings and graded modules

Graded rings appear in many circumstances, both in elementary and advanced areas. Here are two examples.

1 In elementary school when we distribute 10 apples giving 2 apples to each person, we have

$$10 \text{ Apples} : 2 \text{ Apples} = 5 \text{ People}.$$

The psychological problem caused to many kids as to exactly how the word "People" appears in the equation can be overcome by correcting it to

$$10 \text{ Apples} : 2 \text{ Apples / People} = 5 \text{ People}.$$

This shows that already at the level of elementary school arithmetic, children work in a much more sophisticated structure, *i.e.,* the graded ring

$$\mathbb{Z}[x_1^{\pm 1}, \dots, x_n^{\pm 1}]$$

of Laurent polynomial rings! (see the interesting book of Borovik [23, §4.7] on this).

2 If A is a commutative ring generated by a finite number of elements of degree 1, then by the celebrated work of Serre [85], the category of quasicoherent sheaves on the scheme is equivalent to QGr-$A \cong$ Gr-A/ Fdim-A, where Gr-A is the category of graded modules over A and Fdim-A is the Serre subcategory of (direct limits of) finite dimensional submodules. In particular when $A = K[x_0, x_1, \dots, x_n]$, where K is a field, then QCoh-$\mathbb{P}^n$ is equivalent to QGr-$A[x_0, x_1, \dots, x_n]$ (see [85, 9, 79] for more precise statements and relations with noncommutative algebraic geometry).

This book treats graded rings and the category of graded modules over a graded ring. This category is an abelian category (in fact a Grothendieck category). Many of the classical invariants constructed for the category of modules can be constructed, *mutatis mutandis*, starting from the category of graded modules. The general viewpoint of this book is that, once a ring has a natural graded structure, graded invariants capture more information than the non-graded counterparts.

In this chapter we give a concise introduction to the theory of graded rings. We introduce grading on matrices, study graded division rings and introduce gradings on graph algebras that will be the source of many interesting examples.

1.1 Graded rings

1.1.1 Basic definitions and examples

A ring A is called a Γ-*graded ring*, or simply a *graded ring*, if $A = \bigoplus_{\gamma\in\Gamma} A_\gamma$, where Γ is an (abelian) group, each A_γ is an additive subgroup of A and $A_\gamma A_\delta \subseteq A_{\gamma+\delta}$ for all $\gamma, \delta \in \Gamma$.

If A is an algebra over a field K, then A is called a *graded algebra* if A is a graded ring and for any $\gamma \in \Gamma$, A_γ is a K-vector subspace.

The set $A^h = \bigcup_{\gamma\in\Gamma} A_\gamma$ is called the set of *homogeneous elements* of A. The additive group A_γ is called the γ-*component* of A and the nonzero elements of A_γ are called *homogeneous of degree* γ. We write $\deg(a) = \gamma$ if $a \in A_\gamma \backslash \{0\}$. We call the set

$$\boxed{\Gamma_A = \{\gamma \in \Gamma \mid A_\gamma \neq 0\}}$$

the *support* of A. We say the Γ-graded ring A has a *trivial grading*, or A is *concentrated in degree zero*, if the support of A is the trivial group, *i.e.*, $A_0 = A$ and $A_\gamma = 0$ for $\gamma \in \Gamma\backslash\{0\}$.

For Γ-graded rings A and B, a Γ-*graded ring homomorphism* $f : A \to B$ is a ring homomorphism such that $f(A_\gamma) \subseteq B_\gamma$ for all $\gamma \in \Gamma$. A graded homomorphism f is called a *graded isomorphism* if f is bijective and, when such a graded isomorphism exists, we write $A \cong_{\text{gr}} B$. Notice that if f is a graded ring homomorphism which is bijective, then its inverse f^{-1} is also a graded ring homomorphism.

If A is a graded ring and R is a commutative graded ring, then A is called a *graded R-algebra* if A is an R-algebra and the associated algebra homomorphism $\phi : R \to A$ is a graded homomorphism. When R is a field concentrated in degree zero, we retrieve the definition of a graded algebra above.

Proposition 1.1.1 *Let $A = \bigoplus_{\gamma\in\Gamma} A_\gamma$ be a Γ-graded ring. Then*

(1) *1_A is homogeneous of degree 0;*
(2) *A_0 is a subring of A;*
(3) *each A_γ is an A_0-bimodule;*
(4) *for an invertible element $a \in A_\gamma$, its inverse a^{-1} is homogeneous of degree $-\gamma$, i.e., $a^{-1} \in A_{-\gamma}$.*

Proof (1) Suppose $1_A = \sum_{\gamma\in\Gamma} a_\gamma$ for $a_\gamma \in A_\gamma$. Let $b \in A_\delta$, $\delta \in \Gamma$, be an arbitrary nonzero homogeneous element. Then $b = b1_A = \sum_{\gamma\in\Gamma} ba_\gamma$, where $ba_\gamma \in A_{\delta+\gamma}$ for all $\gamma \in \Gamma$. Since the decomposition is unique, $ba_\gamma = 0$ for all $\gamma \in \Gamma$ with $\gamma \neq 0$. But as b was an arbitrary homogeneous element, it follows that $ba_\gamma = 0$ for all $b \in A$ (not necessarily homogeneous), and in particular $1_A a_\gamma = a_\gamma = 0$ if $\gamma \neq 0$. Thus $1_A = a_0 \in A_0$.

(2) This follows since A_0 is an additive subgroup of A with $A_0 A_0 \subseteq A_0$ and $1 \in A_0$.

(3) This is immediate.

(4) Let $b = \sum_{\delta\in\Gamma} b_\delta$, with $\deg(b_\delta) = \delta$, be the inverse of $a \in A_\gamma$, so that $1 = ab = \sum_{\delta\in\Gamma} ab_\delta$, where $ab_\delta \in A_{\gamma+\delta}$. By (1), since 1 is homogeneous of degree 0 and the decomposition is unique, it follows that $ab_\delta = 0$ for all $\delta \neq -\gamma$. Since a is invertible, $b_{-\gamma} \neq 0$, so $b = b_{-\gamma} \in A_{-\gamma}$ as required. □

The ring A_0 is called the *0-component ring* of A and plays a crucial role in the theory of graded rings. The proof of Proposition 1.1.1(4), in fact, shows that if $a \in A_\gamma$ has a left (or right) inverse then that inverse is in $A_{-\gamma}$. In Theorem 1.6.9, we characterise $\mathbb{Z}$-graded rings such that A_1 has a left (or right) invertible element.

Example 1.1.2 GROUP RINGS

For a group Γ, the group ring $\mathbb{Z}[\Gamma]$ has a natural Γ-grading

$$\mathbb{Z}[\Gamma] = \bigoplus_{\gamma\in\Gamma} \mathbb{Z}[\Gamma]_\gamma, \text{ where } \mathbb{Z}[\Gamma]_\gamma = \mathbb{Z}\gamma.$$

In §1.1.4, we construct crossed products which are graded rings and are generalisations of group rings and skew groups rings. A group ring has a natural involution which makes it an involutary graded ring (see §1.9).

In several applications (such as K-theory of rings, Chapter 6) we deal with $\mathbb{Z}$-graded rings with support in $\mathbb{N}$, the so called *positively graded rings*.

Example 1.1.3 TENSOR ALGEBRAS AS POSITIVELY GRADED RINGS

Let A be a commutative ring and M be an A-module. Denote by $T_n(M)$,

$n \geq 1$, the tensor product of n copies of M over A. Set $T_0(M) = A$. Then the natural A-module isomorphism $T_n(M) \otimes_A T_m(M) \to T_{n+m}(M)$, induces a ring structure on

$$T(M) := \bigoplus_{n \in \mathbb{N}} T_n(M).$$

The A-algebra $T(M)$ is called the *tensor algebra* of M. Setting

$$T(M)_n := T_n(M)$$

makes $T(M)$ a $\mathbb{Z}$-graded ring with support $\mathbb{N}$. From the definition, we have $T(M)_0 = A$.

If M is a free A-module, then $T(M)$ is a free algebra over A, generated by a basis of M. Thus free rings are $\mathbb{Z}$-graded rings with the generators being homogeneous elements of degree 1. We will systematically study the grading of free rings in §1.6.1.

Example 1.1.4 FORMAL MATRIX RINGS AS GRADED RINGS

Let R and S be rings, M a R–S-bimodule and N a $S-R$-bimodule. Consider the set

$$T := \left\{ \begin{pmatrix} r & m \\ n & s \end{pmatrix} \,\middle|\, r \in R, s \in S, m \in M, n \in N \right\}.$$

Suppose that there are bimodule homomorphisms $\phi : M \otimes_S N \to R$ and $\psi : N \otimes_R M \to S$ such that $(mn)m' = m(nm')$, where we denote $\phi(m,n) = mn$ and $\psi(n,m) = nm$. One can then check that T with matrix addition and multiplication forms a ring with an identity. The ring T is called the *formal matrix ring* and denoted also by

$$T = \begin{pmatrix} R & M \\ N & S \end{pmatrix}.$$

For example, the Morita ring of a module is a formal matrix ring (see §2.3 and (2.6)).

Considering

$$T_0 = \begin{pmatrix} R & 0 \\ 0 & S \end{pmatrix}, \quad T_1 = \begin{pmatrix} 0 & M \\ N & 0 \end{pmatrix},$$

it is easy to check that T becomes a $\mathbb{Z}_2$-graded ring. In the cases that the images of ϕ and ψ are zero, these rings have been extensively studied (see [57] and references therein).

When $N = 0$, the ring T is called a *formal triangular matrix ring*. In this case there is no need to consider the homomorphisms ϕ and ψ. Setting further $T_i = 0$ for $i \neq 0, 1$ makes T also a $\mathbb{Z}$-graded ring.

One specific example of such a grading on (subrings of) formal triangular matrix rings is used in representation theory. Recall that for a field K, a finite dimensional K-algebra R is called *Frobenius algebra* if $R \cong R^*$ as right R-modules, where $R^* := \operatorname{Hom}_K(R, K)$. Note that R^* has a natural R-bimodule structure.

Starting from a finite dimensional K-algebra R, one constructs the *trivial extension* of R which is a Frobenius algebra and has a natural $\mathbb{Z}$-graded structure as follows. Consider $A := R \bigoplus R^*$, with addition defined component-wise and multiplication defined as

$$(r_1, q_1)(r_2, q_2) = (r_1 r_2, r_1 q_1 + q_2 r_2),$$

where $r_1, r_2 \in R$ and $q_1, q_2 \in R^*$. Clearly A is a Frobenius algebra with identity $(1, 0)$. Moreover, setting

$$\begin{aligned} A_0 &= R \oplus 0, \\ A_1 &= 0 \oplus R^*, \\ A_i &= 0, \text{ otherwise,} \end{aligned}$$

makes A into a $\mathbb{Z}$-graded ring with support $\{0, 1\}$. In fact this ring is a subring of the formal triangular matrix ring

$$T_0 = \begin{pmatrix} R & R^* \\ 0 & R \end{pmatrix},$$

consisting of elements $\begin{pmatrix} a & q \\ 0 & a \end{pmatrix}$.

These rings appear in representation theory (see [46, §2.2]). The graded version of this contraction is carried out in Example 1.2.9.

Example 1.1.5 THE GRADED RING A AS A_0-MODULE

Let A be a Γ-graded ring. Then A can be considered as an A_0-bimodule. In many cases A is a projective A_0-module, for example in the case of group rings (Example 1.1.2) or when A is a strongly graded ring (see §1.1.3 and Theorem 1.5.12). Here is an example that this is not the case in general. Consider the formal matrix ring T

$$T = \begin{pmatrix} R & M \\ 0 & 0 \end{pmatrix},$$

where M is a left R-module which is not a projective R-module. Then by Example 1.1.4, T is a $\mathbb{Z}$-graded ring with $T_0 = R$ and $T_1 = M$. Now T as a T_0-module is $R \oplus M$ as an R-module. Since M is not projective, $R \oplus M$ is not a projective R-module. We also get that T_1 is not a projective T_0-module.

1.1.2 Partitioning graded rings

Let A be a Γ-graded ring and $f : \Gamma \to \Delta$ be a group homomorphism. Then one can assign a natural Δ-graded structure to A as follows: $A = \bigoplus_{\delta \in \Delta} A_\delta$, where

$$A_\delta = \begin{cases} \bigoplus_{\gamma \in f^{-1}(\delta)} A_\gamma & \text{if } f^{-1}(\delta) \neq \varnothing; \\ 0 & \text{otherwise.} \end{cases}$$

In particular, for a subgroup Ω of Γ we have the following constructions.

Subgroup grading The ring $A_\Omega := \bigoplus_{\omega \in \Omega} A_\omega$ forms a Ω-graded ring. In particular, A_0 corresponds to the trivial subgroup of Γ.

Quotient grading Considering

$$A = \bigoplus_{\Omega+\alpha \in \Gamma/\Omega} A_{\Omega+\alpha},$$

where

$$A_{\Omega+\alpha} := \bigoplus_{\omega \in \Omega} A_{\omega+\alpha},$$

makes A a Γ/Ω-graded ring. (Note that if Γ is not abelian, then for this construction, Ω needs to be a normal subgroup.) Notice that with this grading, $A_0 = A_\Omega$. If $\Gamma_A \subseteq \Omega$, then A, considered as a Γ/Ω-graded ring, is concentrated in degree zero.

This construction induces a *forgetful* functor (or with other interpretations, a *block*, or a *coarsening* functor) from the category of Γ-graded rings $\mathcal{R}^\Gamma$ to the category of Γ/Ω-graded rings $\mathcal{R}^{\Gamma/\Omega}$, *i.e.*,

$$U : \mathcal{R}^\Gamma \to \mathcal{R}^{\Gamma/\Omega}.$$

If $\Omega = \Gamma$, this gives the obvious forgetful functor from the category of Γ-graded rings to the category of rings. We give a specific example of this construction in Example 1.1.8 and others in Examples 1.1.20 and 1.6.1.

Example 1.1.6 TENSOR PRODUCT OF GRADED RINGS

Let A be a Γ-graded and B a Ω-graded ring. Then $A \otimes_{\mathbb{Z}} B$ has a natural $\Gamma \times \Omega$-graded ring structure as follows. Since A_γ and B_ω, $\gamma \in \Gamma$, $\omega \in \Omega$, are $\mathbb{Z}$-modules then $A \otimes_{\mathbb{Z}} B$ can be decomposed as a direct sum

$$A \otimes_{\mathbb{Z}} B = \bigoplus_{(\gamma,\omega) \in \Gamma \times \Omega} A_\gamma \otimes B_\omega$$

(to be precise, $A_\gamma \otimes B_\omega$ is the image of $A_\gamma \otimes_{\mathbb{Z}} B_\omega$ in $A \otimes_{\mathbb{Z}} B$).

Now, if $\Omega = \Gamma$ and

$$f : \Gamma \times \Gamma \longrightarrow \Gamma,$$
$$(\gamma_1, \gamma_2) \longmapsto \gamma_1 + \gamma_2,$$

then we get a natural Γ-graded structure on $A \otimes_{\mathbb{Z}} B$. Namely,

$$A \otimes_{\mathbb{Z}} B = \bigoplus_{\gamma \in \Gamma} (A \otimes B)_\gamma,$$

where

$$(A \otimes B)_\gamma = \Big\{ \sum_i a_i \otimes b_i \mid a_i \in A^h, b_i \in B^h, \deg(a_i) + \deg(b_i) = \gamma \Big\}.$$

We give specific examples of this construction in Example 1.1.7. One can replace $\mathbb{Z}$ by a field K, if A and B are K-algebras and A_γ, B_γ are K-modules.

Example 1.1.7 Let A be a ring with identity and Γ be a group. We consider A as a Γ-graded ring concentrated in degree zero. Then, by Example 1.1.6,

$$A[\Gamma] \cong A \otimes_{\mathbb{Z}} \mathbb{Z}[\Gamma]$$

has a Γ-graded structure, *i.e.*, $A[\Gamma] = \bigoplus_{\gamma \in \Gamma} A\gamma$. If A itself is a (nontrivial) Γ-graded ring $A = \bigoplus_{\gamma \in \Gamma} A_\gamma$, then by Example 1.1.6, $A[\Gamma]$ has also a Γ-grading

$$A[\Gamma] = \bigoplus\nolimits_{\gamma \in \Gamma} A^\gamma, \text{ where } A^\gamma = \bigoplus\nolimits_{\gamma = \zeta + \zeta'} A_\zeta \zeta'. \tag{1.1}$$

A specific example is when A is a positively graded $\mathbb{Z}$-graded ring. Then $A[x] \cong A \otimes \mathbb{Z}[x]$ is a $\mathbb{Z}$-graded ring with support $\mathbb{N}$, where

$$A[x]_n = \bigoplus_{i+j=n} A_i x^j.$$

This graded ring will be used in §6.2.4 when we prove the fundamental theorem of K-theory. Such constructions are systematically studied in [72] (see also [75, §6]).

Example 1.1.8 Let A be a $\Gamma \times \Gamma$-graded ring. Define a Γ-grading on A as follows. For $\gamma \in \Gamma$, set

$$A'_\gamma = \sum_{\alpha \in \Gamma} A_{\gamma - \alpha, \alpha}.$$

It is easy to see that $A = \bigoplus_{\gamma \in \Gamma} A'_\gamma$ is a Γ-graded ring. When A is $\mathbb{Z} \times \mathbb{Z}$-graded, then the $\mathbb{Z}$-grading on A is obtained from considering all the homogeneous components on a diagonal together, as Figure 1.1 shows.

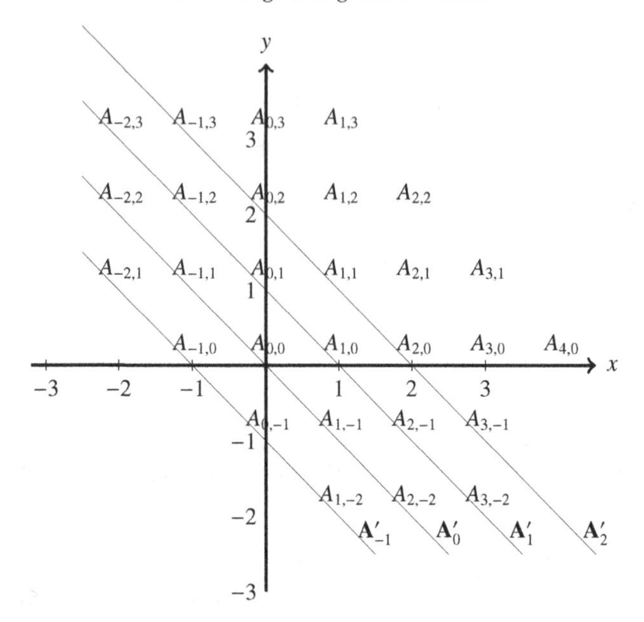

Figure 1.1

In fact this example follows from the general construction given in §1.1.2. Consider the homomorphism $\Gamma\times\Gamma \to \Gamma, (\alpha,\beta) \mapsto \alpha+\beta$. Let Ω be the kernel of this map. Clearly $(\Gamma\times\Gamma)/\Omega \cong \Gamma$. One can check that the $(\Gamma\times\Gamma)/\Omega$-graded ring A gives the graded ring constructed in this example (see also Remark 1.1.26).

Example 1.1.9 THE DIRECT LIMIT OF GRADED RINGS

Let A_i, $i \in I$, be a direct system of Γ-graded rings, *i.e.*, I is a directed partially ordered set and for $i \leq j$ there is a graded homomorphism $\phi_{ij} : A_i \to A_j$ which is compatible with the ordering. Then $A := \varinjlim A_i$ is a Γ-graded ring with homogeneous components $A_\alpha = \varinjlim A_{i\alpha}$. For a detailed construction of such direct limits see [24, II, §11.3, Remark 3].

As an example, the ring $A = \mathbb{Z}[x_i \mid i \in \mathbb{N}]$, where $A = \varinjlim_{i\in\mathbb{N}} \mathbb{Z}[x_1,\dots,x_i]$, with $\deg(x_i) = 1$ is a $\mathbb{Z}$-graded ring with support $\mathbb{N}$. We give another specific example of this construction in Example 1.1.10.

We will study in detail one type of these graded rings, *i.e.*, graded ultramatricial algebras (Chapter 5, Definition 5.2.1) and will show that the graded Grothendieck group (Chapter 3) classifies these graded rings completely.

Example 1.1.10 Let $A = \bigoplus_{\gamma\in\Gamma} A_\gamma$ and $B = \bigoplus_{\gamma\in\Gamma} B_\gamma$ be Γ-graded rings. Then $A \times B$ has a natural grading given by $A \times B = \bigoplus_{\gamma\in\Gamma}(A \times B)_\gamma$ where $(A \times B)_\gamma = A_\gamma \times B_\gamma$.

Example 1.1.11 Localisation of graded rings

Let S be a central multiplicative closed subset of the Γ-graded ring A, consisting of homogeneous elements. Then $S^{-1}A$ has a natural Γ-graded structure. Namely, for $a \in A^h$, define $\deg(a/s) = \deg(a) - \deg(s)$ and for $\gamma \in \Gamma$,

$$(S^{-1}A)_\gamma = \{ a/s \mid a \in A^h, \deg(a/s) = \gamma \}.$$

It is easy to see that this is well-defined and makes $S^{-1}A$ a Γ-graded ring.

Many rings have a "canonical" graded structure, among them are crossed products (group rings, skew group rings, twisted group rings), edge algebras, path algebras, incidence rings, etc. (see [53] for a review of these ring constructions). We will study some of these rings in this book.

Remark 1.1.12 Rings graded by a category

The use of groupoids as a suitable language for structures whose operations are partially defined has now been firmly recognised. There is a generalised notion of groupoid graded rings as follows. Recall that a *groupoid* is a small category with the property that all morphisms are isomorphisms. As an example, let G be a group and I a nonempty set. The set $I \times G \times I$, considered as morphisms, forms a groupoid where the composition is defined by

$$(i, g, j)(j, h, k) = (i, gh, k).$$

One can show that this forms a connected groupoid and any connected groupoid is of this form ([62, Ch. 3.3, Prop. 6]). If $I = \{1, \dots, n\}$, we denote $I \times G \times I$ by $n \times G \times n$.

Let Γ be a groupoid and A be a ring. A is called a Γ-*groupoid graded* ring if $A = \bigoplus_{\gamma \in \Gamma} A_\gamma$, where γ is a morphism of Γ, each A_γ is an additive subgroup of A and $A_\gamma A_\delta \subseteq A_{\gamma\delta}$ if the morphism $\gamma\delta$ is defined and $A_\gamma A_\delta = 0$ otherwise. For a group Γ, considering it as a category with one element and Γ as the set of morphisms, we recover the Γ-group graded ring A (see Example 2.3.1 for an example of a groupoid graded ring).

One can develop the theory of groupoid graded rings in parallel and similarly to the group graded rings. See [65, 66] for this approach. Since adjoining a zero to a groupoid gives a semigroup, a groupoid graded ring is a special case of rings graded by semigroups (see Remark 1.1.13). For a general notion of a ring graded by a category see [1, §2], where it is shown that the category of graded modules (graded by a category) is a Grothendieck category.

Remark 1.1.13 Rings graded by a semigroup

In the definition of a graded ring (§1.1.1), one can replace the group grading with a semigroup. With this setting, the tensor algebras of Example 1.1.3 are $\mathbb{N}$-graded rings. A number of results on group graded rings can also be established in the more general setting of rings graded by cancellative monoids or semigroups (see for example [24, II, §11]). However, in this book we only consider group graded rings.

Remark 1.1.14 Graded rings without identity

For a ring without identity, one defines the concept of the graded ring exactly as when the ring has an identity. The concept of the strongly graded ring is defined similarly. In several occasions in this book we construct graded rings without an identity. For example, Leavitt path algebras arising from infinite graphs are graded rings without an identity, §1.6.4. See also §1.6.1, the graded free rings. The unitisation of a (nonunital) graded ring has a canonical grading. This is studied in relation with graded K_0 of nonunital rings in §3.5 (see (3.25)).

1.1.3 Strongly graded rings

Let A be a Γ-graded ring. By Proposition 1.1.1, $1 \in A_0$. This implies $A_0A_\gamma = A_\gamma$ and $A_\gamma A_0 = A_\gamma$ for any $\gamma \in \Gamma$. If these equalities hold for any two arbitrary elements of Γ, we call the ring a strongly graded ring. Namely, a Γ-graded ring $A = \bigoplus_{\gamma\in\Gamma} A_\gamma$ is called a *strongly graded ring* if $A_\gamma A_\delta = A_{\gamma+\delta}$ for all $\gamma, \delta \in \Gamma$. A graded ring A is called *crossed product* if there is an invertible element in every homogeneous component A_γ of A; that is, $A^* \cap A_\gamma \neq \varnothing$ for all $\gamma \in \Gamma$, where A^* is the group of all invertible elements of A. We define the support of invertible homogeneous elements of A as

$$\boxed{\Gamma_A^* = \{\gamma \in \Gamma \mid A_\gamma^* \neq \varnothing\},} \tag{1.2}$$

where $A_\gamma^* := A^* \cap A_\gamma$. It is easy to see that Γ_A^* is a group and $\Gamma_A^* \subseteq \Gamma_A$ (see Proposition 1.1.1(4)). Clearly A is a crossed product if and only if $\Gamma_A^* = \Gamma$.

Proposition 1.1.15 *Let $A = \bigoplus_{\gamma\in\Gamma} A_\gamma$ be a Γ-graded ring. Then*

(1) *A is strongly graded if and only if $1 \in A_\gamma A_{-\gamma}$ for any $\gamma \in \Gamma$;*
(2) *if A is strongly graded then the support of A is Γ;*
(3) *any crossed product ring is strongly graded;*
(4) *if $f : A \to B$ is a graded homomorphism of graded rings, then B is strongly graded (resp. crossed product) if A is so.*

Proof (1) If A is strongly graded, then $1 \in A_0 = A_\gamma A_{-\gamma}$ for any $\gamma \in \Gamma$. For the converse, the assumption $1 \in A_\gamma A_{-\gamma}$ implies that $A_0 = A_\gamma A_{-\gamma}$ for any $\gamma \in \Gamma$. Then for $\sigma, \delta \in \Gamma$,

$$A_{\sigma+\delta} = A_0 A_{\sigma+\delta} = (A_\sigma A_{-\sigma})A_{\sigma+\delta} = A_\sigma(A_{-\sigma}A_{\sigma+\delta}) \subseteq A_\sigma A_\delta \subseteq A_{\sigma+\delta},$$

proving $A_{\sigma\delta} = A_\sigma A_\delta$, so A is strongly graded.

(2) By (1), $1 \in A_\gamma A_{-\gamma}$ for any $\gamma \in \Gamma$. This implies $A_\gamma \neq 0$ for any γ, *i.e.*, $\Gamma_A = \Gamma$.

(3) Let A be a crossed product ring. By definition, for $\gamma \in \Gamma$, there exists $a \in A^* \cap A_\gamma$. So $a^{-1} \in A_{-\gamma}$ by Proposition 1.1.1(4) and $1 = aa^{-1} \in A_\gamma A_{-\gamma}$. Thus A is strongly graded by (1).

(4) Suppose A is strongly graded. By (1), $1 \in A_\gamma A_{-\gamma}$ for any $\gamma \in \Gamma$. Thus

$$1 \in f(A_\gamma)f(A_{-\gamma}) \subseteq B_\gamma B_{-\gamma}.$$

Again (1) implies B is strongly graded. The case of the crossed product follows easily from the definition. □

The converse of (3) in Proposition 1.1.15 does not hold. One can prove that if A is strongly graded and A_0 is a local ring, then A is a crossed product algebra (see [75, Theorem 3.3.1]). In §1.6 we give examples of a strongly graded algebra A such that A is crossed product but A_0 is not a local ring. We also give an example of a strongly $\mathbb{Z}$-graded ring A such that A_0 is not local and A is not crossed product (Example 1.6.22). Using graph algebras we will produce large classes of strongly graded rings which are not crossed product (see Theorems 1.6.15 and 1.6.16).

If Γ is a finitely generated group, generated by the set $\{\gamma_1, \dots, \gamma_n\}$, then (1) in Proposition 1.1.15 can be simplified to the following: A is strongly graded if and only if $1 \in A_{\gamma_i}A_{-\gamma_i}$ and $1 \in A_{-\gamma_i}A_{\gamma_i}$, where $1 \leq i \leq n$. Thus if $\Gamma = \mathbb{Z}$, in order for A to be strongly graded, we only need to have $1 \in A_1 A_{-1}$ and $1 \in A_{-1}A_1$. This will be used, for example, in Proposition 1.6.6 to show that certain corner skew Laurent polynomial rings (§1.6.2) are strongly graded.

Example 1.1.16 Constructing strongly graded rings via tensor products

Let A and B be Γ-graded rings. Then by Example 1.1.6, $A \otimes_{\mathbb{Z}} B$ is a Γ-graded ring. If one of the rings is strongly graded (resp. crossed product) then $A \otimes_{\mathbb{Z}} B$ is so. Indeed, suppose A is strongly graded (resp. crossed product). Then the claim follows from Proposition 1.1.15(4) and the graded homomorphism $A \to A \otimes_{\mathbb{Z}} B, a \mapsto a \otimes 1$.

As a specific case, suppose A is a $\mathbb{Z}$-graded ring. Then

$$A[x, x^{-1}] = A \otimes \mathbb{Z}[x, x^{-1}]$$

is a strongly graded ring. Notice that with this grading, $A[x, x^{-1}]_0 \cong A$.

Example 1.1.17 STRONGLY GRADED AS A Γ/Ω-GRADED RING

Let A be a Γ-graded ring. Using Proposition 1.1.15, it is easy to see that if A is a strongly Γ-graded ring, then it is also a strongly Γ/Ω-graded ring, where Ω is a subgroup of Γ. However the strongly gradedness is not a "closed" property, i.e, if A is a strongly Γ/Ω-graded ring and A_Ω is a strongly Ω-graded ring, it does not follow that A is strongly Γ-graded.

1.1.4 Crossed products

Natural examples of strongly graded rings are crossed product algebras (see Proposition 1.1.15(3)). They cover, as special cases, the skew group rings and twisted groups rings. We briefly describe the construction here.

Let A be a ring, Γ a group (as usual we use the additive notation), and let $\phi : \Gamma \to \mathrm{Aut}(A)$ and $\psi : \Gamma \times \Gamma \to A^*$ be maps such that for any $\alpha, \beta, \gamma \in \Gamma$ and $a \in A$,

(i) ${}^{\alpha}({}^{\beta}a) = \psi(\alpha,\beta)\, {}^{\alpha+\beta}a\, \psi(\alpha,\beta)^{-1}$,
(ii) $\psi(\alpha,\beta)\psi(\alpha+\beta,\gamma) = {}^{\alpha}\psi(\beta,\gamma)\,\psi(\alpha,\beta+\gamma)$,
(iii) $\psi(\alpha,0) = \psi(0,\alpha) = 1$

Here for $\alpha \in \Gamma$ and $a \in A$, $\phi(\alpha)(a)$ is denoted by ${}^{\alpha}a$. The map ψ is called a *2-cocycle map*. Denote by $A^{\phi}_{\psi}[\Gamma]$ the free left A-module with the basis Γ, and define the multiplication by

$$(a\alpha)(b\beta) = a\, {}^{\alpha}b\, \psi(\alpha,\beta)(\alpha+\beta). \tag{1.3}$$

One can show that with this multiplication, $A^{\phi}_{\psi}[\Gamma]$ is a Γ-graded ring with homogeneous components $A\gamma$, $\gamma \in \Gamma$. In fact $\gamma \in A\gamma$ is invertible, so $A^{\phi}_{\psi}[\Gamma]$ is a crossed product algebra [75, Proposition 1.4.1].

On the other hand, any crossed product algebra is of this form (see [75, §1.4]): for any $\gamma \in \Gamma$ choose $u_\gamma \in A^* \cap A_\gamma$ and define $\phi : \Gamma \to \mathrm{Aut}(A_0)$ by $\phi(\gamma)(a) = u_\gamma a u_\gamma^{-1}$ for $\gamma \in \Gamma$ and $a \in A_0$. Moreover, define the cocycle map

$$\begin{aligned} \psi : \Gamma \times \Gamma &\longrightarrow A_0^*, \\ (\zeta,\eta) &\longmapsto u_\zeta u_\eta u_{\zeta+\eta}^{-1}. \end{aligned}$$

Then

$$A = A_0{}^{\phi}_{\psi}[\Gamma] = \bigoplus\nolimits_{\gamma\in\Gamma} A_0\gamma,$$

with multiplication

$$(a\zeta)(b\eta) = a^{\zeta} b \psi(\zeta,\eta)(\zeta+\eta),$$

where ${}^{\zeta}b$ is defined as $\phi(\zeta)(b)$.

Note that when Γ is cyclic, one can choose $u_i = u_1^i$ for $u_1 \in A^* \cap A_1$ and thus the cocycle map ψ is trivial, ϕ is a homomorphism and the crossed product is a skew group ring. In fact, if $\Gamma = \mathbb{Z}$, then the skew group ring becomes the so-called *skew Laurent polynomial ring*, denoted by $A_0[x, x^{-1}, \phi]$. Moreover, if u_1 is in the centre of A, then ϕ is the identity map and the crossed product ring reduces to the group ring $A_0[\Gamma]$. A variant of this construction, namely corner skew polynomial rings, is studied in §1.6.2.

Skew group rings If $\psi : \Gamma \times \Gamma \to A^*$ is a trivial map, *i.e.*, $\psi(\alpha,\beta) = 1$ for all $\alpha, \beta \in \Gamma$, then Conditions (ii) and (iii) trivially hold, and Condition (i) reduces to ${}^{\alpha}({}^{\beta}a) = {}^{\alpha+\beta}a$ which means that $\phi : \Gamma \to \mathrm{Aut}(A)$ becomes a group homomorphism. In this case $A_{\psi}^{\phi}[\Gamma]$, denoted by $A \star_{\phi} \Gamma$, is a *skew group ring* with multiplication

$$(a\alpha)(b\beta) = a\,{}^{\alpha}b\,(\alpha + \beta). \tag{1.4}$$

Twisted group ring If $\phi : \Gamma \to \mathrm{Aut}(A)$ is trivial, *i.e.*, $\phi(\alpha) = 1_A$ for all $\alpha \in \Gamma$, then Condition (i) implies that $\psi(\alpha,\beta) \in C(A) \cap A^*$ for any $\alpha, \beta \in \Gamma$. Here $C(A)$ stands for the centre of the ring A. In this case $A_{\psi}^{\phi}[\Gamma]$, denoted by $A_{\psi}[\Gamma]$, is a *twisted group ring* with multiplication

$$(a\alpha)(b\beta) = ab\psi(\alpha,\beta)(\alpha + \beta). \tag{1.5}$$

A well-known theorem in the theory of central simple algebras states that if D is a central simple F-algebra with a maximal subfield L such that L/F is a Galois extension and $[A : F] = [L : F]^2$, then D is a crossed product, with $\Gamma = \mathrm{Gal}(L/F)$ and $A = L$ (see [35, §12, Theorem 1]).

Some of the graded rings we treat in this book are of the form $K[x, x^{-1}]$, where K is a field. This is an example of a graded field.

A Γ-graded ring $A = \bigoplus_{\gamma \in \Gamma} A_{\gamma}$ is called a *graded division ring* if every nonzero homogeneous element has a multiplicative inverse. If A is also a commutative ring, then A is called a *graded field*.

Let A be a Γ-graded division ring. It follows from Proposition 1.1.1(4) that Γ_A is a group, so we can write $A = \bigoplus_{\gamma \in \Gamma_A} A_{\gamma}$. Then, as a Γ_A-graded ring, A is a crossed product and it follows from Proposition 1.1.15(3) that A is strongly Γ_A-graded. Note that if $\Gamma_A \neq \Gamma$, then A is not strongly Γ-graded. Also note that if A is a graded division ring, then A_0 is a division ring.

Remark 1.1.18 GRADED DIVISION RINGS AND DIVISION RINGS WHICH ARE GRADED

Note that a graded division ring and a division ring which is graded are different. By definition, A is a graded division ring if and only if $A^h \backslash \{0\}$ is a group. A simple example is the Laurent polynomial ring $D[x, x^{-1}]$, where

D is a division ring (Example 1.1.19). Other examples show that a graded division ring does not need to be a domain (Example 1.1.21). However, if the grade group is totally ordered, then a domain which is also graded has to be concentrated in degree zero. Thus a division ring which is graded by a totally ordered grade group Γ is of the form $A = \bigoplus_{\gamma\in\Gamma} A_\gamma$, where A_0 is a division ring and $A_\gamma = 0$ for $\gamma \neq 0$. This will not be the case if Γ is not totally ordered (see Example 1.1.20).

In the following we give some concrete examples of graded division rings.

Example 1.1.19 THE VERONESE SUBRING

Let $A = \bigoplus_{\gamma\in\Gamma} A_\gamma$ be a Γ-graded ring, where Γ is a torsion-free group. For $n \in \mathbb{Z}\backslash\{0\}$, the nth-*Veronese subring* of A is defined as $A^{(n)} = \bigoplus_{\gamma\in\Gamma} A_{n\gamma}$. This is a Γ-graded ring with $A^{(n)}_\gamma = A_{n\gamma}$. It is easy to see that the support of $A^{(n)}$ is Γ if the support of A is Γ. Note also that if A is strongly graded, so is $A^{(n)}$. Clearly $A^{(1)} = A$ and $A^{(-1)}$ is the graded ring with the components "flipped", *i.e.*, $A^{(-1)}_\gamma = A_{-\gamma}$. For the case of $A^{(-1)}$ we don't need to require the grade group to be torsion-free.

Let D be a division ring and let $A = D[x, x^{-1}]$ be the Laurent polynomial ring. The elements of A consist of finite sums $\sum_{i\in\mathbb{Z}} a_i x^i$, where $a_i \in D$. Then A is a $\mathbb{Z}$-graded division ring with $A = \bigoplus_{i\in\mathbb{Z}} A_i$, where $A_i = \{ax^i \mid a \in D\}$. Consider the nth-Veronese subring $A^{(n)}$ which is the ring $D[x^n, x^{-n}]$. The elements of $A^{(n)}$ consist of finite sums $\sum_{i\in\mathbb{Z}} a_i x^{in}$, where $a_i \in D$. Then $A^{(n)}$ is a $\mathbb{Z}$-graded division ring, with $A^{(n)} = \bigoplus_{i\in\mathbb{Z}} A_{in}$. Here both A and $A^{(n)}$ are strongly graded rings.

There is also another way to consider the $\mathbb{Z}$-graded ring $B = D[x^n, x^{-n}]$ such that it becomes a graded subring of $A = D[x, x^{-1}]$. Namely, we define $B = \bigoplus_{i\in\mathbb{Z}} B_i$, where $B_i = Dx^i$ if $i \in n\mathbb{Z}$ and $B_i = 0$ otherwise. This way B is a graded division ring and a graded subring of A. The support of B is clearly the subgroup $n\mathbb{Z}$ of $\mathbb{Z}$. With this definition, B is not strongly graded.

Example 1.1.20 DIFFERENT GRADINGS ON A GRADED DIVISION RING

Let $\mathbb{H} = \mathbb{R}\oplus\mathbb{R}i\oplus\mathbb{R}j\oplus\mathbb{R}k$ be the real *quaternion algebra*, with multiplication defined by $i^2 = -1$, $j^2 = -1$ and $ij = -ji = k$. It is known that $\mathbb{H}$ is a noncommutative division ring with centre $\mathbb{R}$. We give $\mathbb{H}$ two different graded division ring structures, with grade groups $\mathbb{Z}_2 \times \mathbb{Z}_2$ and $\mathbb{Z}_2$ respectively as follows.

$\mathbb{Z}_2 \times \mathbb{Z}_2$-grading Let $\mathbb{H} = R_{(0,0)} \oplus R_{(1,0)} \oplus R_{(0,1)} \oplus R_{(1,1)}$, where

$$R_{(0,0)} = \mathbb{R}, \quad R_{(1,0)} = \mathbb{R}i, \quad R_{(0,1)} = \mathbb{R}j, \quad R_{(1,1)} = \mathbb{R}k.$$

It is routine to check that $\mathbb{H}$ forms a strongly $\mathbb{Z}_2 \times \mathbb{Z}_2$-graded division ring.

$\mathbb{Z}_2$-grading Let $\mathbb{H} = \mathbb{C}_0 \oplus \mathbb{C}_1$, where $\mathbb{C}_0 = \mathbb{R} \oplus \mathbb{R}i$ and $\mathbb{C}_1 = \mathbb{C}j = \mathbb{R}j \oplus \mathbb{R}k$. One can check that $\mathbb{C}_0\mathbb{C}_0 = \mathbb{C}_0$, $\mathbb{C}_0\mathbb{C}_1 = \mathbb{C}_1\mathbb{C}_0 = \mathbb{C}_1$ and $\mathbb{C}_1\mathbb{C}_1 = \mathbb{C}_0$. This makes $\mathbb{H}$ a strongly $\mathbb{Z}_2$-graded division ring. Note that this grading on $\mathbb{H}$ can be obtained from the first part by considering the quotient grade group $\mathbb{Z}_2 \times \mathbb{Z}_2/0 \times \mathbb{Z}_2$ (§1.1.2). Quaternion algebras are examples of Clifford algebras (see Example 1.1.24).

The following generalises the above example of quaternions as a $\mathbb{Z}_2 \times \mathbb{Z}_2$-graded ring.

Example 1.1.21 SYMBOL ALGEBRAS

Let F be a field, ξ be a primitive nth root of unity and let $a, b \in F^*$. Let

$$A = \bigoplus_{i=0}^{n-1} \bigoplus_{j=0}^{n-1} F x^i y^j$$

be the F-algebra generated by the elements x and y, which are subject to the relations $x^n = a$, $y^n = b$ and $xy = \xi yx$. By [35, Theorem 11.1], A is an n^2-dimensional central simple algebra over F. We will show that A forms a graded division ring. Clearly A can be written as a direct sum

$$A = \bigoplus_{(i,j)\in\mathbb{Z}_n\oplus\mathbb{Z}_n} A_{(i,j)}, \quad \text{where } A_{(i,j)} = F x^i y^j$$

and each $A_{(i,j)}$ is an additive subgroup of A. Using the fact that $\xi^{-kj} x^k y^j = y^j x^k$ for each j, k, with $0 \le j, k \le n-1$, we can show that

$$A_{(i,j)} A_{(k,l)} \subseteq A_{([i+k],[j+l])},$$

for $i, j, k, l \in \mathbb{Z}_n$. A nonzero homogeneous element $f x^i y^j \in A_{(i,j)}$ has an inverse

$$f^{-1} a^{-1} b^{-1} \xi^{-ij} x^{n-i} y^{n-j},$$

proving A is a graded division ring. Clearly the support of A is $\mathbb{Z}_n \times \mathbb{Z}_n$, so A is strongly $\mathbb{Z}_n \times \mathbb{Z}_n$-graded.

These examples can also be obtained from graded free rings (see Example 1.6.3).

Example 1.1.22 A GOOD COUNTER-EXAMPLE

In the theory of graded rings, in many instances it has been established that if the grade group Γ is finite (or in some cases, finitely generated), then a graded property implies the corresponding nongraded property of the ring (*i.e.*, the property is preserved under the forgetful functor). For example, one can prove that if a $\mathbb{Z}$-graded ring is graded Artinian (Noetherian), then the ring is Artinian (Noetherian). One good example which provides counter-examples to such phenomena is the following graded field.

Let K be a field and $A = K[x_1^{\pm 1}, x_2^{\pm 1}, x_3^{\pm 1}, \dots]$ a Laurent polynomial ring in countably many variables. This ring is a graded field with its "canonical" $\bigoplus_\infty \mathbb{Z}$-grading and thus it is graded Artinian and Noetherian. However, A is not Noetherian.

1.1.5 Graded ideals

Let A be a Γ-graded ring. A two-sided ideal I of A is called a *graded ideal* (or *homogeneous ideal*) if

$$I = \bigoplus_{\gamma \in \Gamma} (I \cap A_\gamma). \tag{1.6}$$

Thus I is a graded ideal if and only if for any $x \in I$, $x = \sum x_i$, where $x_i \in A^h$, implies that $x_i \in I$.

The notions of a *graded subring*, a *graded left* and a *graded right ideal* are defined similarly.

Let I be a graded ideal of A. Then the quotient ring A/I forms a graded ring, with

$$\boxed{A/I = \bigoplus_{\gamma \in \Gamma} (A/I)_\gamma, \quad \text{where} \quad (A/I)_\gamma = (A_\gamma + I)/I.} \tag{1.7}$$

With this grading $(A/I)_0 \cong A_0/I_0$, where $I_0 = A_0 \cap I$. From (1.6) it follows that an ideal I of A is a graded ideal if and only if I is generated as a two-sided ideal of A by homogeneous elements. Also, for a two-sided ideal I of A, if (1.7) induces a grading on A/I, then I has to be a graded ideal. By Proposition 1.1.15(4), if A is strongly graded or a crossed product, so is the graded quotient ring A/I.

Example 1.1.23 SYMMETRIC AND EXTERIOR ALGEBRAS AS $\mathbb{Z}$-GRADED RINGS

Recall from Example 1.1.3 that for a commutative ring A and an A-module M, the tensor algebra $T(M)$ is a $\mathbb{Z}$-graded ring with support $\mathbb{N}$. The *symmetric algebra* $S(M)$ is defined as the quotient of $T(M)$ by the ideal generated by elements $x \otimes y - y \otimes x$, $x, y \in M$. Since these elements are homogeneous of degree two, $S(M)$ is a $\mathbb{Z}$-graded commutative ring.

Similarly, the *exterior algebra* of M, denoted by $\bigwedge M$, is defined as the quotient of $T(M)$ by the ideal generated by homogeneous elements $x \otimes x$, $x \in M$. So $\bigwedge M$ is a $\mathbb{Z}$-graded ring.

Let I be a two-sided ideal of a Γ-graded ring A generated by a subset $\{a_i\}$ of not necessarily homogeneous elements of A. If Ω is a subgroup of Γ such that $\{a_i\}$ are homogeneous elements in Γ/Ω-graded ring A (see §1.1.2), then clearly I is a Γ/Ω-graded ideal and consequently A/I is a Γ/Ω-graded ring.

Example 1.1.24 CLIFFORD ALGEBRAS AS $\mathbb{Z}_2$-GRADED RINGS

Let V be a F-vector space and $q : V \to F$ be a quadratic form with its associated nondegenerate symmetric bilinear form $B : V \times V \to F$.

The *Clifford algebra* associated with (V, q) is defined as

$$\mathrm{Cl}(V, q) := T(V)/\langle v \otimes v + q(v) \rangle.$$

Considering $T(V)$ as a $\mathbb{Z}/2\mathbb{Z}$-graded ring (see §1.1.2), the elements $v \otimes v - q(v)$ are homogeneous of degree zero. This induces a $\mathbb{Z}_2$-graded structure on $\mathrm{Cl}(V, q)$. Identifying V with its image in the Clifford algebra $\mathrm{Cl}(V, q)$, V lies in the odd part of the Clifford algebra, *i.e.*, $V \subset \mathrm{Cl}(V, q)_1$.

If $\mathrm{char}(F) \neq 2$, as B is nondegenerate, there exist $x, y \in V$ such that $B(x, y) = 1/2$, and thus

$$xy + yx = 2B(x, y) = 1 \in \mathrm{Cl}(V, q)_1\, \mathrm{Cl}(V, q)_1.$$

Similarly, if $\mathrm{char}(F) = 2$, there exist $x, y \in V$ such that $B(x, y) = 1$, so

$$xy + yx = B(x, y) = 1 \in \mathrm{Cl}(V, q)_1\, \mathrm{Cl}(V, q)_1.$$

It follows from Proposition 1.1.15 that Clifford algebras are strongly $\mathbb{Z}_2$-graded rings.

Recall that for Γ-graded rings A and B, a Γ-graded ring homomorphism $f : A \to B$ is a ring homomorphism such that $f(A_\gamma) \subseteq B_\gamma$ for all $\gamma \in \Gamma$. It can easily be shown that $\ker(f)$ is a graded ideal of A and $\mathrm{im}(f)$ is a graded subring of B. It is also easy to see that f is injective (surjective/bijective) if and only if for any $\gamma \in \Gamma$, the restriction of f on A_γ is injective (surjective/bijective).

Note that if Γ is an abelian group, then the centre of a graded ring A, $C(A)$, is a graded subring of A. More generally, the centraliser of a set of homogeneous elements is a graded subring.

Example 1.1.25 THE CENTRE OF THE GRADED RING

If a group Γ is not abelian, then the centre of a Γ-graded ring may not be a graded subring. For example, let $\Gamma = S_3 = \{e, a, b, c, d, f\}$ be the symmetric group of order 3, where

$$a = (23), \quad b = (13), \quad c = (12), \quad d = (123), \quad f = (132).$$

Let A be a ring, and consider the group ring $R = A[\Gamma]$, which is a Γ-graded ring by Example 1.1.2. Let $x = 1d + 1f \in R$, where $1 = 1_A$, and we note that x is not homogeneous in R. Then $x \in Z(R)$, but the homogeneous components of x are not in the centre of R. As x is expressed uniquely as the sum of homogeneous components, we have $x \notin \bigoplus_{\gamma \in \Gamma}(Z(R) \cap R_\gamma)$.

This example can be generalised by taking a nonabelian finite group Γ with a subgroup Ω which is normal and noncentral. Let A be a ring and consider the group ring $R = A[\Gamma]$ as above. Then $x = \sum_{\omega \in \Omega} 1\omega$ is in the centre of R, but the homogeneous components of x are not all in the centre of R.

Remark 1.1.26 Let Γ and Λ be two groups. Let A be a Γ-graded ring and B be a Λ-graded ring. Suppose $f : A \to B$ is a ring homomorphism and $g : \Gamma \to \Lambda$ a group homomorphism such that for any $\gamma \in \Gamma$, $f(A_\gamma) \subseteq B_{g(\gamma)}$. Then f is called a $\Gamma-\Lambda$-graded homomorphism. In the case $\Gamma = \Lambda$ and $g = \mathrm{id}$, we recover the usual definition of a Γ-graded homomorphism. For example, if Ω is a subgroup of Γ, then the identity map $1_A : A \to A$ is a $\Gamma-\Gamma/\Omega$-graded homomorphism, where A is considered as Γ and Γ/Ω-graded rings, respectively (see §1.1.2).

Throughout this book, we fix a given group Γ and we work with the Γ-graded category and all our considerations are within this category. (See Remark 2.3.14 for references to literature where mixed grading is studied.)

1.1.6 Graded prime and maximal ideals

A graded ideal P of Γ-graded ring A is called a *graded prime ideal* of A if $P \neq A$ and for any two graded ideals I, J, $IJ \subseteq P$, implies $I \subseteq P$ or $J \subseteq P$. If A is commutative, we obtain the familiar formulation that P is a graded prime ideal if and only if for $x, y \in A^h$, $xy \in P$ implies that $x \in P$ or $y \in P$. Note that a graded prime ideal is not necessarily a prime ideal.

A graded ideal P is called a *graded semiprime ideal* if for any graded ideal I in A, $I^2 \subseteq P$, implies $I \subseteq P$. A graded ring A is called a *graded prime (graded semiprime)* ring if the zero ideal is a graded prime (graded semiprime) ideal.

A *graded maximal ideal* of a Γ-graded ring A is defined to be a proper graded ideal of A which is maximal among the set of proper graded ideals of A. Using Zorn's lemma, one can show that graded maximal ideals exist, and it is not difficult to show that a graded maximal ideal is a graded prime. For a graded commutative ring, a graded ideal is maximal if and only if its quotient ring is a graded field. There are similar notions of graded maximal left and right ideals.

Parallel to the nongraded setting, for a Γ-graded ring A, the *graded Jacobson radical*, $J^{\mathrm{gr}}(A)$, is defined as the intersection of all graded left maximal ideals of A. This coincides with the intersection of all graded right maximal ideals and so $J^{gr}(A)$ is a two-sided ideal (see [75, Proposition 2.9.1]). We denote by $J(A)$ the usual Jacobson radical. It is a theorem of G. Bergman that for a $\mathbb{Z}$-graded ring A, $J(A)$ is a graded ideal and $J(A) \subseteq J^{gr}(A)$ (see [19]).

1.1.7 Graded simple rings

A nonzero graded ring A is said to be *graded simple* if the only graded two-sided ideals of A are $\{0\}$ and A. The structure of graded simple Artinian rings are known (see Remark 1.4.8). Following [52] we prove that a graded ring A is simple if and only if A is graded simple and $C(A)$, the centre of A, is a field.

For a Γ-graded ring A, recall the support Γ_A of A, from §1.1.1. For $a \in A$, writing $a = \sum_{\gamma \in \Gamma} a_\gamma$ where $a_\gamma \in A^h$, define the *support* of a to be

$$\Gamma_a = \{\gamma \mid a_\gamma \neq 0\}.$$

We also need the notion of minimal support. A finite set X of Γ is called a *minimal support with respect to an ideal I* if $X = \Gamma_a$ for $0 \neq a \in I$ and there is no $b \in I$ such that $b \neq 0$ and $\Gamma_b \subsetneq \Gamma_a$.

We start with a lemma.

Lemma 1.1.27 *Let A be a Γ-graded simple ring and I an ideal of A. Let $0 \neq a \in I$ with $\Gamma_a = \{\gamma_1, \dots, \gamma_n\}$. Then for any $\alpha \in \Gamma_A$, there is a $0 \neq b \in I$ with $\Gamma_b \subseteq \{\gamma_1 - \gamma_n + \alpha, \dots, \gamma_n - \gamma_n + \alpha\}$.*

Proof Let $0 \neq x \in A_\alpha$, where $\alpha \in \Gamma_A$ and $0 \neq a \in I$ with $\Gamma_a = \{\gamma_1, \dots, \gamma_n\}$. Write $a = \sum_{i=1}^n a_{\gamma_i}$, where $\deg(a_{\gamma_i}) = \gamma_i$. Since A is graded simple,

$$x = \sum_l r_l a_{\gamma_n} s_l, \tag{1.8}$$

where $r_l, s_l \in A^h$. Thus there are $r_k, s_k \in A^h$ such that $r_k a_{\gamma_n} s_k \neq 0$ which implies that $b := r_k a s_k \in I$ is not zero. Comparing the degrees in Equation (1.8), it follows that $\alpha = \deg(r_k) + \deg(s_k) + \gamma_n$, or $\deg(r_k) + \deg(s_k) = \alpha - \gamma_n$. So

$$\Gamma_b \subseteq \Gamma_a + \deg(r_k) + \deg(s_k) = \{\gamma_1 - \gamma_n + \alpha, \dots, \gamma_n - \gamma_n + \alpha\}.$$ □

Theorem 1.1.28 *Let A be a Γ-graded ring. Then A is a simple ring if and only if A is a graded simple ring and $C(A)$ is a field.*

Proof One direction is straightforward.

Suppose A is graded simple and $C(A)$ is a field. We will show that A is a simple ring. Suppose I is a nontrivial ideal of A and $0 \neq a \in I$ with Γ_a a minimal support with respect to I. For any $x \in A^h$, with $\deg(x) = \alpha$ and $\gamma \in \Gamma_a$, we have

$$\Gamma_{axa_\gamma - a_\gamma xa} \subsetneq \Gamma_a + (\gamma + \alpha). \tag{1.9}$$

Set $b = axa_\gamma - a_\gamma xa \in I$. Suppose $b \neq 0$. By (1.9),

$$\Gamma_b \subsetneq \{\gamma_1 + \gamma + \alpha, \dots, \gamma_n + \gamma + \alpha\}.$$

Applying Lemma 1.1.27 with, say, $\gamma_n + \gamma + \alpha \in \Gamma_b$ and $\gamma_n \in \Gamma_A$, we obtain a $0 \neq c \in I$ such that

$$\Gamma_c \subseteq \Gamma_b + (\gamma_n - \gamma_n - \gamma - \alpha) \subsetneq \Gamma_a.$$

This is, however, a contradiction as Γ_a was a minimal support. Thus $b = 0$, *i.e.*, $axa_\gamma = a_\gamma xa$. It follows that for any $\gamma_i \in \Gamma_a$

$$a_{\gamma_i} xa_\gamma = a_\gamma xa_{\gamma_i}. \tag{1.10}$$

Consider the R-bimodule map

$$\phi : R = \langle a_{\gamma_i} \rangle \longrightarrow \langle a_{\gamma_j} \rangle = R,$$
$$\sum_l r_l a_{\gamma_i} s_l \longmapsto \sum_l r_l a_{\gamma_j} s_l.$$

To show that ϕ is well-defined, since $\phi(t+s) = \phi(t) + \phi(s)$, it is enough to show that if $t = 0$ then $\phi(t) = 0$, where $t \in \langle a_{\gamma_i} \rangle$. Suppose $\sum_l r_l a_{\gamma_i} s_l = 0$. Then for any $x \in A^h$, using (1.10) we have

$$0 = a_{\gamma_j} x \big(\sum_l r_l a_{\gamma_i} s_l\big) = \sum_l a_{\gamma_j} x r_l a_{\gamma_i} s_l = \sum_l a_{\gamma_i} x r_l a_{\gamma_j} s_l = a_{\gamma_i} x \big(\sum_l r_l a_{\gamma_j} s_l\big).$$

Since A is graded simple, $\langle a_{\gamma_i} \rangle = 1$. It follows that $\sum_l r_l a_{\gamma_j} s_l = 0$. Thus ϕ is well-defined, injective and also clearly surjective. Then $a_{\gamma_j} = \phi(a_{\gamma_i}) = a_{\gamma_i}\phi(1)$. But $\phi(1) \in C(A)$. Thus $a = \sum_j a_{\gamma_j} = a_{\gamma_i} c$ where $c \in C(A)$. But $C(A)$ is a field, so $a_{\gamma_i} = ac^{-1} \in I$. Again, since R is graded simple, it follows that $I = R$. This finishes the proof. □

Remark 1.1.29 If the grade group is not abelian, in order for Theorem 1.1.28 to be valid, the grade group should be hyper-central; A *hyper-central group* is a group such that any nontrivial quotient has a nontrivial centre. If A is strongly graded, and the grade group is torsion-free hyper-central, then A is simple if and only if A is graded simple and $C(A) \subseteq A_0$ (see [52]).

Remark 1.1.30 Graded simplicity implying simplicity

There are other cases that the graded simplicity of a ring implies that the ring itself is simple. For example, if a ring is graded by an ordered group (such as $\mathbb{Z}$), and has a finite support, then graded simplicity implies the simplicity of the ring [10, Theorem 3].

1.1.8 Graded local rings

Recall that a ring is a *local ring* if the set of noninvertible elements form a two-sided ideal. When A is a commutative ring, then A is local if and only if A has a unique maximal ideal.

A Γ-graded ring A is called a *graded local ring* if the two-sided ideal M generate by noninvertible homogeneous elements is a proper ideal. One can easily observe that the graded ideal M is the unique graded maximal left, right, and graded two-sided ideal of A. When A is a graded commutative ring, then A is graded local if and only if A has a unique graded maximal ideal.

If A is a graded local ring, then the graded ring A/M is a graded division ring. One can further show that A_0 is a local ring with the unique maximal ideal $A_0 \cap M$. In fact we have the following proposition.

Proposition 1.1.31 *Let A be a Γ-graded ring. Then A is a graded local ring if and only if A_0 is a local ring.*

Proof Suppose A is a graded local ring. Then by definition, the two-sided ideal M generated by noninvertible homogeneous elements is a proper ideal. Consider $m = A_0 \cap M$ which is a proper ideal of A_0. Suppose $x \in A_0 \backslash m$. Then x is a homogeneous element which is not in M. Thus x has to be invertible in A and consequently in A_0. This shows that A_0 is a local ring with the unique maximal ideal m.

Conversely, suppose A_0 is a local ring. We first show that any left or right invertible homogeneous element is a two-sided invertible element. Let a be a left invertible homogeneous element. Then there is a homogeneous element b such that $ba = 1$. If ab is not right invertible, then $ab \in m$, where m is the unique maximal ideal of the local ring A_0. Thus $1 - ab \notin m$ which implies that $1 - ab$ is invertible. But $(1 - ab)a = a - aba = a - a = 0$, and since $1 - ab$ is invertible, we get $a = 0$ which is a contradiction to the fact that a has a left inverse. Thus a has a right inverse and so is invertible. A similar argument can be written for right invertible elements. Now let M be the ideal generated by all noninvertible homogeneous elements of A. We will show that M is proper, and thus A is a graded local ring. Suppose M is not proper. Thus $1 = \sum_i r_i a_i s_i$, where a_i are noninvertible homogeneous elements and r_i, s_i are homogeneous elements such that $\deg(r_i a_i s_i) = 0$. If $r_i a_i s_i$ is invertible for some i, using the fact that right and left invertibles are invertibles, it follows that a_i is invertible, which is a contradiction. Thus $r_i a_i s_i$, for all i, are homogeneous elements of degree zero and not invertible. So they are all in m. This implies that $1 \in m$, which is a contradiction. Thus M is a proper ideal of A. □

For more on graded local rings (graded by a cancellative monoid) see [64]. In §3.8 we determine the graded Grothendieck group of these rings.

1.1.9 Graded von Neumann regular rings

The von Neumann regular rings constitute an important class of rings. A unital ring A is *von Neumann regular* if for any $a \in A$, we have $a \in aAa$. There are several equivalent module theoretical definitions, such as A is von Neumann regular if and only if any module over A is flat. This gives a comparison with the class of division rings and semisimple rings. A ring is a division ring if and only if any module is free. A semisimple ring is characterised by the property that any module is projective. Goodearl's book [40] is devoted to the class of von Neumann regular rings. The definition extends to a nonunital ring in an obvious manner.

If a ring has a graded structure, one defines the graded version of regularity in a natural way: the graded ring A is called *graded von Neumann regular* if for any homogeneous element $a \in A$ we have $a \in aAa$. This means, for any homogeneous element $a \in A$, one can find a homogeneous element $b \in A$ such that $a = aba$. As an example, a direct sum of graded division rings is a graded von Neumann regular ring. Many of the module theoretic properties established for von Neumann regular rings can be extended to the graded setting; for example, A is graded regular if and only if any graded module is (graded) flat. We refer the reader to [74, C, I.5] for a treatment of such rings and [11, §2.2] for a concise survey. Several of the graded rings we construct in this book are graded von Neumann regular, such as Leavitt path algebras (Corollary 1.6.17) and corner skew Laurent series (Proposition 1.6.8).

In this section, we briefly give some of the properties of graded von Neumann regular rings. The following proposition is the graded version of [40, Theorem 1.1] which has a similar proof.

Proposition 1.1.32 *Let A be a Γ-graded ring. The following statements are equivalent:*

(1) *A is a graded von Neumann regular ring;*
(2) *any finitely generated right (left) graded ideal of A is generated by a homogeneous idempotent.*

Proof (1) $\Rightarrow$ (2) First we show that any principal graded ideal is generated by a homogeneous idempotent. So consider the principal ideal xA, where $x \in A^h$. By the assumption, there is $y \in A^h$ such that $xyx = x$. This immediately implies $xA = xyA$. Now note that xy is homogeneous idempotent.

Next we will prove the claim for graded ideals generated by two elements. The general case follows by an easy induction. So let $xA+yA$ be a graded ideal generated by two homogeneous elements x, y. By the previous paragraph, $xA = eA$ for a homogeneous idempotent e. Note that $y-ey \in A^h$ and $y-ey \in xA+yA$. Thus

$$xA + yA = eA + (y - ey)A. \tag{1.11}$$

Again, the previous paragraph gives us a homogeneous idempotent f such that $(y - ey)A = fA$. Let $g = f - fe \in A_0$. Notice that $ef = 0$, which implies that e and g are orthogonal idempotents. Moreover, $fg = g$ and $gf = f$. It then follows that $gA = fA = (y - ey)A$. Now from (1.11) we get

$$xA + yA = eA + gA = (e + g)A.$$

(2) $\Rightarrow$ (1) Let $x \in A^h$. Then $xA = eA$ for some homogeneous idempotent e. Thus $x = ea$ and $e = xy$ for some $a, y \in A^h$. Then $x = ea = eea = ex = xyx$. □

Proposition 1.1.33 *Let A be a Γ-graded von Neumann regular ring. Then*

(1) *any graded right (left) ideal of A is idempotent;*
(2) *any graded ideal is graded semiprime;*
(3) *any finitely generated right (left) graded ideal of A is a projective module.*

Moreover, if A is a $\mathbb{Z}$-graded regular ring then

(4) $J(A) = J^{\mathrm{gr}}(A) = 0$.

Proof The proofs of (1)–(3) are similar to the nongraded case [40, Corollary 1.2]. We provide the easy proofs here.

(1) Let I be a graded right ideal. For any homogeneous element $x \in I$ there is $y \in A^h$ such that $x = xyx$. Thus $x = (xy)x \in I^2$. It follows that $I^2 = I$.

(2) This follows immediately from (1).

(3) By Proposition 1.1.32, any finitely generated right ideal is generated by a homogeneous idempotent. However, this latter ideal is a direct summand of the ring, and so is a projective module.

(4) By Bergman's observation, for a $\mathbb{Z}$-graded ring A, $J(A)$ is a graded ideal and $J(A) \subseteq J^{\mathrm{gr}}(A)$ (see [19]). By Proposition 1.1.32, $J^{\mathrm{gr}}(A)$ contains an idempotent, which then forces $J^{\mathrm{gr}}(A) = 0$. □

If the graded ring A is strongly graded then one can show that there is a one-to-one correspondence between the right ideals of A_0 and the graded right ideals of A (similarly for the left ideals) (see Remark 1.5.6). This is always the case for the graded regular rings as the following proposition shows.

Proposition 1.1.34 *Let A be a Γ-graded von Neumann regular ring. Then there is a one-to-one correspondence between the right (left) ideals of A_0 and the graded right (left) ideals of A.*

Proof Consider the following correspondences between the graded right ideals of A and the right ideals of A_0. For a graded right ideal I of A assign I_0 in A_0 and for a right ideal J in A_0 assign the graded right ideal JA in A. Note that $(JA)_0 = J$. We show that $I_0A = I$. It is enough to show that any homogeneous element a of I belongs to I_0A. Since A is graded regular, $axa = a$ for some $x \in A^h$. But $ax \in I_0$ and thus $a = axa \in I_0A$. A similar proof gives the left ideal correspondence. □

In Theorem 1.2.20 we give yet another characterisation of graded von Neumann regular rings based on the concept of divisible modules.

Later, in Corollary 1.5.10, we show that if A is a strongly graded ring, then A is graded von Neumann regular if and only if A_0 is a von Neumann regular ring. The proof uses the equivalence of suitable categories over the rings A and A_0. An element-wise proof of this fact can also be found in [96, Theorem 3].

1.2 Graded modules

1.2.1 Basic definitions

Let A be a Γ-graded ring. A *graded right A-module* M is defined to be a right A-module M with a direct sum decomposition $M = \bigoplus_{\gamma\in\Gamma} M_\gamma$, where each M_γ is an additive subgroup of M such that $M_\lambda A_\gamma \subseteq M_{\lambda+\gamma}$ for all $\gamma, \lambda \in \Gamma$.

For Γ-graded right A-modules M and N, a Γ*-graded module homomorphism* $f : M \to N$ is a module homomorphism such that $f(M_\gamma) \subseteq N_\gamma$ for all $\gamma \in \Gamma$. A graded homomorphism f is called a *graded module isomorphism* if f is bijective and, when such a graded isomorphism exists, we write $M \cong_{\mathrm{gr}} N$. Notice that if f is a graded module homomorphism which is bijective, then its inverse f^{-1} is also a graded module homomorphism.

1.2.2 Shift of modules

Let M be a graded right A-module. For $\delta \in \Gamma$, we define the δ*-suspended* or δ*-shifted* graded right A-module $M(\delta)$ as

$$\boxed{M(\delta) = \bigoplus_{\gamma\in\Gamma} M(\delta)_\gamma, \text{ where } M(\delta)_\gamma = M_{\delta+\gamma}.}$$

This shift plays a pivotal role in the theory of graded rings. For example, if M is a $\mathbb{Z}$-graded A-module, then the following table shows how the shift like "the tick of the clock" moves the homogeneous components of M to the left.

	degrees	**-3**	**-2**	**-1**	**0**	**1**	**2**	**3**
M				M_{-1}	M_0	M_1	M_2	
$M(1)$			M_{-1}	M_0	M_1	M_2		
$M(2)$		M_{-1}	M_0	M_1	M_2			

Let M be a Γ-graded right A-module. A submodule N of M is called a *graded submodule* if

$$N = \bigoplus_{\gamma \in \Gamma} (N \cap M_\gamma).$$

Example 1.2.1 <u>aA AS A GRADED IDEAL AND A GRADED MODULE</u>

Let A be a Γ-graded ring and $a \in A$ a homogeneous element of degree α. Then aA is a graded right A-module with $\gamma \in \Gamma$ homogeneous component defined as

$$(aA)_\gamma := aA_{\gamma-\alpha} \subseteq A_\gamma.$$

With this grading aA is a graded submodule (and graded right ideal) of A. Thus for $\beta \in \Gamma$, a is a homogenous element of the graded A-module $aA(\beta)$ of degree $\alpha - \beta$. This will be used throughout the book, for example in Proposition 1.2.19.

However, note that defining the grading on aA as

$$(aA)_\gamma := aA_\gamma \subseteq A_{\gamma+\alpha}$$

makes aA a graded submodule of $A(\alpha)$, which is the image of the graded homomorphism $A \to A(\alpha)$, $r \mapsto ar$.

There are similar notions of graded left and graded bi-submodules (§1.2.5). When N is a graded submodule of M, the factor module M/N forms a graded A-module, with

$$\boxed{M/N = \bigoplus_{\gamma \in \Gamma} (M/N)_\gamma, \quad \text{where} \quad (M/N)_\gamma = (M_\gamma + N)/N.} \tag{1.12}$$

Example 1.2.2 Let A be a Γ-graded ring. Define a grading on the matrix ring $\mathbb{M}_n(A)$ as follows. For $\alpha \in \Gamma$, $\mathbb{M}_n(A)_\alpha = \mathbb{M}_n(A_\alpha)$ (for a general theory of grading on matrix rings see §1.3). Let $\mathbf{e}_{ii} \in \mathbb{M}_n(A)$, $1 \leq i \leq n$, be a *matrix unit*, *i.e.*, a matrix with 1 in the (i, i) position and zero everywhere else, and consider $\mathbf{e}_{ii}\,\mathbb{M}_n(A)$. By Example 1.2.1, $\mathbf{e}_{ii}\,\mathbb{M}_n(A)$ is a graded right $\mathbb{M}_n(A)$-module and

$$\bigoplus_{i=1}^{n} \mathbf{e}_{ii}\,\mathbb{M}_n(A) = \mathbb{M}_n(A).$$

This shows that the graded module $\mathbf{e}_{ii}\,\mathbb{M}_n(A)$ is a projective module. This is an example of a graded projective module (see §1.2.9).

Example 1.2.3 Let A be a commutative ring. Consider the matrix ring $\mathbb{M}_n(A)$ as a $\mathbb{Z}$-graded ring concentrated in degree zero. Moreover, consider $\mathbb{M}_n(A)$ as a graded $\mathbb{M}_n(A)$-module with the grading defined as follows: $\mathbb{M}_n(A)_i = \mathbf{e}_{ii}\,\mathbb{M}_n(A)$ for $1 \leq i \leq n$ and zero otherwise. Note that all nonzero homogeneous elements of this module are zero-divisors, and thus can't constitute a linear independent set. We will use this example to show that a free module which is graded is not necessarily a graded free module (§1.2.4).

Example 1.2.4 MODULES WITH NO SHIFT

It is easy to construct modules whose shifts don't produce new (nonisomorphic) graded modules. Let M be a graded A-module and consider

$$N = \bigoplus_{\gamma\in\Gamma} M(\gamma).$$

We show that $N \cong_{\mathrm{gr}} N(\alpha)$ for any $\alpha \in \Gamma$. Define the map $f_\alpha : N \to N(\alpha)$ on homogeneous components as follows and extend it to N,

$$N_\beta = \bigoplus_{\gamma\in\Gamma} M_{\gamma+\beta} \longrightarrow \bigoplus_{\gamma\in\Gamma} M_{\gamma+\alpha+\beta} = N(\alpha)_\beta$$
$$\{m_\gamma\} \longmapsto \{m'_\gamma\},$$

where $m'_\gamma = m_{\gamma+\alpha}$ (*i.e.*, shift the sequence α "steps"). It is routine to see that this gives a graded A-module homomorphism with inverse homomorphism $f_{-\alpha}$. For another example, see Corollary 1.3.18.

1.2.3 The Hom groups and the category of graded modules

For graded right A-modules M and N, a *graded A-module homomorphism of degree δ* is an A-module homomorphism $f : M \to N$, such that

$$f(M_\gamma) \subseteq N_{\gamma+\delta}$$

for any $\gamma \in \Gamma$. Let $\operatorname{Hom}_A(M, N)_\delta$ denote the subgroup of $\operatorname{Hom}_A(M, N)$ consisting of all graded A-module homomorphisms of degree δ, *i.e.*,

$$\boxed{\operatorname{Hom}_A(M, N)_\delta = \{f \in \operatorname{Hom}_A(M, N) \mid f(M_\gamma) \subseteq N_{\gamma+\delta}, \gamma \in \Gamma\}.} \tag{1.13}$$

For graded A-modules, M, N and P, under the composition of functions, we then have

$$\mathrm{Hom}_A(N,P)_\gamma \times \mathrm{Hom}_A(M,N)_\delta \longrightarrow \mathrm{Hom}(M,P)_{\gamma+\delta}. \tag{1.14}$$

Clearly a graded module homomorphism defined in §1.2.1 is a graded homomorphism of degree 0.

By Gr-A (or Gr^Γ-A to emphasise the grade group of A), we denote a category that consists of Γ-graded right A-modules as objects and graded homomorphisms as the morphisms. Similarly, A-Gr denotes the category of graded left A-modules. Thus

$$\boxed{\mathrm{Hom}_{\text{Gr-}A}(M,N) = \mathrm{Hom}_A(M,N)_0.}$$

Moreover, for $\alpha \in \Gamma$, as *a set of functions*, one can write

$$\mathrm{Hom}_{\text{Gr-}A}\left(M(-\alpha),N\right) = \mathrm{Hom}_{\text{Gr-}A}\left(M,N(\alpha)\right) = \mathrm{Hom}_A(M,N)_\alpha. \tag{1.15}$$

A full subcategory of Gr-A consisted of all graded finitely generated A-modules is denoted by gr-A.

For $\alpha \in \Gamma$, the *α-suspension functor* or *shift functor*

$$\boxed{\begin{aligned} \mathcal{T}_\alpha : \text{Gr-}A &\longrightarrow \text{Gr-}A, \\ M &\longmapsto M(\alpha), \end{aligned}} \tag{1.16}$$

is an isomorphism with the property $\mathcal{T}_\alpha\mathcal{T}_\beta = \mathcal{T}_{\alpha+\beta}$, where $\alpha, \beta \in \Gamma$.

Remark 1.2.5 Let A be a Γ-graded ring and Ω be a subgroup of Γ such that $\Gamma_A \subseteq \Omega \subseteq \Gamma$. Then the ring A can be considered naturally as a Ω-graded ring. Similarly, if A, B are Γ-graded rings and $f : A \to B$ is a Γ-graded homomorphism and $\Gamma_A, \Gamma_B \subseteq \Omega \subseteq \Gamma$, then the homomorphism f can be naturally considered as a Ω-graded homomorphism. In this case, to make a distinction, we write Gr^Γ-A for the category of Γ-graded A-modules and Gr^Ω-A for the category of Ω-graded A-modules.

Theorem 1.2.6 *For graded right A-modules M and N, such that M is finitely generated, the abelian group* $\mathrm{Hom}_A(M,N)$ *has a natural decomposition*

$$\mathrm{Hom}_A(M,N) = \bigoplus_{\gamma\in\Gamma} \mathrm{Hom}_A(M,N)_\gamma. \tag{1.17}$$

Moreover, the endomorphism ring $\mathrm{Hom}_A(M,M)$ *is Γ-graded.*

Proof Let $f \in \mathrm{Hom}_A(M,N)$ and $\lambda \in \Gamma$. Define a map $f_\lambda : M \to N$ as follows: for $m \in M$,

$$f_\lambda(m) = \sum_{\gamma\in\Gamma} f(m_{\gamma-\lambda})_\gamma, \tag{1.18}$$

where $m = \sum_{\gamma\in\Gamma} m_\gamma$. One can check that $f_\lambda \in \operatorname{Hom}_A(M, N)$.

Now let $m \in M_\alpha$, $\alpha \in \Gamma$. Then (1.18) reduces to

$$f_\lambda(m) = f(m)_{\alpha+\lambda} \subseteq M_{\alpha+\lambda}.$$

This shows that $f_\lambda \in \operatorname{Hom}_A(M, N)_\lambda$. Moreover, $f_\lambda(m)$ is zero for all but a finite number of $\lambda \in \Gamma$ and

$$\sum_\lambda f_\lambda(m) = \sum_\lambda f(m)_{\alpha+\lambda} = f(m).$$

Now since M is finitely generated, there are a finite number of homogeneous elements which generate any element $m \in M$. The above argument shows that only a finite number of the $f_\lambda(m)$ are nonzero and $f = \sum_\lambda f_\lambda$. This in turn shows that

$$\operatorname{Hom}_A(M, N) = \sum_{\gamma\in\Gamma} \operatorname{Hom}_A(M, N)_\gamma.$$

Finally, it is easy to see that $\operatorname{Hom}_A(M, N)_\gamma$, $\gamma \in \Gamma$ constitutes a direct sum.

For the second part, replacing N by M in (1.17), we get

$$\operatorname{Hom}_A(M, M) = \bigoplus_{\gamma\in\Gamma} \operatorname{Hom}_A(M, M)_\gamma.$$

Moreover, by (1.14) if $f \in \operatorname{Hom}_A(M, M)_\gamma$ and $g \in \operatorname{Hom}_A(M, M)_\lambda$ then

$$fg \in \operatorname{Hom}_A(M, M)_{\gamma+\lambda}.$$

This shows that when M is finitely generated $\operatorname{Hom}_A(M, M)$ is a Γ-graded ring. □

Let M be a graded finitely generated right A-module. Then the usual *dual* of M, *i.e.*, $M^* = \operatorname{Hom}_A(M, A)$, is a left A-module. Moreover, using Theorem 1.2.6, one can check that M^* is a graded left A-module. Since

$$\operatorname{Hom}_A(M, N)(\alpha) = \operatorname{Hom}_A(M(-\alpha), N) = \operatorname{Hom}_A(M, N(\alpha)),$$

we have

$$M(\alpha)^* = M^*(-\alpha). \tag{1.19}$$

This should also make sense: the dual of "pushing forward" M by α, is the same as "pulling back" the dual M^* by α.

Note that although $\operatorname{Hom}_A(M, N)$ is defined in the category Mod-A, the graded structures of M and N are intrinsic in the grading defined on $\operatorname{Hom}_A(M, N)$. Thus if M is isomorphic to N as a nongraded A-module, then $\operatorname{End}_A(M)$ is not necessarily graded isomorphic to $\operatorname{End}_A(N)$. However if $M \cong_{\mathrm{gr}} N(\alpha)$, $\alpha \in \Gamma$, then one can observe that $\operatorname{End}_A(M) \cong_{\mathrm{gr}} \operatorname{End}_A(N)$ as graded rings.

When M is a free module, $\mathrm{Hom}_A(M, M)$ can be represented as a matrix ring over A. Next we define graded free modules. In §1.3 we will see that if M is a graded free module, the graded ring $\mathrm{Hom}_A(M, M)$ can be represented as a matrix ring over A with a very concrete grading.

Example 1.2.7 THE VERONESE SUBMODULE

For a Γ-graded ring A, recall the construction of nth-Veronese subring

$$A^{(n)} = \bigoplus_{\gamma \in \Gamma} A_{n\gamma}$$

(Example 1.1.19). In a similar fashion, for a graded A-module M and $n \in \mathbb{Z}$, define the nth-*Veronese module* of M as

$$M^{(n)} = \bigoplus_{\gamma \in \Gamma} M_{n\gamma}.$$

This is a Γ-graded $A^{(n)}$-module. Clearly there is a natural "forgetful" functor

$$U : \text{Gr-}A \longrightarrow \text{Gr-}A^{(n)},$$

which commutes with suspension functors as follows $\mathcal{T}_\alpha U = U\mathcal{T}_{n\alpha}$, *i.e.*,

$$M^{(n)}(\alpha) = M(n\alpha)^{(n)},$$

for $\alpha \in \Gamma$ and $n \in \mathbb{Z}$ (see §1.2.7 for more on forgetful functors).

1.2.4 Graded free modules

A Γ-graded (right) A-module F is called a *graded free A-module* if F is a free right A-module with a homogeneous base. Clearly a graded free module is a free module but the converse is not correct, *i.e.*, a free module which is graded is not necessarily a graded free module. As an example, for $A = \mathbb{R}[x]$ considered as a $\mathbb{Z}$-graded ring, $A \oplus A(1)$ is not a graded free $A \oplus A$-module, whereas $A \oplus A$ is a free $A \oplus A$-module (see also Example 1.2.3). The definition of free given here is consistent with the categorical definition of free objects over a set of homogeneous elements in the category of graded modules ([50, I, §7]).

Consider a Γ-graded A-module $\bigoplus_{i \in I} A(\delta_i)$, where I is an indexing set and $\delta_i \in \Gamma$. Note that for each $i \in I$, the element e_i of the standard basis (*i.e.*, 1 in the ith component and zero elsewhere) is homogeneous of degree $-\delta_i$. The set $\{e_i\}_{i \in I}$ forms a base for $\bigoplus_{i \in I} A(\delta_i)$, which by definition makes this a graded free A-module. On the other hand, a graded free A-module F with a homogeneous base $\{b_i\}_{i \in I}$, where $\deg(b_i) = -\delta_i$ is graded isomorphic to $\bigoplus_{i \in I} A(\delta_i)$. Indeed one can easily observe that the map induced by

$$\varphi : \bigoplus_{i \in I} A(\delta_i) \longrightarrow F \qquad (1.20)$$
$$e_i \longmapsto b_i$$

is a graded A-module isomorphism.

If the indexing set I is finite, say $I = \{1, \dots, n\}$, then

$$\bigoplus_{i \in I} A(\delta_i) = A(\delta_1) \oplus \cdots \oplus A(\delta_n),$$

is also denoted by $A^n(\delta_1, \dots, \delta_n)$ or $A^n(\overline{\delta})$, where $\overline{\delta} = (\delta_1, \dots, \delta_n)$.

In §1.3.4, we consider the situation when the graded free right A-modules $A^n(\overline{\delta})$ and $A^m(\overline{\alpha})$, where $\overline{\delta} = (\delta_1, \dots, \delta_n)$ and $\overline{\alpha} = (\alpha_1, \dots, \alpha_m)$, are isomorphic. In §1.7, we will also consider the concept of graded rings with the graded invariant basis numbers.

1.2.5 Graded bimodules

The notion of graded *left* A-modules is developed similarly. The category of graded left A-modules with graded homomorphisms is denoted by A-Gr. In a similar manner for Γ-graded rings A and B, we can consider the *graded $A-B$-bimodule M*. That is, M is a $A-B$-bimodule and additionally $M = \bigoplus_{\gamma \in \Gamma} M_\gamma$ is a graded left A-module and a graded right B-module, *i.e.*,

$$A_\alpha M_\gamma B_\beta \subseteq M_{\alpha+\gamma+\beta},$$

where $\alpha, \gamma, \beta \in \Gamma$. The category of graded A-bimodules is denoted by Gr-A-Gr.

Remark 1.2.8 Shift of nonabelian group graded modules

If the grade group Γ is not abelian, then in order that the shift of components matches, for a graded left A-module M one needs to define

$$M(\delta)_\gamma = M_{\gamma\delta},$$

whereas for the graded right M-module A, shift is defined by

$$M(\delta)_\gamma = M_{\delta\gamma}.$$

With these definitions, for $\mathcal{T}_\alpha, \mathcal{T}_\beta$: Gr-A $\rightarrow$ Gr-A, we have $\mathcal{T}_\alpha\mathcal{T}_\beta = \mathcal{T}_{\beta\alpha}$, whereas for $\mathcal{T}_\alpha, \mathcal{T}_\beta$: A-Gr $\rightarrow$ A-Gr, we have $\mathcal{T}_\alpha\mathcal{T}_\beta = \mathcal{T}_{\alpha\beta}$. For this reason, in the nonabelian grade group setting, several books choose to work with the graded left modules as opposed to the graded right modules we have adopted in this book.

1.2.6 Tensor product of graded modules

Let A be a Γ-graded ring and M_i, $i \in I$, be a direct system of Γ-graded A-modules, *i.e.*, I is a directed partially ordered set and for $i \leq j$, there is a graded A-homomorphism $\phi_{ij} : M_i \to M_j$ which is compatible with the ordering. Then $M := \varinjlim M_i$ is a Γ-graded A-module with homogeneous components $M_\alpha = \varinjlim M_{i\alpha}$ (see Example 1.1.9 for the similar construction for rings).

In particular, let $\{M_i \mid i \in I\}$ be Γ-graded right A-modules. Then $\bigoplus_{i\in I} M_i$ has a natural graded A-module given by $(\bigoplus_{i\in I} M_i)_\alpha = \bigoplus_{i\in I} M_{i\alpha}$, $\alpha \in \Gamma$.

Let M be a graded right A-module and N be a graded left A-module. We will observe that the tensor product $M \otimes_A N$ has a natural Γ-graded $\mathbb{Z}$-module structure. Since each of M_γ, $\gamma \in \Gamma$, is a right A_0-module and similarly N_γ, $\gamma \in \Gamma$, is a left A_0-module, then $M \otimes_{A_0} N$ can be decomposed as a direct sum

$$\boxed{M \otimes_{A_0} N = \bigoplus_{\gamma\in\Gamma} (M \otimes N)_\gamma,}$$

where

$$\boxed{(M \otimes N)_\gamma = \Big\{ \sum_i m_i \otimes n_i \mid m_i \in M^h, n_i \in N^h, \deg(m_i) + \deg(n_i) = \gamma \Big\}.}$$

Now note that $M \otimes_A N \cong (M \otimes_{A_0} N)/J$, where J is a subgroup of $M \otimes_{A_0} N$ generated by the homogeneous elements

$$\{ma \otimes n - m \otimes an \mid m \in M^h, n \in N^h, a \in A^h\}.$$

This shows that $M \otimes_A N$ is also a graded module. It is easy to check that, for example, if N is a graded A-bimodule, then $M \otimes_A N$ is a graded right A-module. It follows from the definition that

$$M \otimes N(\alpha) = M(\alpha) \otimes N = (M \otimes N)(\alpha). \tag{1.21}$$

Observe that for a graded right A-module M, the map

$$\begin{aligned} M \otimes_A A(\alpha) &\longrightarrow M(\alpha), \\ m \otimes a &\longmapsto ma, \end{aligned} \tag{1.22}$$

is a graded isomorphism. In particular, for any $\alpha, \beta \in \Gamma$, there is a graded A-bimodule isomorphism

$$\boxed{A(\alpha) \otimes_A A(\beta) \cong_{\text{gr}} A(\alpha + \beta).} \tag{1.23}$$

Example 1.2.9 GRADED FORMAL MATRIX RINGS

The construction of formal matrix rings (Example 1.1.4) can be carried over to the graded setting as follows. Let R and S be Γ-graded rings, M be a graded R–S-bimodule and N be a graded S–R-bimodule. Suppose that there are graded bimodule homomorphisms $\phi : M \otimes_S N \to R$ and $\psi : N \otimes_R M \to S$ such that $(mn)m' = n(nm')$, where we denote $\phi(m,n) = mn$ and $\psi(n,m) = nm$. Consider the ring

$$T = \begin{pmatrix} R & M \\ N & S \end{pmatrix},$$

and define, for any $\gamma \in \Gamma$,

$$T_\gamma = \begin{pmatrix} R_\gamma & M_\gamma \\ N_\gamma & S_\gamma \end{pmatrix}.$$

One checks that T is a Γ-graded ring, called a *graded formal matrix ring*. One specific type of such rings is a Morita ring which appears in graded Morita theory (§2.3).

1.2.7 Forgetting the grading

Most forgetful functors in algebra tend to have left adjoints, which have a "free" construction. One such example is the forgetful functor from the category of abelian groups to abelian monoids that we will study in Chapter 3 in relation to Grothendieck groups. However, some of the forgetful functors in the graded setting naturally have right adjoints, as we will see below.

Consider the *forgetful* functor

$$U : \text{Gr-}A \longrightarrow \text{Mod-}A, \tag{1.24}$$

which simply assigns to any graded module M in Gr-A its underlying module M in Mod-A, ignoring the grading. Similarly, the graded homomorphisms are sent to the same homomorphisms, disregarding their graded compatibilities.

There is a functor $F : \text{Mod-}A \to \text{Gr-}A$ which is a right adjoint to U. The construction is as follows: let M be an A-module. Consider the abelian group $F(M) := \bigoplus_{\gamma\in\Gamma} M_\gamma$, where M_γ is a copy of M. Moreover, for $a \in A_\alpha$ and $m \in M_\gamma$ define $m.a = ma \in M_{\alpha+\gamma}$. This defines a graded A-module structure on $F(M)$ and makes F an exact functor from Mod-A to Gr-A. One can prove that for any $M \in \text{Gr-}A$ and $N \in \text{Mod-}A$, we have a bijective map

$$\begin{aligned} \operatorname{Hom}_{\text{Mod-}A}(U(M), N) &\stackrel{\phi}{\longrightarrow} \operatorname{Hom}_{\text{Gr-}A}(M, F(N)), \\ f &\longmapsto \phi_f, \end{aligned}$$

where $\phi_f(m_\alpha) = f(m_\alpha) \in N_\alpha$.

Remark 1.2.10 It is not difficult to observe that for any $M \in \text{Gr-}A$,

$$FU(M) \cong_{\text{gr}} \bigoplus_{\gamma \in \Gamma} M(\gamma).$$

By Example 1.2.4, we have $FU(M) \cong_{\text{gr}} FU(M)(\alpha)$ for any $\alpha \in \Gamma$. We also note that if Γ is finite, then F is also a left adjoint functor of U. Further, if U has a left adjoint functor, then one can prove that Γ is finite (see [75, §2.5] for details).

1.2.8 Partitioning graded modules

Let $f : \Gamma \to \Delta$ be a group homomorphism. Recall from §1.1.2 that there is a functor from the category of Γ-graded rings to the category of Δ-graded rings which gives the natural forgetful functor when $\Delta = 0$. This functor has a right adjoint functor (see [75, Proposition 1.2.2] for the case of $\Delta = 0$). The homomorphism f induces a forgetful functor on the level of module categories. We describe this here.

Let A be a Γ-graded ring and consider the corresponding Δ-graded structure induced by the homomorphism $f : \Gamma \to \Delta$ (§1.1.2). Then one can construct a functor $U_f : \text{Gr}^{\Gamma}\text{-}A \to \text{Gr}^{\Delta}\text{-}A$ which has a right adjoint. In particular, for a subgroup Ω of Γ, we have the following canonical "forgetful" functor (a *block* functor or a *coarsening* functor)

$$U : \text{Gr}^{\Gamma}\text{-}A \to \text{Gr}^{\Gamma/\Omega}\text{-}A,$$

such that when $\Omega = \Gamma$, it gives the functor (1.24). The construction is as follows. Let $M = \bigoplus_{\alpha \in \Gamma} M_\alpha$ be a Γ-graded A-module. Write

$$M = \bigoplus_{\Omega + \alpha \in \Gamma/\Omega} M_{\Omega+\alpha}, \tag{1.25}$$

where

$$M_{\Omega+\alpha} := \bigoplus_{\omega \in \Omega} M_{\omega+\alpha}. \tag{1.26}$$

One can easily check that M is a Γ/Ω-graded A-module. Moreover,

$$U(M(\alpha)) = M(\Omega + \alpha).$$

We will use this functor to relate the Grothendieck groups of these categories in Examples 3.1.10 and 3.1.11.

In a similar manner we have the following functor

$$(-)_\Omega : \text{Gr}^{\Gamma}\text{-}A \longrightarrow \text{Gr}^{\Omega}\text{-}A_\Omega,$$

where Gr^{Ω}-A is the category of Ω-graded (right) A_Ω-modules, where $\Omega = \ker(f)$.

The above construction motivates the following which will establish a relation between the categories Gr^{Γ}-A and Gr^{Ω}-A_Ω.

Consider the quotient group Γ/Ω and fix a complete set of coset representatives $\{\alpha_i\}_{i\in I}$. Let $\beta \in \Gamma$ and consider the permutation map ρ_β:

$$\begin{aligned}\rho_\beta : \Gamma/\Omega &\longrightarrow \Gamma/\Omega,\\ \Omega + \alpha_i &\longmapsto \Omega + \alpha_i + \beta = \Omega + \alpha_j.\end{aligned}$$

This defines a bijective map (called ρ_β again) $\rho_\beta : \{\alpha_i\}_{i\in I} \to \{\alpha_i\}_{i\in I}$. Moreover, for any α_i, since

$$\Omega + \alpha_i + \beta = \Omega + \alpha_j = \Omega + \rho_\beta(\alpha_i),$$

there is a unique $w_i \in \Omega$ such that

$$\alpha_i + \beta = \omega_i + \rho_\beta(\alpha_i). \tag{1.27}$$

Recall that if $\mathcal{C}$ is an additive category, $\bigoplus_I \mathcal{C}$, where I is a nonempty index set, is defined in the obvious manner, with objects $\bigoplus_{i\in I} M_i$, where M_i are objects of $\mathcal{C}$ and morphisms accordingly.

Define the functor

$$\boxed{\begin{aligned}\mathcal{P} : \mathrm{Gr}^{\Gamma}\text{-}A &\longrightarrow \bigoplus_{\Gamma/\Omega} \mathrm{Gr}^{\Omega}\text{-}A_\Omega,\\ M &\longmapsto \bigoplus_{\Omega+\alpha_i\in\Gamma/\Omega} M_{\Omega+\alpha_i},\end{aligned}} \tag{1.28}$$

where

$$\boxed{M_{\Omega+\alpha_i} = \bigoplus_{\omega\in\Omega} M_{\omega+\alpha_i}.}$$

Since $M_{\Omega+\alpha}$, $\alpha \in \Gamma$, as defined in (1.26), can be naturally considered as an Ω-graded A_Ω-module, where

$$(M_{\Omega+\alpha})_\omega = M_{\omega+\alpha}, \tag{1.29}$$

it follows that the functor $\mathcal{P}$ defined in (1.28) is well-defined. Note that the homogeneous components defined in (1.29) depend on the coset representation, thus choosing another complete set of coset representatives gives a different functor between these categories.

For any $\beta \in \Gamma$, define a shift functor

$$\overline{\rho}_\beta : \bigoplus_{\Gamma/\Omega} \mathrm{Gr}^{\Omega}\text{-}A_\Omega \longrightarrow \bigoplus_{\Gamma/\Omega} \mathrm{Gr}^{\Omega}\text{-}A_\Omega,$$

$$\bigoplus_{\Omega+\alpha_i \in \Gamma/\Omega} M_{\Omega+\alpha_i} \longmapsto \bigoplus_{\Omega+\alpha_i \in \Gamma/\Omega} M(\omega_i)_{\Omega+\rho_\beta(\alpha_i)}, \tag{1.30}$$

where $\rho_\beta(\alpha_i)$ and ω_i are defined in (1.27). The action of $\overline{\rho}_\beta$ on morphisms are defined accordingly. Note that in the left hand side of (1.30) the graded A_Ω-module which appears in $\Omega + \alpha_i$ component is denoted by $M_{\Omega+\alpha_i}$. When M is a Γ-graded module, then $M_{\Omega+\alpha_i}$ has a Ω-structure as described in (1.26).

We are in a position to prove the next theorem.

Theorem 1.2.11 *Let A be a Γ-graded ring and Ω a subgroup of Γ. Then for any $\beta \in \Gamma$, the following diagram is commutative,*

$$\begin{array}{ccc} \mathrm{Gr}^{\Gamma}\text{-}A & \xrightarrow{\ \mathcal{P}\ } & \bigoplus_{\Gamma/\Omega} \mathrm{Gr}^{\Omega}\text{-}A_\Omega \\ \downarrow{\scriptstyle \mathcal{T}_\beta} & & \downarrow{\scriptstyle \overline{\rho}_\beta} \\ \mathrm{Gr}^{\Gamma}\text{-}A & \xrightarrow{\ \mathcal{P}\ } & \bigoplus_{\Gamma/\Omega} \mathrm{Gr}^{\Omega}\text{-}A_\Omega, \end{array} \tag{1.31}$$

where the functors $\mathcal{P}$ and $\overline{\rho}_\beta$ are defined in (1.28) *and* (1.30), *respectively. Moreover, if $\Gamma_A \subseteq \Omega$, then the functor $\mathcal{P}$ induces an equivalence of categories.*

Proof We first show that Diagram (1.31) is commutative. Let $\beta \in \Gamma$ and M be a Γ-graded A-module. As in (1.27), let $\{\alpha_i\}$ be a fixed complete set of coset representative and

$$\alpha_i + \beta = \omega_i + \rho_\beta(\alpha_i).$$

Then

$$\mathcal{P}(\mathcal{T}_\beta(M)) = \mathcal{P}(M(\beta)) = \bigoplus_{\Omega+\alpha_i \in \Gamma/\Omega} M(\beta)_{\Omega+\alpha_i}. \tag{1.32}$$

But

$$M(\beta)_{\Omega+\alpha_i} = \bigoplus_{\omega \in \Omega} M(\beta)_{\omega+\alpha_i} = \bigoplus_{\omega \in \Omega} M_{\omega+\alpha_i+\beta} = \bigoplus_{\omega \in \Omega} M_{\omega+\omega_i+\rho_\beta(\alpha_i)} =$$
$$\bigoplus_{\omega \in \Omega} M(\omega_i)_{\omega+\rho_\beta(\alpha_i)} = M(\omega_i)_{\Omega+\rho_\beta(\alpha_i)}.$$

Replacing this into Equation (1.32) we have

$$\mathcal{P}(\mathcal{T}_\beta(M)) = \bigoplus_{\Omega+\alpha_i \in \Gamma/\Omega} M(\omega_i)_{\Omega+\rho_\beta(\alpha_i)}. \tag{1.33}$$

On the other hand, by (1.30),

$$\overline{\rho}_\beta \mathcal{P}(M) = \overline{\rho}_\beta\Big(\bigoplus_{\Omega+\alpha_i \in \Gamma/\Omega} M_{\Omega+\alpha_i}\Big) = \bigoplus_{\Omega+\alpha_i \in \Gamma/\Omega} M(\omega_i)_{\Omega+\rho_\beta(\alpha_i)}. \tag{1.34}$$

Comparing (1.33) and (1.34) shows that Diagram (1.31) is commutative.

For the last part of the theorem, suppose $\Gamma_A \subseteq \Omega$. We construct a functor

$$\mathcal{P}' : \bigoplus_{\Gamma/\Omega} \text{Gr}^{\Omega}\text{-}A_\Omega \longrightarrow \text{Gr}^{\Gamma}\text{-}A,$$

which depends on the coset representative $\{\alpha_i\}_{i\in I}$ of Γ/Ω. First note that any $\alpha \in \Gamma$ can be written uniquely as $\alpha = \alpha_i + \omega$, for some $i \in I$ and $\omega \in \Omega$. Now let

$$\bigoplus_{\Omega+\alpha_i \in \Gamma/\Omega} N_{\Omega+\alpha_i} \in \bigoplus_{\Gamma/\Omega} \text{Gr}^{\Omega}\text{-}A_\Omega,$$

where $N_{\Omega+\alpha_i}$ is an Ω-graded A_Ω-module. Define a Γ-graded A-module N as follows: $N = \bigoplus_{\alpha\in\Gamma} N_\alpha$, where $N_\alpha := (N_{\Omega+\alpha_i})_\omega$ and $\alpha = \alpha_i + \omega$. We check that N is a Γ-graded A-module, *i.e.*, $N_\alpha A_\gamma \subseteq N_{\alpha+\gamma}$, for $\alpha, \gamma \in \Gamma$. If $\gamma \notin \Omega$, since $\Gamma_A \subseteq \Omega$, $A_\gamma = 0$ and thus $0 = N_\alpha A_\gamma \subseteq N_{\alpha+\gamma}$. Let $\gamma \in \Omega$. Then

$$N_\alpha A_\gamma = (N_{\Omega+\alpha_i})_\omega A_\gamma \subseteq (N_{\Omega+\alpha_i})_{\omega+\gamma} = N_{\alpha+\gamma},$$

as $\alpha + \gamma = \alpha_i + \omega + \gamma$.

We define $\mathcal{P}'(\bigoplus_{\Omega+\alpha_i\in\Gamma/\Omega} N_{\Omega+\alpha_i}) = N$ for the objects and similarly for the morphisms. It is now not difficult to check that $\mathcal{P}'$ is an inverse of the functor $\mathcal{P}$. This finishes the proof. □

The above theorem will be used to compare the graded K-theories with respect to Γ and Ω (see Example 3.1.11).

Corollary 1.2.12 *Let A be a Γ-graded ring concentrated in degree zero. Then*

$$\text{Gr-}A \approx \bigoplus_{\Gamma} \text{Mod-}A.$$

The action of Γ on $\bigoplus_\Gamma$ Mod-A described in (1.30) *reduces to the following: for $\beta \in \Gamma$,*

$$\overline{\rho}_\beta\Big(\bigoplus_{\alpha\in\Gamma} M_\alpha\Big) = \bigoplus_{\alpha\in\Gamma} M(\beta)_\alpha = \bigoplus_{\alpha\in\Gamma} M_{\alpha+\beta}. \tag{1.35}$$

Proof This follows by replacing Ω by a trivial group in Theorem 1.2.11. □

The following corollary, which is a more general case of Corollary 1.2.12 with a similar proof, will be used in the proof of Lemma 6.1.6.

Corollary 1.2.13 *Let A be a $\Gamma \times \Omega$ graded ring which is concentrated in Ω. Then*

$$\mathrm{Gr}^{\Gamma\times\Omega}\text{-}A \cong \bigoplus_{\Gamma} \mathrm{Gr}^{\Omega}\text{-}A.$$

The action of $\Gamma\times\Omega$ on $\bigoplus_{\Gamma} \mathrm{Gr}^{\Omega}\text{-}A$ described in (1.30) *reduces to the following: for $(\beta, \omega) \in \Gamma \times \Omega$,*

$$\overline{\rho}_{(\beta,\omega)}\Big(\bigoplus_{\alpha\in\Gamma} M_\alpha\Big) = \bigoplus_{\alpha\in\Gamma} M_{\alpha+\beta}(\omega).$$

1.2.9 Graded projective modules

Graded projective modules play a crucial role in this book. They will appear in the graded Morita theory in Chapter 2 and will be used to define the graded Grothendieck groups in Chapter 3. Moreover, the graded higher K-theory is constructed from the exact category consisting of graded finitely generated projective modules (see Chapter 6). In this section we define the graded projective modules and give several equivalent criteria for a module to be graded projective. As before, unless stated otherwise, we work in the category of (graded) right modules.

A graded A-module P is called a *graded projective module* if it is a projective object in the abelian category Gr-A. More concretely, P is graded projective if for any diagram of graded modules and graded A-module homomorphisms

$$\begin{array}{ccccc} & & P & & \\ & \overset{h}{\swarrow} & \downarrow{\scriptstyle j} & & \\ M & \xrightarrow[g]{} & N & \longrightarrow & 0, \end{array} \tag{1.36}$$

there is a graded A-module homomorphism $h : P \to M$ with $gh = j$.

In Proposition 1.2.15 we give some equivalent characterisations of graded projective modules, including the one that shows an A-module is graded projective if and only if it is graded and projective as an A-module. By Pgr-A (or Pgr^{Γ}-A to emphasise the grade group of A) we denote a full subcategory of Gr-A, consisting of graded finitely generated projective right A-modules. This is the primary category we are interested in. The graded Grothendieck group (Chapter 3) and higher K-groups (Chapter 6) are constructed from this exact category (see Definition 3.12.1).

We need the following lemma, which says if a graded map factors into two maps, with one being graded, then we can replace the other one with a graded map as well.

Lemma 1.2.14 *Let P, M, N be graded A-modules, with A-module homomorphisms f, g, h*

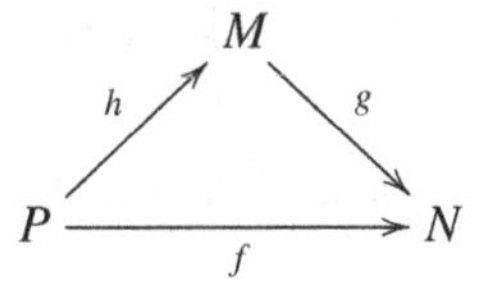

such that $f = gh$, where f is a graded A-module homomorphism. If g (resp. h) is a graded A-homomorphism then there exists a graded homomorphism $h' : P \to M$ (resp. $g' : M \to N$) such that $f = gh'$ (resp. $f = g'h$).

Proof Suppose $g : M \to N$ is a graded A-module homomorphism. Define $h' : P \to M$ as follows: for $p \in P_\alpha$, $\alpha \in \Gamma$, let $h'(p) = h(p)_\alpha$ and extend this linearly to all elements of P, *i.e.*, for $p \in P$ with $p = \sum_{\alpha \in \Gamma} p_\alpha$,

$$h(p) = \sum_{\alpha \in \Gamma} h(p_\alpha)_\alpha.$$

One can easily see that $h' : P \to M$ is a graded A-module homomorphism. Moreover, for $p \in P_\alpha$, $\alpha \in \Gamma$, we have

$$f(p) = gh(p) = g\Big(\sum_{\gamma \in \Gamma} h(p)_\gamma\Big) = \sum_{\gamma \in \Gamma} g(h(p)_\gamma).$$

Since f and g are graded homomorphisms, comparing the degrees of the homogeneous elements of each side of the equation, we get

$$f(p) = g(h(p)_\alpha) = gh'(p).$$

Using the linearity of f, g, h' it follows that $f = gh'$. This proves the lemma for the case g. The other case is similar. □

We are in a position to give equivalent characterisations of graded projective modules.

Proposition 1.2.15 *Let A be a Γ-graded ring and P be a graded A-module. Then the following are equivalent:*

(1) *P is graded and projective;*
(2) *P is graded projective;*
(3) $\operatorname{Hom}_{\text{Gr-}A}(P, -)$ *is an exact functor in* Gr-A;
(4) *every short exact sequence of graded A-module homomorphisms*

$$0 \longrightarrow L \stackrel{f}{\longrightarrow} M \stackrel{g}{\longrightarrow} P \longrightarrow 0$$

splits via a (graded) map;

(5) *P is graded isomorphic to a direct summand of a graded free A-module.*

Proof (1) ⇒ (2) Consider the diagram

$$\begin{array}{ccccc} & & P & & \\ & & \downarrow{\scriptstyle j} & & \\ M & \xrightarrow[g]{} & N & \longrightarrow & 0, \end{array}$$

where M and N are graded modules, g and j are graded homomorphisms and g is surjective. Since P is projective, there is an A-module homomorphism $h : P \to M$ with $gh = j$. By Lemma 1.2.14, there is a graded A-module homomorphism $h' : P \to M$ with $gh' = j$. This gives that P is a graded projective module.

(2) ⇒ (3) In exactly the same way as the nongraded setting, we can show (with no assumption on P) that $\operatorname{Hom}_{\text{Gr-}A}(P, -)$ is left exact (see [50, §IV, Theorem 4.2]). The right exactness follows immediately from the definition of graded projective modules that any diagram of the form (1.36) can be completed.

(3) ⇒ (4) Let

$$0 \longrightarrow L \xrightarrow{f} M \xrightarrow{g} P \longrightarrow 0 \tag{1.37}$$

be a short exact sequence. Since $\operatorname{Hom}_{\text{Gr-}A}(P, -)$ is exact,

$$\begin{aligned} \operatorname{Hom}_{\text{Gr-}A}(P, M) &\longrightarrow \operatorname{Hom}_{\text{Gr-}A}(P, P) \\ h &\longmapsto gh \end{aligned}$$

is an epimorphism. In particular, there is a graded homomorphism h such that $gh = 1$, *i.e.*, the short exact sequence (1.37) is spilt.

(4) ⇒ (5) First note that P is a homomorphic image of a graded free A-module as follows: Let $\{p_i\}_{i \in I}$ be a homogeneous generating set for P, where $\deg(p_i) = \delta_i$. Let $\bigoplus_{i \in I} A(-\delta_i)$ be the graded free A-module with standard homogeneous basis $\{e_i\}_{i \in I}$ where $\deg(e_i) = \delta_i$. Then there is an exact sequence

$$0 \longrightarrow \ker(g) \xrightarrow{\subseteq} \bigoplus_{i \in I} A(-\delta_i) \xrightarrow{g} P \longrightarrow 0, \tag{1.38}$$

as the map

$$\begin{aligned} g : \bigoplus_{i \in I} A(-\delta_i) &\longrightarrow P, \\ e_i &\longmapsto p_i, \end{aligned}$$

is a surjective graded A-module homomorphism. By the assumption, there is a A-module homomorphism $h : P \to \bigoplus_{i\in I} A(-\delta_i)$ such that $gh = \mathrm{id}_P$. By Lemma 1.2.14 one can assume h is a graded homomorphism.

Since the exact sequence (1.38) is, in particular, a split exact sequence of A-modules, we know from the nongraded setting [67, Proposition 2.5] that there is an A-module isomorphism

$$\begin{aligned} \theta : P \oplus \ker(g) &\longrightarrow \bigoplus_{i\in I} A(-\delta_i) \\ (p, q) &\longmapsto h(p) + q. \end{aligned}$$

Clearly this map is also a graded A-module homomorphism, so

$$P \oplus \ker(g) \cong_{\mathrm{gr}} \bigoplus_{i\in I} A(-\delta_i).$$

(5) $\Rightarrow$ (1) Graded free modules are free, so P is isomorphic to a direct summand of a free A-module. From the nongraded setting, we know that P is projective. □

The proof of Proposition 1.2.15 (see in particular (4) $\Rightarrow$ (5) and (5) $\Rightarrow$ (1)) shows that a graded A-module P is a graded finitely generated projective A-module if and only if

$$\boxed{P \oplus Q \cong_{\mathrm{gr}} A^n(\overline{\alpha})} \tag{1.39}$$

for some $\overline{\alpha} = (\alpha_1, \dots, \alpha_n)$, $\alpha_i \in \Gamma$. This fact will be used frequently throughout this book.

Remark 1.2.16 Recall the functor $\mathcal{P}$ from (1.28). It is easy to see that if M is a Γ-graded projective A-module, then $M_{\Omega+\alpha}$ is a Ω-graded projective A_Ω-graded module. Thus the functor $\mathcal{P}$ restricts to

$$\boxed{\begin{aligned} \mathcal{P} : \mathrm{Pgr}^{\Gamma}\text{-}A &\longrightarrow \bigoplus_{\Gamma/\Omega} \mathrm{Pgr}^{\Omega}\text{-}A_\Omega, \\ M &\longmapsto \bigoplus_{\Omega+\alpha_i\in\Gamma/\Omega} M_{\Omega+\alpha_i}. \end{aligned}} \tag{1.40}$$

This will be used later in Examples 3.1.5, 3.1.11 and Lemma 6.1.6.

Theorem 1.2.17 (The dual basis lemma) *Let A be a Γ-graded ring and P be a graded A-module. Then P is graded projective if and only if there exists $p_i \in P^h$ with $\deg(p_i) = \delta_i$ and $f_i \in \mathrm{Hom}_{\mathrm{Gr}\text{-}A}(P, A(-\delta_i))$, for some indexing set I, such that*

(1) *for every $p \in P$, $f_i(p) = 0$ for all but a finite subset of $i \in I$,*
(2) *for every $p \in P$, $\sum_{i\in I} f_i(p)p_i = p$.*

Proof Since P is graded projective, by Proposition 1.2.15(5), there is a graded module Q such that $P \oplus Q \cong_{\mathrm{gr}} \bigoplus_i A(-\delta_i)$. This gives two graded maps

$$\phi : P \to \bigoplus_i A(-\delta_i) \text{ and } \pi : \bigoplus_i A(-\delta_i) \to P,$$

such that $\pi\phi = 1_P$. Let

$$\pi_i : \bigoplus_i A(-\delta_i) \longrightarrow A(-\delta_i),$$
$$\{a_i\}_{i\in I} \longmapsto a_i$$

be the projection on the ith component. So if

$$a = \{a_i\}_{i\in I} \in \bigoplus_i A(-\delta_i),$$

then

$$\sum_i \pi_i(a)e_i = a,$$

where $\{e_i\}_{i\in I}$ is the standard homogeneous basis of $\bigoplus_i A(-\delta_i)$. Now let $p_i = \pi(e_i)$ and $f_i = \pi_i\phi$. Note that $\deg(p_i) = \delta_i$ and

$$f_i \in \mathrm{Hom}_{\mathrm{Gr}\text{-}A}(P, A(-\delta_i)).$$

Clearly $f_i(p) = \pi_i\phi(p)$ is zero for all but a finite number of $i \in I$. This gives (1). Moreover,

$$\sum_i p_i f_i(p) = \sum_i p_i \pi_i\phi(p) = \sum \pi(e_i)\pi_i\phi(p) = \pi(\sum_i e_i\pi_i\phi(p)) = \pi\phi(p) = p.$$

This gives (2).

Conversely, suppose that there exists a dual basis $\{p_i, f_i \mid i \in I\}$. Consider the maps

$$\phi : P \longrightarrow \bigoplus_i A(-\delta_i),$$
$$p \longmapsto \{f_i(p)\}_{i\in I}$$

and

$$\pi : \bigoplus_i A(-\delta_i) \longrightarrow P,$$
$$\{a_i\}_{i\in I} \longmapsto \sum_i p_i a_i.$$

One sees easily that ϕ and π are graded right A-module homomorphisms and $\pi\phi = 1_P$. Therefore the exact sequence

$$0 \longrightarrow \ker(\pi) \longrightarrow \bigoplus_i A(-\delta_i) \stackrel{\pi}{\longrightarrow} P \longrightarrow 0$$

splits. Thus P is a direct summand of the graded free module $\bigoplus_i A(-\delta_i)$. By Proposition 1.2.15, P is a graded projective. □

Remark 1.2.18 GRADED INJECTIVE MODULES

Proposition 1.2.15 shows that an A-module P is graded projective if and only if P is graded and projective. However, the similar statement is not valid for graded injective modules. A graded A-module I is called a *graded injective module* if for any diagram of graded modules and graded A-module homomorphisms

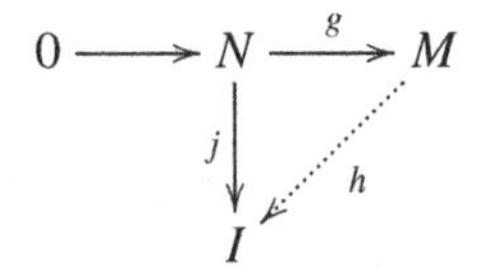

there is a graded A-module homomorphism $h : M \to I$ with $hg = j$.

Using Lemma 1.2.14 one can show that if a graded module is injective, then it is also graded injective. However, a graded injective module is not necessarily injective. The reason for this difference between projective and injective behaviour is that the forgetful functor U : Gr-$A \to$ Mod-A is a left adjoint functor (see Remark 1.2.7). In detail, consider a graded projective module P and the diagram

$$\begin{array}{ccccc} & & P & & \\ & & \downarrow{\scriptstyle j} & & \\ M & \xrightarrow[g]{} & N & \longrightarrow & 0, \end{array}$$

where M and N are A-modules. Since the diagram below is commutative

$$\begin{array}{ccc} \operatorname{Hom}_{\text{Mod-}A}(U(P), M) & \xrightarrow{\cong} & \operatorname{Hom}_{\text{Gr-}A}(P, F(M)) \\ \downarrow & & \downarrow \\ \operatorname{Hom}_{\text{Mod-}A}(U(P), N) & \xrightarrow{\cong} & \operatorname{Hom}_{\text{Gr-}A}(P, F(N)) \end{array}$$

and there is a graded homomorphism $h' : P \to F(M)$ such that the diagram

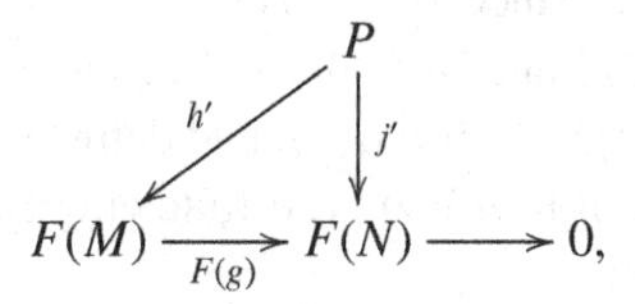

is commutative, there is a homomorphism $h : P \to M$ such that $gh = j$. So P is projective (see Proposition 1.2.15 for another proof).

If the grade group is finite, then the forgetful functor is right adjoint as well (see Remark 1.2.7) and a similar argument as above shows that a graded injective module is injective.

1.2.10 Graded divisible modules

Here we define the notion of graded divisible modules and we give yet another characterisation of graded von Neumann regular rings (see §1.1.9).

Let A be a Γ-graded ring and M a graded right A-module. We say $m \in M^h$ (a homogeneous element of M) is divisible by $a \in A^h$ if $m \in Ma$, *i.e.,* there is a homogeneous element $n \in M$ such that $m = na$. We say that M is a *graded divisible module* if for any $m \in M^h$ and any $a \in A^h$, where $\operatorname{ann}_r(a) \subseteq \operatorname{ann}(m)$, we have that m is divisible by a. Note that for $m \in M^h$, *the annihilator* of m,

$$\operatorname{ann}(m) := \{ a \in A \mid ma = 0 \}$$

is a graded ideal of A.

Proposition 1.2.19 *Let A be a Γ-graded ring and M a graded right A-module. Then the following are equivalent:*

(1) *M is a graded divisible module;*
(2) *for any $a \in A^h$, $\gamma \in \Gamma$, and any graded A-module homomorphism $f : aA(\gamma) \to M$, the following diagram can be completed:*

$$\begin{array}{ccccc} 0 & \longrightarrow & aA(\gamma) & \xrightarrow{\subseteq} & A(\gamma) \\ & & \downarrow{\scriptstyle f} & \swarrow{\scriptstyle \overline{f}} & \\ & & M. & & \end{array}$$

Proof (1) $\Rightarrow$ (2) Let $f : aA(\gamma) \to M$ be a graded A-homomorphism. Set $m := f(a)$. Note that $\deg(m) = \alpha - \gamma$, where $\deg(a) = \alpha$ (see Example 1.2.1). If $x \in \operatorname{ann}_r(a)$ then $0 = f(ax) = f(a)x = mx$, thus $x \in \operatorname{ann}(m)$. Since M is graded divisible, m is divisible by a, *i.e.,* $m = na$ for some $n \in M$. It follows that $\deg(n) = -\gamma$. Define $\overline{f} : A(\gamma) \to M$ by $\overline{f}(1) = n$ and extend it to $A(\gamma)$. Thisis a graded A-module. Since $\overline{f}(a) = \overline{f}(1)a = na = m$, $\overline{f}$ extends f.

(2) $\Rightarrow$ (1) Let $m \in M^h$ and $a \in A^h$, where $\operatorname{ann}_r(a) \subseteq \operatorname{ann}(m)$. Suppose $\deg(a) = \alpha$ and $\deg(m) = \beta$. Let $\gamma = \alpha - \beta$ and define the map $f : aA(\gamma) \to M$, by $f(a) = m$ and extend it to $aA(\gamma)$. The following shows that f is in fact a graded A-homomorphism:

$$f\big((aA(\gamma))_\delta\big) \subseteq f((aA)_{\gamma+\delta}) \subseteq f(aA_{\gamma+\delta-\alpha}) = f(aA_{\delta-\beta}) = mA_{\delta-\beta} \subseteq M_\delta.$$

Thus there is an $\overline{f} : A(\gamma) \to M$ which extends f. So $m = f(a) = \overline{f}(a) = \overline{f}(1)a = na$, where $f(1) = n$. This means m is divisible by a and the proof is complete. □

Theorem 1.2.20 *Let A be a Γ-graded ring. Then A is graded von Neumann regular if and only if any graded right A-module is divisible.*

Proof Let A be a graded regular ring. Consider the exact sequence of graded right A-modules

$$0 \longrightarrow aA(\gamma) \overset{\subseteq}{\longrightarrow} A(\gamma) \longrightarrow A(\gamma)/aA(\gamma) \longrightarrow 0.$$

Define

$$f : A(\gamma) \longrightarrow aA(\gamma),$$
$$x \longmapsto abx.$$

Since $\deg(ab) = 0$ this gives a split graded homomorphism for the exact sequence above. Thus $aA(\gamma)$ is a direct summand of $A(\gamma)$. This shows that any graded A-module homomorphism $f : aA(\gamma) \to M$ can be extend to $A(\gamma)$.

Conversely, consider the diagram

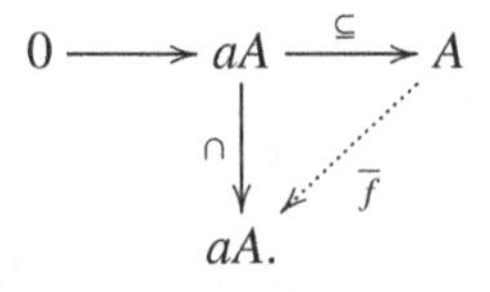

Since aA is divisible, there is an $\overline{f}$ which completes this diagram. Set $\overline{f}(1) = ab$, where $b \in A^h$. We then have $a = \overline{f}(a) = \overline{f}(1)a = aba$. Thus A is a graded regular ring. □

Example 1.2.21 Graded rings associated with filter rings

A ring A with identity is called a *filtered ring* if there is an ascending family $\{A_i \mid i \in \mathbb{Z}\}$ of additive subgroups of A such that $1 \in A_0$ and $A_iA_j \subseteq A_{i+j}$, for all

$i, j \in \mathbb{Z}$. Let A be a filtered ring and M be a right A-module. M is called a *filtered module* if there is an ascending family $\{M_i \mid i \in \mathbb{Z}\}$ of additive subgroups of M such that $M_i A_j \subseteq M_{i+j}$, for all $i, j \in \mathbb{Z}$. An A-module homomorphism $f : M \to N$ of filtered modules M and N is called a *filtered homomorphism* if $f(M_i) \subseteq N_i$ for $i \in \mathbb{Z}$. A category consisting of filtered right A-modules for objects and filtered homomorphisms as morphisms is denoted by Filt-A. If M is a filtered A-module then

$$\operatorname{gr}(M) := \bigoplus_{i \in \mathbb{Z}} M_i/M_{i-1}$$

is a $\mathbb{Z}$-graded $\operatorname{gr}(A) := \bigoplus_{i \in \mathbb{Z}} A_i/A_{i-1}$-module. The operations here are defined naturally. This gives a functor $\operatorname{gr} : \text{Filt-}A \to \text{Gr-gr}(A)$. In Example 1.4.7 we use a variation of this construction to associate a graded division algebra with a valued division algebra.

In the theory of filtered rings, one defines the concepts of filtered free and projective modules and under certain conditions the functor gr sends these objects to the corresponding objects in the category Gr-gr(A). For a comprehensive study of filtered rings see [73].

1.3 Grading on matrices

Let A be an arbitrary ring and Γ an arbitrary group. Then one can consider Γ-gradings on the matrix ring $\mathbb{M}_n(A)$ which, at first glance, might look somewhat artificial. However, these types of gradings on matrices appear quite naturally in the graded rings arising from graphs. In this section we study the grading on matrices. We then include a section on graph algebras (including path algebras and Leavitt path algebras, §1.6). These algebras give us a wealth of examples of graded rings and graded matrix rings.

For a free right A-module V of dimension n, there is a ring isomorphism $\operatorname{End}_A(V) \cong \mathbb{M}_n(A)$. When A is a Γ-graded ring and V is a graded free module of finite rank, by Theorem 1.2.6, $\operatorname{End}_A(V)$ has a natural Γ-grading. This induces a graded structure on the matrix ring $\mathbb{M}_n(A)$. In this section we study this grading on matrices. For an n-tuple $(\delta_1, \dots, \delta_n)$, $\delta_i \in \Gamma$, we construct a grading on the matrix ring $\mathbb{M}_n(A)$, denoted by $\mathbb{M}_n(A)(\delta_1, \dots, \delta_n)$, and we show that

$$\operatorname{End}_A \left(A(-\delta_1) \oplus A(-\delta_2) \oplus \cdots \oplus A(-\delta_n)\right) \cong_{\operatorname{gr}} \mathbb{M}_n(A)(\delta_1, \dots, \delta_n).$$

We will see that these graded structures on matrices appear very naturally when studying the graded structure of path algebras in §1.6.

1.3.1 Graded calculus on matrices

Let A be a Γ-graded ring and let $M = M_1 \oplus \cdots \oplus M_n$, where M_i are graded finitely generated right A-modules. Then M is also a graded right A-module (see §1.2.6). Let

$$\big(\operatorname{Hom}(M_j, M_i)\big)_{1\le i,j\le n} := \begin{pmatrix} \operatorname{Hom}_A(M_1,M_1) & \operatorname{Hom}_A(M_2,M_1) & \cdots & \operatorname{Hom}_A(M_n,M_1) \\ \operatorname{Hom}_A(M_1,M_2) & \operatorname{Hom}_A(M_2,M_2) & \cdots & \operatorname{Hom}_A(M_n,M_2) \\ \vdots & \vdots & \ddots & \vdots \\ \operatorname{Hom}_A(M_1,M_n) & \operatorname{Hom}_A(M_2,M_n) & \cdots & \operatorname{Hom}_A(M_n,M_n) \end{pmatrix}.$$

It is easy to observe that $\big(\operatorname{Hom}(M_j, M_i)\big)_{1\le i,j\le n}$ forms a ring with component-wise addition and matrix multiplication. Moreover, for $\lambda \in \Gamma$, assigning the additive subgroup

$$\begin{pmatrix} \operatorname{Hom}_A(M_1,M_1)_\lambda & \operatorname{Hom}_A(M_2,M_1)_\lambda & \cdots & \operatorname{Hom}_A(M_n,M_1)_\lambda \\ \operatorname{Hom}_A(M_1,M_2)_\lambda & \operatorname{Hom}_A(M_2,M_2)_\lambda & \cdots & \operatorname{Hom}_A(M_n,M_2)_\lambda \\ \vdots & \vdots & \ddots & \vdots \\ \operatorname{Hom}_A(M_1,M_n)_\lambda & \operatorname{Hom}_A(M_2,M_n)_\lambda & \cdots & \operatorname{Hom}_A(M_n,M_n)_\lambda \end{pmatrix} \tag{1.41}$$

as a λ-homogeneous component of $\big(\operatorname{Hom}(M_j, M_i)\big)_{1\le i,j\le n}$, using Theorem 1.2.6 and (1.14), it follows that $\big(\operatorname{Hom}(M_j, M_i)\big)_{1\le i,j\le n}$ is a Γ-graded ring.

Let $\pi_j : M \to M_j$ and $\kappa_j : M_j \to M$ be the (graded) projection and injection homomorphisms. For the next theorem, we need the following identities:

$$\sum_{i=1}^{n} \kappa_i \pi_i = \mathrm{id}_M \quad \text{and} \quad \pi_i \kappa_j = \delta_{ij}\, \mathrm{id}_{M_j}, \tag{1.42}$$

where δ_{ij} is the Kronecker delta.

Theorem 1.3.1 *Let A be a Γ-graded ring and $M = M_1 \oplus \cdots \oplus M_n$, where M_i are graded finitely generated right A-modules. Then there is a graded ring isomorphism*

$$\Phi : \operatorname{End}_A(M) \longrightarrow \big(\operatorname{Hom}(M_j, M_i)\big)_{1\le i,j\le n}$$

defined by $f \mapsto (\pi_i f \kappa_j)$, $1 \le i, j \le n$.

Proof The map Φ is clearly well-defined. Since for $f, g \in \operatorname{End}_A(M)$,

$$\begin{aligned}\Phi(f+g) = (\pi_i(f+g)\kappa_j)_{1\le i,j\le n} &= (\pi_i f \kappa_j + \pi_i g \kappa_j)_{1\le i,j\le n} \\ &= (\pi_i f \kappa_j)_{1\le i,j\le n} + (\pi_i g \kappa_j)_{1\le i,j\le n} = \Phi(f) + \Phi(g)\end{aligned}$$

and

$$\Phi(fg) = (\pi_i f g \kappa_j)_{1\leq i,j\leq n} = \Big(\pi_i f(\sum_{l=1}^{n} \kappa_l \pi_l) g \kappa_j\Big)_{1\leq i,j\leq n}$$

$$= \Big(\sum_{l=1}^{n} (\pi_i f \kappa_l)(\pi_l g \kappa_j)\Big)_{1\leq i,j\leq n} = \Phi(f)\Phi(g),$$

Φ is a ring homomorphism. Moreover, if $f \in \operatorname{End}_A(M)_\lambda$, $\lambda \in \Gamma$, then

$$\pi_i f \kappa_j \in \operatorname{Hom}_A(M_j, M_i)_\lambda,$$

for $1 \leq i, j \leq n$. This (see (1.41)) shows that Φ is a graded ring homomorphism. Define the map

$$\Psi : \big(\operatorname{Hom}(M_j, M_i)\big)_{1\leq i,j\leq n} \longrightarrow \operatorname{End}_A(M),$$

$$(g_{ij})_{1\leq i,j\leq n} \longmapsto \sum_{1\leq i,j\leq n} \kappa_i g_{ij} \pi_j.$$

Using the identities (1.42), one can observe that the compositions $\Psi\Phi$ and $\Phi\Psi$ give the identity maps of the corresponding rings. Thus Φ is an isomorphism. □

For a graded ring A, consider $A(\delta_i)$, $1 \leq i \leq n$, as graded right A-modules and observe that

$$\Phi_{\delta_j,\delta_i} : \operatorname{Hom}_A\big(A(\delta_i), A(\delta_j)\big) \cong_{\mathrm{gr}} A(\delta_j - \delta_i), \tag{1.43}$$

as graded left A-modules such that

$$\Phi_{\delta_k,\delta_i}(gf) = \Phi_{\delta_k,\delta_j}(g)\Phi_{\delta_j,\delta_i}(f),$$

where $f \in \operatorname{Hom}(A(\delta_i), A(\delta_j))$ and $g \in \operatorname{Hom}(A(\delta_j), A(\delta_k))$ (see (1.17)). If

$$V = A(-\delta_1) \oplus A(-\delta_2) \oplus \cdots \oplus A(-\delta_n),$$

then by Theorem 1.3.1,

$$\operatorname{End}_A(V) \cong_{\mathrm{gr}} \Big(\operatorname{Hom}\big(A(-\delta_j), A(-\delta_i)\big)\Big)_{1\leq i,j\leq n}.$$

Applying Φ_{δ_j,δ_i} defined in (1.43) to each entry, we have

$$\operatorname{End}_A(V) \cong_{\mathrm{gr}} \Big(\operatorname{Hom}\big(A(-\delta_j), A(-\delta_i)\big)\Big)_{1\leq i,j\leq n} \cong_{\mathrm{gr}} \big(A(\delta_j - \delta_i)\big)_{1\leq i,j\leq n}.$$

Denoting the graded matrix ring $(A(\delta_j - \delta_i))_{1\leq i,j\leq n}$ by $\mathbb{M}_n(A)(\delta_1, \ldots, \delta_n)$, we have

$$\mathbb{M}_n(A)(\delta_1,\dots,\delta_n)=\begin{pmatrix} A(\delta_1-\delta_1) & A(\delta_2-\delta_1) & \cdots & A(\delta_n-\delta_1)\\ A(\delta_1-\delta_2) & A(\delta_2-\delta_2) & \cdots & A(\delta_n-\delta_2)\\ \vdots & \vdots & \ddots & \vdots\\ A(\delta_1-\delta_n) & A(\delta_2-\delta_n) & \cdots & A(\delta_n-\delta_n)\end{pmatrix}. \tag{1.44}$$

By (1.41), $\mathbb{M}_n(A)(\delta_1,\dots,\delta_n)_\lambda$, the λ-homogeneous elements, are the $n \times n$-matrices over A with the degree shifted (suspended) as follows:

$$\boxed{\mathbb{M}_n(A)(\delta_1,\dots,\delta_n)_\lambda=\begin{pmatrix} A_{\lambda+\delta_1-\delta_1} & A_{\lambda+\delta_2-\delta_1} & \cdots & A_{\lambda+\delta_n-\delta_1}\\ A_{\lambda+\delta_1-\delta_2} & A_{\lambda+\delta_2-\delta_2} & \cdots & A_{\lambda+\delta_n-\delta_2}\\ \vdots & \vdots & \ddots & \vdots\\ A_{\lambda+\delta_1-\delta_n} & A_{\lambda+\delta_2-\delta_n} & \cdots & A_{\lambda+\delta_n-\delta_n}\end{pmatrix}.} \tag{1.45}$$

This also shows that for $x \in A^h$,

$$\deg(\mathbf{e}_{ij}(x)) = \deg(x) + \delta_i - \delta_j, \tag{1.46}$$

where $\mathbf{e}_{ij}(x)$ is a matrix with x in the ij-position and zero elsewhere.

In particular the zero homogeneous component (which is a ring) is of the form

$$\boxed{\mathbb{M}_n(A)(\delta_1,\dots,\delta_n)_0=\begin{pmatrix} A_0 & A_{\delta_2-\delta_1} & \cdots & A_{\delta_n-\delta_1}\\ A_{\delta_1-\delta_2} & A_0 & \cdots & A_{\delta_n-\delta_2}\\ \vdots & \vdots & \ddots & \vdots\\ A_{\delta_1-\delta_n} & A_{\delta_2-\delta_n} & \cdots & A_0\end{pmatrix}.} \tag{1.47}$$

Setting $\overline{\delta} = (\delta_1,\dots,\delta_n) \in \Gamma^n$, one denotes the graded matrix ring (1.44) by $\mathbb{M}_n(A)(\overline{\delta})$. To summarise, we have shown that there is a graded ring isomorphism

$$\boxed{\operatorname{End}_A\big(A(-\delta_1)\oplus A(-\delta_2)\oplus\cdots\oplus A(-\delta_n)\big) \cong_{\mathrm{gr}} \mathbb{M}_n(A)(\delta_1,\dots,\delta_n).} \tag{1.48}$$

Remark 1.3.2 MATRIX RINGS OF A NONABELIAN GROUP GRADING

If the grade group Γ is nonabelian, the homogeneous components of the matrix ring take the following form:

$$\mathbb{M}_n(A)(\delta_1\dots,\delta_n)_\lambda=\begin{pmatrix} A_{\delta_1\lambda\delta_1^{-1}} & A_{\delta_1\lambda\delta_2^{-1}} & \cdots & A_{\delta_1\lambda\delta_n^{-1}}\\ A_{\delta_2\lambda\delta_1^{-1}} & A_{\delta_2\lambda\delta_2^{-1}} & \cdots & A_{\delta_2\lambda\delta_n^{-1}}\\ \vdots & \vdots & \ddots & \vdots\\ A_{\delta_n\lambda\delta_1^{-1}} & A_{\delta_n\lambda\delta_2^{-1}} & \cdots & A_{\delta_n\lambda\delta_n^{-1}}\end{pmatrix}.$$

Consider the graded A-bimodule $A^n(\overline{\delta}) = A(\delta_1) \oplus \cdots \oplus A(\delta_n)$. Then one can check that $A^n(\overline{\delta})$ is a graded right $\mathbb{M}_n(A)(\overline{\delta})$-module and $A^n(-\overline{\delta})$ is a graded left $\mathbb{M}_n(A)(\overline{\delta})$-module, where $-\overline{\delta} = (-\delta_1, \dots, -\delta_n)$. These will be used in the graded Morita theory (see Proposition 2.1.1).

One can easily check the graded ring $R = \mathbb{M}_n(A)(\overline{\delta})$, where $\overline{\delta} = (\delta_1, \dots \delta_n)$, $\delta_i \in \Gamma$, has the support

$$\Gamma_R = \bigcup_{1 \leq i,j \leq n} \Gamma_A + \delta_i - \delta_j. \tag{1.49}$$

One can rearrange the shift, without changing the graded matrix ring, as the following theorem shows (see also [75, pp. 60–61]).

Theorem 1.3.3 *Let A be a Γ-graded ring and $\delta_i \in \Gamma$, $1 \leq i \leq n$.*

(1) *If $\alpha \in \Gamma$, and $\pi \in S_n$ is a permutation then*

$$\mathbb{M}_n(A)(\delta_1, \dots, \delta_n) \cong_{\mathrm{gr}} \mathbb{M}_n(A)(\delta_{\pi(1)} + \alpha, \dots, \delta_{\pi(n)} + \alpha). \tag{1.50}$$

(2) *If $\tau_1, \dots, \tau_n \in \Gamma_A^*$, then*

$$\mathbb{M}_n(A)(\delta_1, \dots, \delta_n) \cong_{\mathrm{gr}} \mathbb{M}_n(A)(\delta_1 + \tau_1, \dots, \delta_n + \tau_n). \tag{1.51}$$

Proof (1) Let V be a graded free module over A with a homogeneous basis $v_1, \dots, v_n$ of degree $\lambda_1, \dots, \lambda_n$, respectively. It is easy to see that ((1.20))

$$V \cong_{\mathrm{gr}} A(-\lambda_1) \oplus \cdots \oplus A(-\lambda_n),$$

and thus $\mathrm{End}_A(V) \cong_{\mathrm{gr}} \mathbb{M}_n(A)(\lambda_1, \dots, \lambda_n)$ (see (1.48)). Now let $\pi \in S_n$. Rearranging the homogeneous basis as $v_{\pi(1)}, \dots, v_{\pi(n)}$ and defining the A-graded isomorphism $\phi : V \to V$ by $\phi(v_i) = v_{\pi^{-1}(i)}$, we get a graded isomorphism in the level of endomorphism rings, called ϕ again

$$\mathbb{M}_n(A)(\lambda_1, \dots, \lambda_n) \cong_{\mathrm{gr}} \mathrm{End}_A(V) \xrightarrow{\phi} \mathrm{End}_A(V) \cong_{\mathrm{gr}} \mathbb{M}_n(A)(\lambda_{\pi(1)}, \dots, \lambda_{\pi(n)}). \tag{1.52}$$

Moreover, (1.45) shows that it does not make any difference adding a fixed $\alpha \in \Gamma$ to each of the entries in the shift. This gives us (1.50).

In fact, the isomorphism ϕ in (1.52) is defined as $\phi(M) = P_\pi M P_\pi^{-1}$, where P_π is the $n \times n$ permutation matrix with entries at $(i, \pi(i))$, $1 \leq i \leq n$, being 1 and zero elsewhere.

(2) For (1.51), let $\tau_i \in \Gamma_A^*$, $1 \leq i \leq n$, that is, $\tau_i = \deg(u_i)$ for some units $u_i \in A^h$. Consider the basis $v_i u_i$, $1 \leq i \leq n$ for V. With this basis,

$$\mathrm{End}_A(V) \cong_{\mathrm{gr}} \mathbb{M}_n(A)(\delta_1 + \tau_1, \dots, \delta_n + \tau_n).$$

Consider the A-graded isomorphism $\mathrm{id} : V \to V$, by $\mathrm{id}(v_i) = (v_i u_i) u_i^{-1}$. A similar argument as Part (1) now gives (1.51). The isomorphism is given by $\phi(M) = P^{-1}MP$, where $P = D[u_1, \dots, u_n]$ is the diagonal matrix. □

Note that if A has a trivial Γ-grading, *i.e.*, $A = \bigoplus_{\gamma\in\Gamma} A_\gamma$, where $A_0 = A$ and $A_\gamma = 0$, for $0 \neq \gamma \in \Gamma$, this construction induces a *good grading* on $\mathbb{M}_n(A)$. By definition, this is a grading on $\mathbb{M}_n(A)$ such that the matrix unit $\mathbf{e}_{ij}$, the matrix with 1 in the ij-position and zero everywhere else, is homogeneous, for $1 \leq i, j \leq n$. This particular group gradings on matrix rings have been studied by Dăscălescu et al. [34] (see Remark 1.3.9). Therefore, for $x \in A$,

$$\deg(\mathbf{e}_{ij}(x)) = \delta_i - \delta_j. \tag{1.53}$$

One can easily check that for a ring A with trivial Γ-grading, the graded ring $\mathbb{M}_n(A)(\overline{\delta})$, where $\overline{\delta} = (\delta_1, \dots \delta_n)$, $\delta_i \in \Gamma$, has the support $\{\, \delta_i - \delta_j \mid 1 \leq i, j \leq n \,\}$. (This follows also immediately from (1.49).)

The grading on matrices appears quite naturally in the graded rings arising from graphs. We will show that the graded structure Leavitt path algebras of acyclic and comet graphs are in effect the graded matrix rings as constructed above (see §1.6, in particular, Theorems 1.6.19 and 1.6.21).

Example 1.3.4 Let A be a ring, Γ a group and A graded trivially by Γ, *i.e.*, A is concentrated in degree zero (see §1.1.1). Consider the Γ-graded matrix ring

$$R = \mathbb{M}_n(A)(0, -1, \dots, -n+1),$$

where $n \in \mathbb{N}$. By (1.49) the support of R is the set $\{-n+1, -n+2, \dots, n-2, n-1\}$. By (1.45), for $k \in \mathbb{Z}$ we have the following arrangements for the homogeneous elements of R:

$$R_k = \begin{pmatrix} A_k & A_{k-1} & \dots & A_{k+1-n} \\ A_{k+1} & A_k & \dots & A_{k+2-n} \\ \vdots & \vdots & \ddots & \vdots \\ A_{k+n-1} & A_{k+n-2} & \dots & A_k \end{pmatrix}.$$

Thus the 0-component ring is

$$R_0 = \begin{pmatrix} A & 0 & \dots & 0 \\ 0 & A & \dots & 0 \\ \vdots & \vdots & \ddots & \vdots \\ 0 & 0 & \dots & A \end{pmatrix}$$

and

$$R_{-1} = \begin{pmatrix} 0 & 0 & \dots & 0 \\ A & 0 & \dots & 0 \\ \vdots & \ddots & \ddots & \vdots \\ 0 & \dots & A & 0 \end{pmatrix}, \dots, R_{-n+1} = \begin{pmatrix} 0 & 0 & \dots & 0 \\ 0 & 0 & \dots & 0 \\ \vdots & \vdots & \ddots & \vdots \\ A & 0 & \dots & 0 \end{pmatrix}$$

$$R_{1} = \begin{pmatrix} 0 & A & \dots & 0 \\ 0 & 0 & \ddots & \vdots \\ \vdots & \vdots & \ddots & A \\ 0 & 0 & \dots & 0 \end{pmatrix}, \dots, \quad R_{n-1} = \begin{pmatrix} 0 & 0 & \dots & A \\ 0 & 0 & \dots & 0 \\ \vdots & \vdots & \ddots & \vdots \\ 0 & 0 & \dots & 0 \end{pmatrix}.$$

In Chapter 2, we will see that R is graded Morita equivalent to the trivially graded ring A.

Example 1.3.5 Let S be a ring, $S[x, x^{-1}]$ the $\mathbb{Z}$-graded Laurent polynomial ring and $A = S[x^3, x^{-3}]$ the $\mathbb{Z}$-graded subring with support $3\mathbb{Z}$ (see Example 1.1.19). Consider the $\mathbb{Z}$-graded matrix ring

$$\mathbb{M}_6(A)(0, 1, 1, 2, 2, 3).$$

By (1.45), the homogeneous elements of degree 1 have the form

$$\begin{pmatrix} A_1 & A_0 & A_0 & A_{-1} & A_{-1} & A_{-2} \\ A_2 & A_1 & A_1 & A_0 & A_0 & A_{-1} \\ A_2 & A_1 & A_1 & A_0 & A_0 & A_{-1} \\ A_3 & A_2 & A_2 & A_1 & A_1 & A_0 \\ A_3 & A_2 & A_2 & A_1 & A_1 & A_0 \\ A_4 & A_3 & A_3 & A_2 & A_2 & A_1 \end{pmatrix} = \begin{pmatrix} 0 & S & S & 0 & 0 & 0 \\ 0 & 0 & 0 & S & S & 0 \\ 0 & 0 & 0 & S & S & 0 \\ Sx^3 & 0 & 0 & 0 & 0 & S \\ Sx^3 & 0 & 0 & 0 & 0 & S \\ 0 & Sx^3 & Sx^3 & 0 & 0 & 0 \end{pmatrix}.$$

Example 1.3.6 Let K be a field. Consider the $\mathbb{Z}$-graded ring

$$R = \mathbb{M}_5(K)(0, 1, 2, 2, 3).$$

Then the support of this ring is $\{0, \pm 1, \pm 2\}$ and by (1.47) the zero homogeneous component (which is a ring) is

$$R_0 = \begin{pmatrix} K & 0 & 0 & 0 & 0 \\ 0 & K & 0 & 0 & 0 \\ 0 & 0 & K & K & 0 \\ 0 & 0 & K & K & 0 \\ 0 & 0 & 0 & 0 & K \end{pmatrix} \cong K \oplus K \oplus \mathbb{M}_2(K) \oplus K.$$

Example 1.3.7 <u>$\mathbb{M}_n(A)(\delta_1, \dots, \delta_n)$ WITH $\Gamma = \{\delta_1, \dots, \delta_n\}$ IS A SKEW GROUP RING</u>

Let A be a Γ-graded ring, where $\Gamma = \{\delta_1, \dots, \delta_n\}$ is a finite group. Consider $\mathbb{M}_n(A)(\delta_1, \dots, \delta_n)$, which is a Γ-graded ring with its homogeneous components described by (1.45). We will show that this graded ring is the skew group ring $\mathbb{M}_n(A)_0 \star \Gamma$. In particular, by Proposition 1.1.15(3), it is a strongly graded ring. Consider the matrix $u_\alpha \in \mathbb{M}_n(A)(\delta_1, \dots, \delta_n)_\alpha$, where in each row i, we have 1 in the (i, j) position, where $\delta_j - \delta_i + \alpha = 0$, and zero everywhere else. One can easily see that u_α is a permutation matrix with exactly one 1 in each row and column. Moreover, for $\alpha, \beta \in \Gamma$, $u_\alpha u_\beta = u_{\alpha+\beta}$. Indeed, consider the ith row of u_α, with the only 1 in the jth column where $\delta_j - \delta_i + \alpha = 0$. Now, consider the jth row of u_β with a kth column such that $\delta_k - \delta_j + \beta = 0$ and so with 1 in the (j, k) row. Thus, multiplying u_α with u_β, we have zero everywhere in the ith row except in the (i, k)th position. On the other hand, since $\delta_k - \delta_i + \alpha + \beta = 0$, in the ith row of $u_{\alpha+\beta}$ we have zero except in the (i, k)th position. Repeating this argument for each row of u_α shows that $u_\alpha u_\beta = u_{\alpha+\beta}$.

Now defining $\phi : \Gamma \to \operatorname{Aut}(\mathbb{M}_n(A)_0)$ by $\phi(\alpha)(a) = u_\alpha a {u_\alpha}^{-1}$, and setting the 2-cocycle ψ trivial, by §1.1.4, $R = \mathbb{M}_n(A)_0 \star_\phi \Gamma$.

This was observed in [76], where it was proved that $\mathbb{M}_n(A)(\delta_1, \dots, \delta_n)_0$ is isomorphic to the smash product of Cohen and Montgomery [29] (see Remark 2.3.13).

Example 1.3.8 The following examples (from [34, Example 1.3]) provide two instances of $\mathbb{Z}_2$-grading on $\mathbb{M}_2(K)$, where K is a field. The first grading is a good grading, whereas the second one is not a good grading.

1 Let $R = \mathbb{M}_2(K)$ with the $\mathbb{Z}_2$-grading defined by

$$R_0 = \left\{ \begin{pmatrix} a & 0 \\ 0 & b \end{pmatrix} \mid a, b \in K \right\} \quad \text{and} \quad R_1 = \left\{ \begin{pmatrix} 0 & c \\ d & 0 \end{pmatrix} \mid c, d \in K \right\}.$$

Since $e_{11}, e_{22} \in R_0$ and $e_{12}, e_{21} \in R_1$, by definition, this is a good grading. Note that $R = \mathbb{M}_2(K)(0, 1)$.

2 Let $S = \mathbb{M}_2(K)$ with the $\mathbb{Z}_2$-grading defined by

$$S_0 = \left\{ \begin{pmatrix} a & b-a \\ 0 & b \end{pmatrix} \mid a, b \in K \right\} \quad \text{and} \quad S_1 = \left\{ \begin{pmatrix} d & c \\ d & -d \end{pmatrix} \mid c, d \in K \right\}.$$

Then S is a graded ring, such that the $\mathbb{Z}_2$-grading is not a good grading, since e_{11} is not homogeneous. Moreover, comparing S_0 with (1.47), shows that the grading on S does not come from the construction given by (1.44).

Consider the map

$$f : R \longrightarrow S; \quad \begin{pmatrix} a & b \\ c & d \end{pmatrix} \longmapsto \begin{pmatrix} a+c & b+d-a-c \\ c & d-c \end{pmatrix}.$$

This map is in fact a graded ring isomorphism, and so $R \cong_{\mathrm{gr}} S$. This shows that the good grading is not preserved under graded isomorphisms.

Remark 1.3.9 GOOD GRADINGS ON MATRIX ALGEBRAS

Let K be a field and Γ be an abelian group. One can put a Γ-grading on the ring $\mathbb{M}_n(K)$, by assigning a degree (an element of the group Γ) to each matrix unit $\mathbf{e}_{ij}$, $1 \leq i, j \leq n$. This is called a *good grading* or an *elementary grading*. This grading has been studied in [34]. In particular it has been shown that a grading on $\mathbb{M}_n(K)$ is good if and only if it can be described as $\mathbb{M}_n(K)(\delta_1, \dots, \delta_n)$ for some $\delta_i \in \Gamma$. Moreover, any grading on $\mathbb{M}_n(K)$ is a good grading if Γ is torsion free. It has also been shown that if $R = \mathbb{M}_n(K)$ has a Γ-grading such that $\mathbf{e}_{ij}$ is a homogeneous for some $1 \leq i, j \leq n$, then there exists a good grading on $S = \mathbb{M}_n(K)$ with a graded isomorphism $R \cong S$. It is shown that if Γ is finite, then the number of good gradings on $\mathbb{M}_n(K)$ is $|\Gamma|^{n-1}$. Moreover, (for a finite Γ) the class of strongly graded and crossed product good gradings of $\mathbb{M}_n(K)$ have been classified.

Remark 1.3.10 Let A be a Γ-graded ring and Ω a subgroup of Γ. Then A can be considered as Γ/Ω-graded ring. Recall that this gives the forgetful functor $U : \mathcal{R}^{\Gamma} \to \mathcal{R}^{\Gamma/\Omega}$ (§1.1.2). Similarly, on the level of modules, one has (again) the forgetful functor $U : \mathrm{Gr}^{\Gamma}\text{-}A \to \mathrm{Gr}^{\Gamma/\Omega}\text{-}A$ (§1.2.8).

If M is a finitely generated A-module, then by Theorem 1.2.6, $\mathrm{End}(M)$ is a Γ-graded ring. One can observe that

$$U(\mathrm{End}(M)) = \mathrm{End}(U(M)).$$

In particular, $\mathbb{M}_n(A)(\delta_1, \dots, \delta_n)$ as a Γ/Ω-graded ring coincides with

$$\mathbb{M}_n(A)(\Omega + \delta_1, \dots, \Omega + \delta_n),$$

where A here in the latter case is considered as a Γ/Ω ring.

Remark 1.3.11 GRADING ON MATRIX RINGS WITH INFINITE ROWS AND COLUMNS

Let A be a Γ-graded ring and I an index set which can be uncountable. Denote by $\mathbb{M}_I(A)$ the matrix ring with entries indexed by $I \times I$, namely, $a_{ij} \in A$, where $i, j \in I$, which are all but a finite number nonzero. For $i \in I$, choose $\delta_i \in \Gamma$ and following the grading on usual matrix rings (see (1.46)) for $a \in A^h$, define

$$\deg(a_{ij}) = \deg(a) + \delta_i - \delta_j. \tag{1.54}$$

This makes $\mathbb{M}_I(A)$ a Γ-graded ring. Clearly if I is finite, then this graded ring coincides with $\mathbb{M}_n(A)(\delta_1, \dots, \delta_n)$, where $|I| = n$.

1.3.2 Homogeneous idempotents calculus

The idempotents naturally arise in relation to the decomposition of rings and modules. The following facts about idempotents are well known in the non-graded setting and one can check that they translate into the graded setting with similar proofs (see [60, §21]). Let P_i, $1 \leq i \leq l$, be graded right ideals of A such that $A = P_1 \oplus \cdots \oplus P_l$. Then there are homogeneous orthogonal idempotents e_i (hence of degree zero) such that $1 = e_1 + \cdots + e_l$ and $e_i A = P_i$.

Let e and f be homogeneous idempotent elements in the graded ring A. (Note that, in general, there are nonhomogeneous idempotents in a graded ring.) Let $\theta : eA \to fA$ be a right A-module homomorphism. Then $\theta(e) = \theta(e^2) = \theta(e)e = fae$ for some $a \in A$ and for $b \in eA$, $\theta(b) = \theta(eb) = \theta(e)b$. This shows that there is a map

$$\begin{aligned} \operatorname{Hom}_A(eA, fA) &\to fAe, \\ \theta &\mapsto \theta(e) \end{aligned} \tag{1.55}$$

and one can easily check this is a group isomorphism. We have

$$fAe = \bigoplus_{\gamma \in \Gamma} fA_\gamma e$$

and by Theorem 1.2.6,

$$\operatorname{Hom}_A(eA, fA) \cong \bigoplus_{\gamma \in \Gamma} \operatorname{Hom}_A(eA, fA)_\gamma.$$

Then one can see that the homomorphism (1.55) respects the graded decomposition.

Now if $\theta : eA \to fA(\alpha)$, where $\alpha \in \Gamma$, is a graded A-isomorphism, then $x = \theta(e) \in fA_\alpha e$ and $y = \theta^{-1}(f) \in eA_{-\alpha} f$, where x and y are homogeneous of degrees α and $-\alpha$, respectively, such that $yx = e$ and $xy = f$.

Finally, for $f = 1$, the map (1.55) gives that

$$\operatorname{Hom}_A(eA, A) \to Ae$$

is a graded left A-module isomorphism and for $f = e$,

$$\operatorname{End}_A(eA) \to eAe$$

is a graded ring isomorphism. In particular, we have a ring isomorphism

$$\operatorname{End}_A(eA)_0 = \operatorname{End}_{\text{Gr-}A}(eA) \cong eA_0 e.$$

These facts will be used later in Theorem 5.1.3.

1.3.3 Graded matrix units

Let A be a Γ-graded ring. Modelling on the properties of the matrix units $\mathbf{e}_{ij}$, we call a set of homogeneous elements $\{ e_{ij} \in A \mid 1 \leq i, j \leq n \}$, a set of *graded matrix units* if

$$e_{ij}e_{kl} = \delta_{jk}e_{il}, \tag{1.56}$$

where δ_{jk} are the Kronecker deltas. Let $\deg(e_{i1}) = \delta_i$. From (1.56) it follows that $\deg(e_{ii}) = 0$, $\deg(e_{1i}) = -\delta_i$ and

$$\deg(e_{ij}) = \delta_i - \delta_j. \tag{1.57}$$

The above set is called a *full set of graded matrix units* if $\sum_{i=1}^{n} e_{ii} = 1$. If a graded ring contains a full set of graded matrix units, then the ring is of the form of a matrix ring over an appropriate graded ring (Lemma 1.3.12). We can use this to characterise the two-sided ideals of graded matrix rings (Corollary 1.3.14). For this we adopt Lam's presentation [61, §17A] to the graded setting.

Lemma 1.3.12 *Let R be a Γ-graded ring. Then $R = \mathbb{M}_n(A)(\delta_1, \dots, \delta_n)$ for some graded ring A if and only if R has a full set of graded matrix units $\{ e_{ij} \in R \mid 1 \leq i, j \leq n \}$.*

Proof One direction is obvious. Suppose $\{e_{ij} \in R \mid 1 \leq i, j \leq n\}$ is a full set of graded matrix units in R and A is its centraliser in R which is a graded subring of R. We show that R is a graded free A-module with the basis $\{e_{ij}\}$. Let $x \in R$ and set

$$a_{ij} = \sum_{k=1}^{n} e_{ki}xe_{jk} \in R.$$

Since $a_{ij}e_{uv} = e_{ui}xe_{jv} = e_{uv}a_{ij}$, it follows that $a_{ij} \in A$. Let $u = i, v = j$. Then $a_{ij}e_{ij} = e_{ii}xe_{jj}$, and since $\{e_{ij}\}$ is full, $\sum_{i,j} a_{ij}e_{ij} = \sum_{ij} e_{ii}xe_{jj} = x$. This shows that $\{e_{ij}\}$ generates R as an A-module. It is easy to see that $\{e_{ij} \mid 1 \leq i, j \leq n\}$ is linearly independent as well. Let $\deg(e_{i1}) = \delta_i$. Then (1.57) shows that the map

$$\begin{aligned} R &\longrightarrow \mathbb{M}_n(A)(\delta_1, \dots, \delta_n), \\ ae_{ij} &\longmapsto \mathbf{e}_{ij}(a) \end{aligned}$$

induces a graded isomorphism. □

Corollary 1.3.13 *Let A, R and S be Γ-graded rings. Suppose*

$$R = \mathbb{M}_n(A)(\delta_1, \dots, \delta_n)$$

and there is a graded ring homomorphism $f : R \to S$. Then

$$S = \mathbb{M}_n(B)(\delta_1, \dots, \delta_n),$$

for a graded ring B and f is induced by a graded homomorphism $f_0 : A \to B$.

Proof Consider the standard full graded matrix units $\{\mathbf{e}_{ij} \mid 1 \leq i, j \leq n\}$ in R. Then $\{f(\mathbf{e}_{ij})\}$ is a set of full graded matrix units in S. Since f is a graded homomorphism, $\deg(f(\mathbf{e}_{ij})) = \deg(e_{ij}) = \delta_i - \delta_j$. Let B be the centraliser of this set in S. By Lemma 1.3.12 (and its proof),

$$S = \mathbb{M}_n(B)(0, \delta_2 - \delta_1, \dots, \delta_n - \delta_1) = \mathbb{M}_n(B)(\delta_1, \delta_2, \dots, \delta_n).$$

Since A is the centraliser of $\{\mathbf{e}_{ij}\}$, f sends A to B and thus induces the map on the matrix algebras. □

Corollary 1.3.14 *Let A be a Γ-graded ring, $R = \mathbb{M}_n(A)(\delta_1, \dots, \delta_n)$ and I be a graded ideal of R. Then $I = \mathbb{M}_n(I_0)(\delta_1, \dots, \delta_n)$, where I_0 is a graded ideal of A.*

Proof Consider the canonical graded quotient homomorphism $f : R \to R/I$. Set $I_0 = \ker(f|_A)$. One can easily see $\mathbb{M}_n(I_0)(\delta_1, \dots, \delta_n) \subseteq I$. By Lemma 1.3.12, $R/I = \mathbb{M}_n(B)(\delta_1, \dots, \delta_n)$, where B is the centraliser of the set $\{f(\mathbf{e}_{ij})\}$. Since A is the centraliser of $\{\mathbf{e}_{ij}\}$, $f(A) \subseteq B$. Now for $x \in I$, write $x = \sum_{i,j} a_{ij}\mathbf{e}_{ij}$, $a_{ij} \in A$. Then $0 = f(x) = \sum_{i,j} f(a_{ij})f(\mathbf{e}_{ij})$, which implies $f(a_{ij}) = 0$ as $f(\mathbf{e}_{ij})$ are linear independent (see the proof of Lemma 1.3.12). Thus $a_{i,j} \in I_0$. This shows $I \subseteq \mathbb{M}_n(I_0)(\delta_1, \dots, \delta_n)$, which finishes the proof. □

Corollary 1.3.14 shows that there is a one-to-one inclusion preserving correspondence between the graded ideals of A and the graded ideals of $\mathbb{M}_n(A)(\overline{\delta})$, where $\overline{\delta} = (\delta_1, \dots, \delta_n)$.

1.3.4 Mixed shift

For a Γ-graded ring A, $\overline{\alpha} = (\alpha_1, \dots \alpha_m) \in \Gamma^m$ and $\overline{\delta} = (\delta_1, \dots, \delta_n) \in \Gamma^n$, set

$$\boxed{\mathbb{M}_{m\times n}(A)[\overline{\alpha}][\overline{\delta}] := \begin{pmatrix} A_{\alpha_1-\delta_1} & A_{\alpha_1-\delta_2} & \cdots & A_{\alpha_1-\delta_n} \\ A_{\alpha_2-\delta_1} & A_{\alpha_2-\delta_2} & \cdots & A_{\alpha_2-\delta_n} \\ \vdots & \vdots & \ddots & \vdots \\ A_{\alpha_m-\delta_1} & A_{\alpha_m-\delta_2} & \cdots & A_{\alpha_m-\delta_n} \end{pmatrix}}.$$

So $\mathbb{M}_{m\times n}(A)[\overline{\alpha}][\overline{\delta}]$ consists of matrices with the ij-entry in $A_{\alpha_i-\delta_j}$.

If $a \in \mathbb{M}_{m\times n}(A)[\overline{\alpha}][\overline{\delta}]$, then one can easily check that multiplying a from the left induces a graded right A-module homomorphism

$$\phi_a : A^n(\overline{\delta}) \longrightarrow A^m(\overline{\alpha}), \tag{1.58}$$

$$\begin{pmatrix} a_1 \\ a_2 \\ \vdots \\ a_n \end{pmatrix} \longmapsto a \begin{pmatrix} a_1 \\ a_2 \\ \vdots \\ a_n \end{pmatrix}.$$

Conversely, suppose $\phi : A^n(\overline{\delta}) \to A^m(\overline{\alpha})$ is graded right A-module homomorphism. Let e_j denote the standard basis element of $A^n(\overline{\delta})$ with 1 in the j-th entry and zeros elsewhere. Let $\phi(e_j) = (a_{1j}, a_{2j}, \dots, a_{mj})$, $1 \leq j \leq n$. Since ϕ is a graded map, comparing the grading of both sides, one can observe that $\deg(a_{ij}) = \alpha_i - \delta_j$. So that the map ϕ is represented by the left multiplication with the matrix $a = (a_{ij})_{m\times n} \in \mathbb{M}_{m\times n}(A)[\overline{\alpha}][\overline{\delta}]$.

In particular $\mathbb{M}_{m\times m}(A)[\overline{\alpha}][\overline{\alpha}]$ represents $\operatorname{End}(A^m(\alpha), A^m(\alpha))_0$. Combining this with (1.48), we get

$$\mathbb{M}_{m\times m}(A)[\overline{\alpha}][\overline{\alpha}] = \mathbb{M}_m(A)(-\overline{\alpha})_0. \tag{1.59}$$

The mixed shift will be used in §3.2 to describe graded Grothendieck groups by idempotent matrices. The following simple lemma comes in handy.

Lemma 1.3.15 *Let $a \in \mathbb{M}_{m\times n}(A)[\overline{\alpha}][\overline{\delta}]$ and $b \in \mathbb{M}_{n\times k}(A)[\overline{\delta}][\overline{\beta}]$. Then $ab \in \mathbb{M}_{m\times k}(A)[\overline{\alpha}][\overline{\beta}]$.*

Proof Let $\phi_a : A^n(\overline{\delta}) \to A^m(\overline{\alpha})$ and $\phi_b : A^k(\overline{\beta}) \to A^n(\overline{\delta})$ be the graded right A-module homomorphisms induced by multiplications with a and b, respectively (see 1.58). Then

$$\phi_{ab} = \phi_a\phi_b : A^k(\overline{\beta}) \longrightarrow A^m(\overline{\alpha}).$$

This shows that $ab \in \mathbb{M}_{m\times k}(A)[\overline{\alpha}][\overline{\beta}]$. (This can also be checked directly, by multiplying the matrices a and b and taking into account the shift arrangements.) □

Proposition 1.3.16 *Let A be a Γ-graded ring and let $\overline{\alpha} = (\alpha_1, \dots, \alpha_m) \in \Gamma^m$, $\overline{\delta} = (\delta_1, \dots, \delta_n) \in \Gamma^n$. Then the following are equivalent:*

(1) *$A^m(\overline{\alpha}) \cong_{\mathrm{gr}} A^n(\overline{\delta})$ as graded right A-modules;*
(2) *$A^m(-\overline{\alpha}) \cong_{\mathrm{gr}} A^n(-\overline{\delta})$ as graded left A-modules;*
(3) *there exist $a = (a_{ij}) \in \mathbb{M}_{n\times m}(A)[\overline{\delta}][\overline{\alpha}]$ and $b = (b_{ij}) \in \mathbb{M}_{m\times n}(A)[\overline{\alpha}][\overline{\delta}]$ such that $ab = \mathbb{I}_n$ and $ba = \mathbb{I}_m$.*

Proof (1) ⇒ (3) Let $\phi : A^m(\overline{\alpha}) \to A^n(\overline{\delta})$ and $\psi : A^n(\overline{\delta}) \to A^m(\overline{\alpha})$ be graded right A-module isomorphisms such that $\phi\psi = 1$ and $\psi\phi = 1$. The paragraph prior to Lemma 1.3.15 shows that the map ϕ is represented by the left multiplication with a matrix $a = (a_{ij})_{n\times m} \in \mathbb{M}_{n\times m}(A)[\overline{\delta}][\overline{\alpha}]$. In the same way one can construct $b \in \mathbb{M}_{m\times n}(A)[\overline{\alpha}][\overline{\delta}]$ which induces ψ. Now $\phi\psi = 1$ and $\psi\phi = 1$ translate to $ab = \mathbb{I}_n$ and $ba = \mathbb{I}_m$.

(3) ⇒ (1) If $a \in \mathbb{M}_{n\times m}(A)[\overline{\delta}][\overline{\alpha}]$, then multiplication from the left induces a graded right A-module homomorphism $\phi_a : A^m(\overline{\alpha}) \longrightarrow A^n(\overline{\delta})$. Similarly b induces $\psi_b : A^n(\overline{\delta}) \longrightarrow A^m(\overline{\alpha})$. Now $ab = \mathbb{I}_n$ and $ba = \mathbb{I}_m$ translate to $\phi_a\psi_b = 1$ and $\psi_b\phi_a = 1$.

(2) ⟺ (3) This part is similar to the previous cases by considering the matrix multiplication from the right. Specifically, the graded left A-module homomorphism $\phi : A^m(-\overline{\alpha}) \to A^n(-\overline{\delta})$ represented by a matrix multiplication from the right of the form $\mathbb{M}_{m\times n}(A)[\overline{\alpha}][\overline{\delta}]$ and similarly ψ gives a matrix in $\mathbb{M}_{n\times m}(A)[\overline{\delta}][\overline{\alpha}]$. The rest follows easily. □

The following corollary shows that $A(\alpha) \cong_{\text{gr}} A$ as graded right A-modules if and only if $\alpha \in \Gamma_A^*$. In fact, replacing $m = n = 1$ in Proposition 1.3.16 we obtain the following.

Corollary 1.3.17 *Let A be a Γ-graded ring and $\alpha \in \Gamma$. Then the following are equivalent:*

(1) *$A(\alpha) \cong_{\text{gr}} A$ as graded right A-modules;*
(2) *$A(-\alpha) \cong_{\text{gr}} A$ as graded right A-modules;*
(3) *$A(\alpha) \cong_{\text{gr}} A$ as graded left A-modules;*
(4) *$A(-\alpha) \cong_{\text{gr}} A$ as graded left A-modules;*
(5) *there is an invertible homogeneous element of degree α;*
(6) *there is an invertible homogeneous element of degree $-\alpha$.*

Proof This follows from Proposition 1.3.16. □

Corollary 1.3.18 *Let A be a Γ-graded ring. Then the following are equivalent:*

(1) *A is crossed product;*
(2) *$A(\alpha) \cong_{\text{gr}} A$, as graded right A-modules, for all $\alpha \in \Gamma$;*
(3) *$A(\alpha) \cong_{\text{gr}} A$, as graded left A-modules, for all $\alpha \in \Gamma$;*
(4) *the shift functor $\mathcal{T}_\alpha : \text{Gr-}A \to \text{Gr-}A$ is isomorphic to identity functor, for all $\alpha \in \Gamma$.*

Proof This follows from Corollary 1.3.17, (1.22) and the definition of the crossed product rings (§1.1.3). □

The Corollary above will be used to show that the action of Γ on the graded Grothendieck group of a crossed product algebra is trivial (see Example 3.1.9).

Example 1.3.19 THE LEAVITT ALGEBRA $\mathcal{L}(1,n)$

In [63] Leavitt considered the free associative ring A with coefficient in $\mathbb{Z}$ generated by symbols $\{x_i, y_i \mid 1 \leq i \leq n\}$ subject to the relations

$$x_i y_j = \delta_{ij}, \text{ for all } 1 \leq i, j \leq n, \tag{1.60}$$
$$\sum_{i=1}^{n} y_i x_i = 1,$$

where $n \geq 2$ and δ_{ij} is the Kronecker delta. The relations guarantee the right A-module homomorphism

$$\phi : A \longrightarrow A^n \tag{1.61}$$
$$a \mapsto (x_1 a, x_2 a, \dots, x_n a)$$

has an inverse

$$\psi : A^n \longrightarrow A \tag{1.62}$$
$$(a_1, \dots, a_n) \mapsto y_1 a_1 + \cdots + y_n a_n,$$

so $A \cong A^n$ as right A-modules. He showed that A is universal with respect to this property, of type $(1, n-1)$ (see §1.7) and it is a simple ring.

Leavitt's algebra constructed in (1.60) has a natural grading; assigning 1 to y_i and -1 to x_i, $1 \leq i \leq n$, since the relations are homogeneous (of degree zero), the ring A is a $\mathbb{Z}$-graded ring (see §1.6.1 for a general construction of graded rings from free algebras). The isomorphism (1.61) induces a graded isomorphism

$$\phi : A \longrightarrow A(-1)^n \tag{1.63}$$
$$a \longmapsto (x_1 a, x_2 a, \dots, x_n a),$$

where $A(-1)$ is the suspension of A by -1. In fact, letting

$$y = (y_1, \dots, y_n) \text{ and}$$
$$x = \begin{pmatrix} x_1 \\ x_2 \\ \vdots \\ x_n \end{pmatrix},$$

we have $y \in \mathbb{M}_{1\times n}(A)[\overline{\alpha}][\overline{\delta}]$ and $x \in \mathbb{M}_{n\times 1}(A)[\overline{\delta}][\overline{\alpha}]$, where $\overline{\alpha} = (0)$ and $\overline{\delta} = (-1, \dots, -1)$. Thus by Proposition 1.3.16, $A \cong_{\mathrm{gr}} A(-1)^n$.

Motivated by this algebra, the Leavitt path algebras were introduced in [2, 5], which associate with a direct graph a certain algebra. When the graph has one vertex and n loops, the algebra corresponds to this graph is the Leavitt algebra constructed in (1.60) and is denoted by $\mathcal{L}(1, n)$ or $\mathcal{L}_n$. The Leavitt path algebras will provide a vast array of examples of graded algebras. We will study these algebras in §1.6.4.

1.4 Graded division rings

Graded fields and their noncommutative version, *i.e.,* graded division rings, are among the simplest graded rings. With a little effort, we can completely compute the invariants of these algebras which we are interested in, namely, the graded Grothendieck groups (§3.7) and the graded Picard groups (Chapter 4).

Recall from §1.1.4 that a Γ-graded ring $A = \bigoplus_{\gamma\in\Gamma} A_\gamma$ is called a graded division ring if every nonzero homogeneous element has a multiplicative inverse. Throughout this section we consider graded right modules over graded division rings. Note that we work with the abelian grade groups, however, all the results are valid for nonabelian grading as well. We first show that for graded modules over a graded division ring, there is well-defined notion of dimension. The proofs follow the standard proofs in the nongraded setting (see [50, §IV, Theorem 2.4, 2.7, 2.13]), or the graded setting (see [75, Proposition 4.6.1], [90, Chapter 2]).

Proposition 1.4.1 *Let A be a Γ-graded ring. Then A is a graded division ring if and only if any graded A-module is graded free. If M is a graded module over graded division ring A, then any linearly independent subset of M consisting of homogeneous elements can be extended to a homogeneous basis of M.*

Proof Suppose any graded (right) module is graded free. Let I be a right ideal of A. Consider A/I as a right A-module, which is graded free by assumption. Thus $I = \operatorname{ann}(A/I) = 0$. This shows that the only graded right ideal of A is the zero ideal. This gives that A is a graded division ring.

For the converse, note that if A is a graded division ring (*i.e.,* all homogeneous elements are invertible), then for any $m \in M^h$, $\{m\}$ is a linearly independent subset of M. This immediately gives the converse of the statement of the theorem as a consequence of the second part of the theorem.

Fix a linearly independent subset X of M consisting of homogeneous elements. Let

$$F = \{Q \subseteq M^h \mid X \subseteq Q \text{ and } Q \text{ is } A\text{-linearly independent}\}.$$

This is a nonempty partially ordered set with inclusion, and every chain $Q_1 \subseteq Q_2 \subseteq \dots$ in F has an upper bound $\bigcup Q_i \in F$. By Zorn's lemma, F has a maximal element, which we denote by P. If $\langle P \rangle \neq M$, then there is a homogeneous element $m \in M^h \setminus \langle P \rangle$. We will show that $P \cup \{m\}$ is a linearly independent set containing X, contradicting the maximality of P.

Suppose $ma + \sum p_i a_i = 0$, where $a, a_i \in A$, $p_i \in P$ with $a \neq 0$. Then there is a homogeneous component of a, say a_λ, which is also nonzero. Considering the $\lambda + \deg(m)$-homogeneous component of this sum, we have

$$m = m a_\lambda a_\lambda^{-1} = -\sum p_i a'_i a_\lambda^{-1}$$

for a'_i homogeneous, which contradicts the assumption $m \in M^h \setminus \langle P \rangle$. Hence $a = 0$, which implies each $a_i = 0$. This gives the required contradiction, so $M = \langle P \rangle$, completing the proof. □

The following proposition shows in particular that a graded division ring has graded invariant basis number (we discuss this type of ring in §1.7).

Proposition 1.4.2 *Let A be a Γ-graded division ring and M be a graded A-module. Then any two homogeneous bases of M over A have the same cardinality.*

Proof By [50, §IV, Theorem 2.6], if a module M has an infinite basis over a ring, then any other basis of M has the same cardinality. This proves the proposition in the case where the homogeneous basis is infinite.

Now suppose that M has two finite homogeneous bases X and Y. Then $X = \{x_1, \dots, x_n\}$ and $Y = \{y_1, \dots, y_m\}$, for $x_i, y_i \in M^h \setminus 0$. As X is a basis for M, we can write

$$y_m = x_1 a_1 + \cdots + x_n a_n,$$

for some $a_i \in A^h$, where $\deg(y_m) = \deg(a_i) + \deg(x_i)$ for each $1 \leq i \leq n$. Since $y_m \neq 0$, we have at least one $a_i \neq 0$. Let a_k be the first nonzero a_i, and we note that a_k is invertible as it is nonzero and homogeneous in A. Then

$$x_k = y_m a_k^{-1} - x_{k+1} a_{k+1} a_k^{-1} - \cdots - x_n a_n a_k^{-1},$$

and the set $X' = \{y_m, x_1, \dots, x_{k-1}, x_{k+1}, \dots, x_n\}$ spans M since X spans M. So

$$y_{m-1} = y_m b_m + x_1 c_1 + \cdots + x_{k-1} c_{k-1} + x_{k+1} c_{k+1} + \cdots + x_n c_n,$$

for $b_m, c_i \in A^h$. There is at least one nonzero c_i, since if all the c_i are zero, then either y_m and y_{m-1} are linearly dependent or y_{m-1} is zero, which are not the case. Let c_j denote the first nonzero c_i. Then x_j can be written as a linear combination of y_{m-1}, y_m and those x_i with $i \neq j, k$. Therefore the set

$$X'' = \{ y_{m-1}, y_m \} \cup \{ x_i : i \neq j, k \}$$

spans M since X' spans M.

Continuing this process of adding a y and removing an x gives, after the kth step, a set which spans M consisting of $y_m, y_{m-1}, \dots, y_{m-k+1}$ and $n-k$ of the x_i. If $n < m$, then after the nth step, we would have that the set $\{y_m, \dots, y_{m-n+1}\}$ spans M. But if $n < m$, then $m - n + 1 \geq 2$, so this set does not contain y_1, and therefore y_1 can be written as a linear combination of the elements of this set. This contradicts the linear independence of Y, so we must have $m \leq n$. Repeating a similar argument with X and Y interchanged gives $n \leq m$, so $n = m$. □

The Propositions 1.4.1 and 1.4.2 above show that for a graded module M over a graded division ring A, M has a homogeneous basis and any two homogeneous bases of M have the same cardinality. The cardinal number of any homogeneous basis of M is called the *dimension* of M over A, and it is denoted by $\dim_A(M)$ or $[M : A]$.

Proposition 1.4.3 *Let A be a Γ-graded division ring and M be a graded A-module. If N is a graded submodule of M, then*

$$\dim_A(N) + \dim_A(M/N) = \dim_A(M).$$

Proof By Proposition 1.4.1, the submodule N is a graded free A-module with a homogeneous basis Y which can be extended to a homogeneous basis X of M. We will show that $U = \{x + N \mid x \in X \setminus Y\}$ is a homogeneous basis of M/N. Note that by (1.12), U consists of homogeneous elements. Let $t \in (M/N)^h$. Again by (1.12), $t = m + N$, where $m \in M^h$ and $m = \sum x_i a_i + \sum y_j b_j$ where a_i, $b_j \in A$, $y_j \in Y$ and $x_i \in X \setminus Y$. So $m + N = \sum (x_i + N) a_i$, which shows that U spans M/N. If $\sum (x_i + N) a_i = 0$, for $a_i \in A$, $x_i \in X \setminus Y$, then $\sum x_i a_i \in N$ which implies that $\sum x_i a_i = \sum y_k b_k$ for $b_k \in A$ and $y_k \in Y$, which implies that $a_i = 0$ and $b_k = 0$ for all i, k. Therefore U is a homogeneous basis for M/N and as we can construct a bijective map $X \setminus Y \to U$, we have $|U| = |X \setminus Y|$. Then

$$\dim_A M = |X| = |Y| + |X \setminus Y| = |Y| + |U| = \dim_A N + \dim_A(M/N). \qquad \square$$

The following statement is the graded version of a similar statement on simple rings (see [50, §IX.1]). This is required for the proof of Theorem 1.4.5.

Proposition 1.4.4 *Let A and B be Γ-graded division rings. If*

$$\mathbb{M}_n(A)(\lambda_1, \dots, \lambda_n) \cong_{\mathrm{gr}} \mathbb{M}_m(B)(\gamma_1, \dots, \gamma_m)$$

as graded rings, where $\lambda_i, \gamma_j \in \Gamma$, $1 \leq i \leq n$, $1 \leq j \leq m$, then $n = m$ and $A \cong_{\mathrm{gr}} B$.

Proof The proof follows the nongraded case (see [50, §IX.1]) with an extra attention given to the grading. We refer the reader to [71, §4.3] for a detailed proof. □

We can further determine the relations between the shift $(\lambda_1, \dots, \lambda_n)$ and $(\gamma_1, \dots, \gamma_m)$ in the above proposition. For this we need to extend [27, Theorem 2.1] (see also [75, Theorem 9.2.18]) from fields (with trivial grading) to graded division algebras. The following theorem states that two graded matrix algebras over a graded division ring with two shifts are isomorphic if and only if one can obtain one shift from the other by applying (1.50) and (1.51).

Theorem 1.4.5 *Let A be a Γ-graded division ring. Then for $\lambda_i, \gamma_j \in \Gamma$, $1 \leq i \leq n$, $1 \leq j \leq m$,*

$$\mathbb{M}_n(A)(\lambda_1, \dots, \lambda_n) \cong_{\mathrm{gr}} \mathbb{M}_m(A)(\gamma_1, \dots, \gamma_m) \tag{1.64}$$

if and only if $n = m$ and for a suitable permutation $\pi \in S_n$, we have $\lambda_i = \gamma_{\pi(i)} + \tau_i + \sigma$, $1 \leq i \leq n$, where $\tau_i \in \Gamma_A$ and a fixed $\sigma \in \Gamma$, i.e., $(\lambda_1, \dots, \lambda_n)$ is obtained from $(\gamma_1, \dots, \gamma_m)$ by applying (1.50) and (1.51).

Proof One direction is Theorem 1.3.3, noting that since A is a graded division ring, $\Gamma_A = \Gamma_A^*$.

We now prove the converse. That $n = m$ follows from Proposition 1.4.4. By (1.4.1) one can find $\epsilon = (\varepsilon_1, \dots, \varepsilon_1, \varepsilon_2, \dots, \varepsilon_2, \dots, \varepsilon_k, \dots, \varepsilon_k)$ in Γ such that $\mathbb{M}_n(A)(\lambda_1, \dots, \lambda_n) \cong_{\mathrm{gr}} \mathbb{M}_n(A)(\epsilon)$ as in (1.68). Now set

$$V = A(-\varepsilon_1) \times \cdots \times A(-\varepsilon_1) \times \cdots \times A(-\varepsilon_k) \times \cdots \times A(-\varepsilon_k)$$

and pick the (standard) homogeneous basis e_i, $1 \leq i \leq n$ and define $E_{ij} \in \mathrm{End}_A(V)$ by $E_{ij}(e_t) = \delta_{j,t} e_i$, $1 \leq i, j, t \leq n$. One can easily see that E_{ij} is a A-graded homomorphism of degree $\varepsilon_{s_i} - \varepsilon_{s_j}$ where ε_{s_i} and ε_{s_j} are ith and jth elements in ϵ. Moreover, $\mathrm{End}_A(V) \cong_{\mathrm{gr}} \mathbb{M}_n(A)(\epsilon)$ and E_{ij} corresponds to the matrix e_{ij} in $\mathbb{M}_n(A)(\epsilon)$. In a similar manner, one can find

$$\epsilon' = (\varepsilon'_1, \dots, \varepsilon'_1, \varepsilon'_2, \dots, \varepsilon'_2, \dots, \varepsilon'_{k'}, \dots, \varepsilon'_{k'})$$

and a graded A-vector space W such that

$$\mathbb{M}_n(A)(\gamma_1, \dots, \gamma_n) \cong_{\mathrm{gr}} \mathbb{M}_n(A)(\epsilon'),$$

and $\mathrm{End}_A(W) \cong_{\mathrm{gr}} \mathbb{M}_n(A)(\epsilon')$. Therefore (1.64) provides a graded ring isomorphism

$$\theta : \mathrm{End}_A(V) \to \mathrm{End}_A(W).$$

Define $E'_{ij} := \theta(E_{ij})$ and $E'_{ii}(W) = Q_i$, for $1 \leq i, j \leq n$. Since $\{E_{ii} \mid 1 \leq i \leq n\}$ is a complete system of orthogonal idempotents, so is $\{E'_{ii} \mid 1 \leq i \leq n\}$. It follows that

$$W \cong_{\mathrm{gr}} \bigoplus\nolimits_{1\leq j\leq n} Q_j.$$

Moreover, $E'_{ij}E'_{tr} = \delta_{j,t}E'_{ir}$ and E'_{ii} acts as identity on Q_i. These relations show that restricting E'_{ij} on Q_j induces an A-graded isomorphism $E'_{ij} : Q_j \to Q_i$ of degree $\varepsilon_{s_i} - \varepsilon_{s_j}$ (same degree as E_{ij}). So $Q_j \cong_{\mathrm{gr}} Q_1(\varepsilon_{s_1} - \varepsilon_{s_j})$ for any $1 \leq j \leq n$. Therefore

$$W \cong_{\mathrm{gr}} \bigoplus\nolimits_{1\leq j\leq n} Q_1(\varepsilon_{s_1} - \varepsilon_{s_j}).$$

By dimension count (see Proposition 1.4.3), it follows that $\dim_A Q_1 = 1$.

A similar argument for the identity map $\mathrm{id} : \mathrm{End}_A(V) \to \mathrm{End}_A(V)$ produces

$$V \cong_{\mathrm{gr}} \bigoplus\nolimits_{1\leq j\leq n} P_1(\varepsilon_{s_1} - \varepsilon_{s_j}),$$

where $P_1 = E_{11}(V)$, with $\dim_A P_1 = 1$.

Since P_1 and Q_1 are A-graded vector spaces of dimension 1, there is $\sigma \in \Gamma$, such that $Q_1 \cong_{\mathrm{gr}} P_1(\sigma)$. Using the fact that for an A-graded module P and $\alpha, \beta \in \Gamma$, $P(\alpha)(\beta) = P(\alpha + \beta) = P(\beta)(\alpha)$, we have

$$\begin{aligned} W \cong_{\mathrm{gr}} \bigoplus\nolimits_{1\leq j\leq n} Q_1(\varepsilon_{s_1} - \varepsilon_{s_j}) &\cong_{\mathrm{gr}} \bigoplus\nolimits_{1\leq j\leq n} P_1(\sigma)(\varepsilon_{s_1} - \varepsilon_{s_j}) \\ &\cong_{\mathrm{gr}} \bigoplus\nolimits_{1\leq j\leq n} P_1(\varepsilon_{s_1} - \varepsilon_{s_j})(\sigma) \cong_{\mathrm{gr}} V(\sigma). \end{aligned} \tag{1.65}$$

We denote this A-graded isomorphism with $\phi : W \to V(\sigma)$. Let e'_i, $1 \leq i \leq n$ be a (standard) homogeneous basis of degree ε'_{s_i} in W. Then $\phi(e'_i) = \sum_{1\leq j\leq n} e_j a_j$, where $a_j \in A^h$ and e_j are homogeneous of degree $\varepsilon_{s_j} - \sigma$ in $V(\sigma)$. Since $\deg(\phi(e'_i)) = \varepsilon'_{s_i}$, any e_j with nonzero a_j in the sum has the same degree. For if $\varepsilon_{s_j} - \sigma = \deg(e_j) \neq \deg(e_l) = \varepsilon_{s_l} - \sigma$, then since $\deg(e_j a_j) = \deg(e_l a_l) = \varepsilon'_{s_i}$ it follows that $\varepsilon_{s_j} - \varepsilon_{s_l} \in \Gamma_A$ which is a contradiction as $\Gamma_A + \varepsilon_{s_j}$ and $\Gamma_A + \varepsilon_{s_l}$ are distinct. Thus $\varepsilon'_{s_i} = \varepsilon_{s_j} + \tau_j - \sigma$, where $\tau_j = \deg(a_j) \in \Gamma_A$. In the same manner one can show that $\varepsilon'_{s_i} = \varepsilon'_{s_{i'}}$ in ϵ' if and only if ε_{s_j} and $\varepsilon_{s_{j'}}$ assigned to them by the previous argument are also equal. This shows that ϵ' can be obtained from ϵ by applying (1.50) and (1.51). Since ϵ' and ϵ are also obtained from $\gamma_1, \dots, \gamma_n$ and $\lambda_1, \dots, \lambda_n$, respectively, by applying (1.50) and (1.51), putting these together shows that $\lambda_1, \dots, \lambda_n$ and $\gamma_1, \dots, \gamma_n$ have similar relations, *i.e.*, $\lambda_i = \gamma_{\pi(i)} + \tau_i + \sigma$, $1 \leq i \leq n$, where $\tau_i \in \Gamma_A$ and a fixed $\sigma \in \Gamma$. □

A *graded division algebra* A is defined to be a graded division ring with centre R such that $[A : R] < \infty$. Note that since R is a graded field, by Propositions 1.4.1 and 1.4.2, A has a well-defined dimension over R. A graded division algebra A over its centre R is said to be *unramified* if $\Gamma_A = \Gamma_R$ and *totally ramified* if $A_0 = R_0$.

Let A be a graded division ring and let R be a graded subfield of A which is contained in the centre of A. It is clear that $R_0 = R \cap A_0$ is a field and A_0 is a division ring. The group of invertible homogeneous elements of A is denoted by A^{h*}, which is equal to $A^h \setminus 0$. Considering A as a graded R-module, since R is a graded field, there is a uniquely defined dimension $[A : R]$ by Theorem 1.4.1. The proposition below has been proven in [90, Chapter 5] for two graded fields $R \subseteq S$ with a torsion-free abelian grade group.

Proposition 1.4.6 *Let A be a graded division ring and let R be a graded subfield of A which is contained in the centre of A. Then*

$$[A : R] = [A_0 : R_0]|\Gamma_A : \Gamma_R|.$$

Proof Since A is a graded division ring, A_0 is a division ring. Moreover, R_0 is a field. Let $\{x_i\}_{i\in I}$ be a basis for A_0 over R_0. Consider the cosets of Γ_A over Γ_R and let $\{\delta_j\}_{j\in J}$ be a coset representative, where $\delta_j \in \Gamma_A$. Take $\{y_j\}_{j\in J} \subseteq A^{h*}$ such that $\deg(y_j) = \delta_j$ for each j. We will show that $\{x_i y_j\}$ is a basis for A over F.

Consider the map

$$\begin{aligned} \psi : A^{h*} &\longrightarrow \Gamma_A/\Gamma_R, \\ a &\longmapsto \deg(a) + \Gamma_R. \end{aligned}$$

This is a group homomorphism with kernel $A_0 R^{h*}$, since for any $a \in \ker(\psi)$ there is some $r \in R^{h*}$ with $ar^{-1} \in A_0$. For $a \in A$, $a = \sum_{\gamma\in\Gamma} a_\gamma$, where $a_\gamma \in A_\gamma$ and $\psi(a_\gamma) = \gamma + \Gamma_R = \delta_j + \Gamma_F$ for some δ_j in the coset representative of Γ_A over Γ_R. Then there is some y_j with $\deg(y_j) = \delta_j$ and $a_\gamma y_j^{-1} \in \ker(\psi) = A_0 R^{h*}$. So

$$a_\gamma y_j^{-1} = (\sum_i x_i r_i) g$$

for $g \in R^{h*}$ and $r_i \in R_0$. Since R is in the centre of A, it follows that $a_\gamma = \sum_i x_i y_j r_i g$. So a can be written as an R-linear combination of the elements of $\{x_i y_j\}$.

To show linear independence, suppose

$$\sum_{i=1}^{n} x_i y_i r_i = 0, \tag{1.66}$$

for $r_i \in R$. Write r_i as the sum of its homogeneous components, and then consider a homogeneous component of the sum (1.66), say $\sum_{k=1}^{m} x_k y_k r'_k$, where $\deg(x_k y_k r'_k) = \alpha$. Since $x_k \in A_0$, $\deg(r'_k) + \deg(y_k) = \alpha$ for all k, so all of the y_k

are the same. This implies that $\sum_k x_k r'_k = 0$, where all of the r'_k have the same degree. If $r'_k = 0$ for all k then $r_i = 0$ for all i we are done. Otherwise, for some $r'_l \neq 0$, we have $\sum_k x_k(r'_k {r'}_l^{-1}) = 0$. Since $\{x_i\}$ forms a basis for A_0 over R_0, this implies $r'_k = 0$ for all k and thus $r_i = 0$ for all $1 \leq i \leq n$. □

Example 1.4.7 GRADED DIVISION ALGEBRAS FROM VALUED DIVISION ALGEBRAS

Let D be a division algebra with a valuation. To this one associates a graded division algebra $\mathrm{gr}(D) = \bigoplus_{\gamma\in\Gamma_D} \mathrm{gr}(D)_\gamma$, where Γ_D is the value group of D and the summands $\mathrm{gr}(D)_\gamma$ arise from the filtration on D induced by the valuation (see details below and also Example 1.2.21). As is illustrated in [90], even though computations in the graded setting are often easier than working directly with D, it seems that not much is lost in passage from D to its corresponding graded division algebra $\mathrm{gr}(D)$. This has provided motivation to systematically study this correspondence, notably by Hwang, Tignol and Wadsworth [90], and to compare certain functors defined on these objects, notably the Brauer group [90, Chapter 6] and the reduced Whitehead group SK_1 [90, Chapter 11]. We introduce this correspondence here and in Chapter 3 we calculate their graded Grothendieck groups (Example 3.7.5).

Let D be a division algebra finite dimensional over its centre F, with a *valuation* $v : D^* \to \Gamma$. So Γ is a totally ordered abelian group, and for any $a, b \in D^*$, v satisfies the following conditions:

(i) $v(ab) = v(a) + v(b)$;
(ii) $v(a + b) \geq \min\{ v(a), v(b)\} \quad (b \neq -a)$.

Let

$V_D = \{ a \in D^* : v(a) \geq 0 \} \cup \{0\}$, the valuation ring of v;
$M_D = \{a \in D^* : v(a) > 0 \} \cup \{0\}$, the unique maximal left and right ideal of V_D;
$\overline{D} = V_D/M_D$, the residue division ring of v on D; and
$\Gamma_D = \mathrm{im}(v)$, the value group of the valuation.

For background on valued division algebras, see [90, Chapter 1]. One associates to D a graded division algebra as follows. For each $\gamma \in \Gamma_D$, let

$D^{\geq\gamma} = \{ d \in D^* : v(d) \geq \gamma \} \cup \{0\}$, an additive subgroup of D ;
$D^{>\gamma} = \{ d \in D^* : v(d) > \gamma \} \cup \{0\}$, a subgroup of $D^{\geq\gamma}$; and
$\mathrm{gr}(D)_\gamma = D^{\geq\gamma}/D^{>\gamma}$.

Then define

$$\mathrm{gr}(D) = \bigoplus_{\gamma\in\Gamma_D} \mathrm{gr}(D)_\gamma.$$

Because $D^{>\gamma}D^{\geq\delta} + D^{\geq\gamma}D^{>\delta} \subseteq D^{>(\gamma+\delta)}$ for all $\gamma, \delta \in \Gamma_D$, the multiplication on $\operatorname{gr}(D)$ induced by multiplication on D is well-defined, giving that $\operatorname{gr}(D)$ is a Γ-graded ring, called the *associated graded ring* of D. The multiplicative property (i) of the valuation v implies that $\operatorname{gr}(D)$ is a graded division ring. Clearly, we have $\operatorname{gr}(D)_0 = \overline{D}$, and $\Gamma_{\operatorname{gr}(D)} = \Gamma_D$. For $d \in D^*$, we write $\widetilde{d}$ for the image $d + D^{>v(d)}$ of d in $\operatorname{gr}(D)_{v(d)}$. Thus, the map given by $d \mapsto \widetilde{d}$ is a group epimorphism $D^* \to \operatorname{gr}(D)^*$ with kernel $1 + M_D$.

The restriction $v|_F$ of the valuation on D to its centre F is a valuation on F, which induces a corresponding graded field $\operatorname{gr}(F)$. Then it is clear that $\operatorname{gr}(D)$ is a graded $\operatorname{gr}(F)$-algebra, and one can prove that for

$$[\operatorname{gr}(D) : \operatorname{gr}(F)] = [\overline{D} : \overline{F}]\,|\Gamma_D : \Gamma_F| \leq [D : F] < \infty.$$

Now let F be a field with a *henselian* valuation v, *i.e.,* the valuation v has a unique extension to any algebraic extension of F. It was proved that (see [90, Chapter 1]) the valuation v extends uniquely to D as well. With respect to this valuation, D is said to be *tame* if $Z(\overline{D})$ is separable over $\overline{F}$ and

$$\operatorname{char}(\overline{F}) \nmid \operatorname{ind}(D)/(\operatorname{ind}(\overline{D})[Z(\overline{D}) : \overline{F}]).$$

It is known ([90, Chapter 8]) that D is tame if and only if

$$[\operatorname{gr}(D) : \operatorname{gr}(F)] = [D : F]$$

and $Z(\operatorname{gr}(D)) = \operatorname{gr}(F)$.

We compute the graded Grothendieck group and the graded Picard group of these division algebras in Examples 3.7.5 and 4.2.6.

1.4.1 The zero component ring of a graded central simple ring

Let A be a Γ-graded division ring and $\mathbb{M}_n(A)(\lambda_1, \dots, \lambda_n)$ be a graded simple ring, where $\lambda_i \in \Gamma$, $1 \leq i \leq n$. Since A is a graded division ring, Γ_A is a subgroup of Γ. Consider the quotient group Γ/Γ_A and let $\Gamma_A + \varepsilon_1, \dots, \Gamma_A + \varepsilon_k$ be the distinct elements in Γ/Γ_A representing the cosets $\Gamma_A + \lambda_i$, $1 \leq i \leq n$, and for each ε_l, let r_l be the number of i with $\Gamma_A + \lambda_i = \Gamma_A + \varepsilon_l$. It was observed in [90, Chapter 2] that

$$\boxed{\mathbb{M}_n(A)(\lambda_1, \dots, \lambda_n)_0 \cong \mathbb{M}_{r_1}(A_0) \times \cdots \times \mathbb{M}_{r_k}(A_0)} \tag{1.67}$$

and in particular, $\mathbb{M}_n(A)(\lambda_1, \dots, \lambda_n)_0$ is a simple ring if and only if $k = 1$. Indeed, using (1.50) and (1.51) we get

$$\mathbb{M}_n(A)(\lambda_1, \dots, \lambda_n) \cong_{\operatorname{gr}} \mathbb{M}_n(A)(\varepsilon_1, \dots, \varepsilon_1, \varepsilon_2, \dots, \varepsilon_2, \dots, \varepsilon_k, \dots, \varepsilon_k), \tag{1.68}$$

with each ε_l occurring r_l times. Now (1.45) for $\lambda = 0$ and

$$(\delta_1, \dots, \delta_n) = (\varepsilon_1, \dots \varepsilon_1, \varepsilon_2, \dots, \varepsilon_2, \dots, \varepsilon_k, \dots, \varepsilon_k)$$

immediately gives (1.67).

Remark 1.4.8 THE GRADED ARTIN–WEDDERBURN STRUCTURE THEOREM

The Artin–Wedderburn theorem shows that division rings are the basic "building blocks" of ring theory, *i.e.*, if a ring A satisfies some finite condition, for example A is right Artinian, then $A/J(A)$ is isomorphic to a finite product of matrix rings over division rings. A graded version of the Artin–Wedderburn structure theorem also holds. We state the statement here without proof. We refer the reader to [75, 90] for proofs of these statements.

A Γ-graded ring B is isomorphic to $\mathbb{M}_n(A)(\lambda_1, \dots, \lambda_n)$, where A is a Γ-graded division ring and $\lambda_i \in \Gamma$, $1 \leq i \leq n$, if and only if B is *graded right Artinian* (*i.e.*, a decreasing chain of graded right ideals becomes stationary) and graded simple.

A Γ-graded ring B is isomorphic to a finite product of matrix rings overs graded division rings (with suitable shifts) if and only if B is graded right Artinian and *graded primitive* (*i.e.*, $J^{\text{gr}}(B) = 0$).

1.5 Strongly graded rings and Dade's theorem

Let A be a Γ-graded ring and Ω be a subgroup of Γ. Recall from §1.1.2 that A has a natural Γ/Ω-graded structure and $A_\Omega = \bigoplus_{\gamma \in \Omega} A_\gamma$ is a Ω-graded ring. If A is a Γ/Ω-strongly graded ring, then one can show that the category of Γ-graded A-modules, $\text{Gr}^\Gamma\text{-}A$, is equivalent to the category of Ω-graded A_Ω-modules, $\text{Gr}^\Omega\text{-}A_\Omega$. In fact, the equivalence

$$\text{Gr}^\Gamma\text{-}A \approx \text{Gr}^\Omega\text{-}A_\Omega,$$

under the given natural functors (see Theorem 1.5.7) implies that A is a Γ/Ω-strongly graded ring. This was first proved by Dade [32] in the case of $\Omega = 0$, *i.e.*, when $\text{Gr}^\Gamma\text{-}A \approx \text{Mod-}A_0$. We prove Dade's theorem (Theorem 1.5.1) and then state this more general case in Theorem 1.5.7.

Let A be a Γ-graded ring. For any right A_0-module N and any $\gamma \in \Gamma$, we identify the right A_0-module $N \otimes_{A_0} A_\gamma$ with its image in $N \otimes_{A_0} A$. Since $A = \bigoplus_{\gamma \in \Gamma} A_\gamma$ and A_γ are A_0-bimodules, $N \otimes_{A_0} A$ is a Γ-graded right A-module, with

$$N \otimes_{A_0} A = \bigoplus_{\gamma \in \Gamma} (N \otimes_{A_0} A_\gamma). \tag{1.69}$$

Consider the *restriction functor*

$$\boxed{\begin{aligned} \mathcal{G} := (-)_0 : \text{Gr-}A &\longrightarrow \text{Mod-}A_0 \\ M &\longmapsto M_0 \\ \psi &\longmapsto \psi|_{M_0} \end{aligned}}$$

and the *induction functor* defined by

$$\boxed{\begin{aligned} \mathcal{I} := - \otimes_{A_0} A : \text{Mod-}A_0 &\longrightarrow \text{Gr-}A \\ N &\longmapsto N \otimes_{A_0} A \\ \phi &\longmapsto \phi \otimes \mathrm{id}_A . \end{aligned}}$$

One can easily check that $\mathcal{G} \circ \mathcal{I} \cong \mathrm{id}_{A_0}$ with the natural transformation

$$\begin{aligned} \mathcal{G}\mathcal{I}(N) = \mathcal{G}(N \otimes_{A_0} A) = N \otimes_{A_0} A_0 &\longrightarrow N, \\ n \otimes a &\mapsto na. \end{aligned} \tag{1.70}$$

On the other hand, there is a natural transformation

$$\begin{aligned} \mathcal{I}\mathcal{G}(M) = \mathcal{I}(M_0) = M_0 \otimes_{A_0} A &\longrightarrow M, \\ m \otimes a &\mapsto ma. \end{aligned} \tag{1.71}$$

The theorem below shows that $\mathcal{I} \circ \mathcal{G} \cong \mathrm{id}_A$ (under (1.71)), if and only if A is a strongly graded ring. Theorem 1.5.1 was proved by Dade [32, Theorem 2.8] (see also [75, Theorem 3.1.1]).

Theorem 1.5.1 (Dade's theorem) *Let A be a Γ-graded ring. Then A is strongly graded if and only if the functors*

$$(-)_0 : \text{Gr-}A \to \text{Mod-}A_0$$

and

$$- \otimes_{A_0} A : \text{Mod-}A_0 \to \text{Gr-}A$$

form mutually inverse equivalences of categories.

Proof One can easily check that (without using the assumption that A is strongly graded) $\mathcal{G} \circ \mathcal{I} \cong \mathrm{id}_{A_0}$ (see (1.70)). Suppose A is strongly graded. We show that $\mathcal{I} \circ \mathcal{G} \cong \mathrm{id}_A$.

For a graded A-module M, we have $\mathcal{I} \circ \mathcal{G}(M) = M_0 \otimes_{A_0} A$. We show that the natural homomorphism

$$\begin{aligned} \phi : M_0 \otimes_{A_0} A &\to M, \\ m \otimes a &\mapsto ma, \end{aligned}$$

is a Γ-graded A-module isomorphism. The map ϕ is clearly graded (see (1.69)). Since A is strongly graded, it follows that for $\gamma, \delta \in \Gamma$,

$$M_{\gamma+\delta} = M_{\gamma+\delta}A_0 = M_{\gamma+\delta}A_{-\gamma}A_\gamma \subseteq M_\delta A_\gamma \subseteq M_{\gamma+\delta}. \tag{1.72}$$

Thus $M_\delta A_\gamma = M_{\gamma+\delta}$. Therefore, $\phi(M_0 \otimes_{A_0} A_\gamma) = M_0 A_\gamma = M_\gamma$, which implies that ϕ is surjective.

Let $N = \ker(\phi)$, which is a graded A-submodule of $M_0 \otimes_{A_0} A$, so $N_0 = N \cap (M_0 \otimes_{A_0} A_0)$. However, the restriction of ϕ to $M_0 \otimes_{A_0} A_0 \to M_0$ is the canonical isomorphism, so $N_0 = 0$. Since N is a graded A-module, a similar argument as (1.72) shows $N_\gamma = N_0 A_\gamma = 0$ for all $\gamma \in \Gamma$. It follows that ϕ is injective. Thus $\mathcal{J} \circ \mathcal{G}(M) = M_0 \otimes_{A_0} A \cong M$. Since all the homomorphisms involved are natural, this shows that $\mathcal{J} \circ \mathcal{G} \cong \mathrm{id}_A$.

For the converse, suppose $\mathcal{J}$ and $\mathcal{G}$ are mutually inverse (under (1.71) and (1.70)). For any graded A-module M, $\mathcal{J} \circ \mathcal{G}(M) \cong_{\mathrm{gr}} M$, which gives that the map

$$M_0 \otimes_{A_0} A_\alpha \longrightarrow M_\alpha,$$
$$m \otimes a \mapsto ma$$

is bijective, where $\alpha \in \Gamma$. This immediately implies

$$M_0 A_\alpha = M_\alpha. \tag{1.73}$$

Now for any $\beta \in \Gamma$, consider the graded A-module $A(\beta)$. Replacing M by $A(\beta)$ in (1.73), we get $A(\beta)_0 A_\alpha = A(\beta)_\alpha$, *i.e.*, $A_\beta A_\alpha = A_{\beta+\alpha}$. This shows that A is strongly graded. □

Corollary 1.5.2 *Let A be a Γ-graded ring and Ω a subgroup of Γ such that A is a Γ/Ω-strongly graded ring. Then the functors*

$$(-)_0 : \mathrm{Gr}^{\Gamma/\Omega}\text{-}A \longrightarrow \mathrm{Mod}\text{-}A_\Omega$$

and

$$- \otimes_{A_\Omega} A : \mathrm{Mod}\text{-}A_\Omega \longrightarrow \mathrm{Gr}^{\Gamma/\Omega}\text{-}A$$

form mutually inverse equivalences of categories.

Proof The result follows from Theorem 1.5.1. □

Remark 1.5.3 Recall that gr-A denotes the category of graded finitely generated right A-modules and Pgr-A denotes the category of graded finitely generated projective right A-modules. Note that in general the restriction functor $(-)_0 : \mathrm{Gr}\text{-}A \to \mathrm{Mod}\text{-}A_0$ does not induce a functor $(-)_0 : \mathrm{Pgr}\text{-}A \to \mathrm{Pr}\text{-}A_0$. In fact, one can easily produce a graded finitely generated projective A-module

P such that P_0 is not a projective A_0-module. As an example, consider the $\mathbb{Z}$-graded ring T of Example 1.1.5. Then $T(1)$ is clearly a graded finitely generated projective T-module. However $T(1)_0 = M$ is not $T_0 = R$-module.

Remark 1.5.4 The proof of Theorem 1.5.1 also shows that A is strongly graded if and only if gr-$A \cong$ mod-A_0, if and only if Pgr-$A \cong$ Pr-A_0, (see Remark 1.5.3) via the same functors $(-)_0$ and $- \otimes_{A_0} A$ of the Theorem 1.5.1.

Remark 1.5.5 STRONGLY GRADED MODULES

Let A be a Γ-graded ring and M be a graded A-module. Then M is called a *strongly graded* A-module if

$$M_\alpha A_\beta = M_{\alpha+\beta}, \tag{1.74}$$

for any $\alpha, \beta \in \Gamma$. The proof of Theorem 1.5.1 shows that A is strongly graded if and only if any graded A-module is strongly graded. Indeed, if A is strongly graded then (1.72) shows that any graded A-module is strongly graded. Conversely, if any graded module is strongly graded, then considering A as a graded A-module, (1.74) for $M = A$, shows that $A_\alpha A_\beta = A_{\alpha+\beta}$ for any $\alpha, \beta \in \Gamma$.

Remark 1.5.6 IDEALS CORRESPONDENCE BETWEEN A_0 AND A

The proof of Theorem 1.5.1 shows that there is a one-to-one correspondence between the right ideals of A_0 and the graded right ideals of A (similarly for the left ideals). However, this correspondence does not hold between two-sided ideals. As an example, $A = \mathbb{M}_2(K[x^2, x^{-2}])(0, 1)$, where K is a field, is a strongly $\mathbb{Z}$-graded simple ring, whereas $A_0 \cong K \otimes K$ is not a simple ring. (See §1.4.1. Also see Proposition 4.2.9 for a relation between simplicity of A_0 and A.)

In the same way, the equivalence Gr-$A \approx$ Mod-A_0 of Theorem 1.5.1 gives a correspondence between several (one-sided) properties of graded objects in A with objects over A_0. For example, one can easily show that A is graded right (left) Noetherian if and only if A_0 is right (left) Noetherian (see also Corollary 1.5.10).

Using Theorem 1.5.1, we will see that the graded Grothendieck group of a strongly graded ring coincides with the (classical) Grothendieck group of its 0-component ring (see §3.1.3).

We need a more general version of grading defined in (1.69) in order to extend Dade's theorem. Let A be a Γ-graded ring and Ω a subgroup of Γ. Let N be a Ω-graded right A_Ω-module. Then $N \otimes_{A_\Omega} A$ is a Γ-graded right A-module, with the grading defined by

$$(N \otimes_{A_\Omega} A)_\gamma = \Big\{ \sum_i n_i \otimes a_i \mid n_i \in N^h, a_i \in A^h, \deg(n_i) + \deg(a_i) = \gamma \Big\}.$$

A similar argument as in §1.2.6 for tensor products will show that this grading is well-defined. Note that with this grading,

$$(N \otimes_{A_\Omega} A)_\Omega = N \otimes_{A_\Omega} A_\Omega \cong N,$$

as graded right A_Ω-modules.

Theorem 1.5.7 *Let A be a Γ-graded ring and Ω be a subgroup of Γ. Consider A as a Γ/Ω-graded ring. Then A is a Γ/Ω-strongly graded ring if and only if*

$$\begin{aligned} (-)_\Omega : \mathrm{Gr}^\Gamma\text{-}A &\longrightarrow \mathrm{Gr}^\Omega\text{-}A_\Omega \\ M &\longmapsto M_\Omega \\ \psi &\longmapsto \psi|_{M_\Omega} \end{aligned}$$

and

$$\begin{aligned} - \otimes_{A_0} A : \mathrm{Gr}^\Omega\text{-}A_\Omega &\longrightarrow \mathrm{Gr}^\Gamma\text{-}A \\ N &\longmapsto N \otimes_{A_\Omega} A \\ \phi &\longmapsto \phi \otimes \mathrm{id}_A \end{aligned}$$

form mutually inverse equivalences of categories.

Proof The proof is similar to the proof of Theorem 1.5.1 and it is omitted. □

Remark 1.5.8 Compare Theorem 1.5.7, with the following statement. Let A be a Γ-graded ring and Ω be a subgroup of Γ. Then A is Γ-strongly graded ring if and only if

$$\begin{aligned} (-)_\Omega : \mathrm{Gr}^\Gamma\text{-}A &\longrightarrow \mathrm{Gr}^\Omega\text{-}A_\Omega \\ M &\longmapsto M_\Omega \\ \psi &\longmapsto \psi|_{M_\Omega} \end{aligned}$$

and

$$\begin{aligned} - \otimes_{A_0} A : \mathrm{Gr}^\Omega\text{-}A_\Omega &\longrightarrow \mathrm{Gr}^\Gamma\text{-}A \\ N &\longmapsto N_0 \otimes_{A_0} A \\ \phi &\longmapsto \phi_0 \otimes \mathrm{id}_A \end{aligned}$$

form mutually inverse equivalences of categories.

Example 1.5.9 Let A be a $\Gamma \times \Omega$-graded ring such that $1 \in A_{(\alpha,\Omega)}A_{(-\alpha,\Omega)}$ for any $\alpha \in \Gamma$, where $A_{(\alpha,\Omega)} = \bigoplus_{\omega \in \Omega} A_{(\alpha,\omega)}$. Then by Theorem 1.5.7

$$\mathrm{Gr}^{\Gamma \times \Omega}\text{-}A \approx \mathrm{Gr}^{\Omega}\text{-}A_{(0,\Omega)}.$$

This example will be used in §6.4. Compare this also with Corollary 1.2.13.

Another application of Theorem 1.5.1 is to provide a condition when a strongly graded ring is a graded von Neumann ring (§1.1.9). This will be used later in Corollary 1.6.17 to show that the Leavitt path algebras are von Neumann regular rings.

Corollary 1.5.10 *Let A be a strongly graded ring. Then A a is graded von Neumann regular ring if and only if A_0 is a von Neumann regular ring.*

Sketch of proof Since any (graded) flat module is a direct limit of (graded) projective modules, from the equivalence of categories $\mathrm{Gr}\text{-}A \approx_{\mathrm{gr}} \mathrm{Mod}\text{-}A_0$ (Theorem 1.5.1), it follows that A is graded von Neumann regular if and only if A_0 is von Neumann regular. □

Remark 1.5.11 An element-wise proof of Corollary 1.5.10 can also be found in [96, Theorem 3].

For a Γ-graded ring A, and $\alpha, \beta \in \Gamma$, one has an A_0-bimodule homomorphism

$$\begin{aligned} \phi_{\alpha,\beta} : A_\alpha \otimes_{A_0} A_\beta &\longrightarrow A_{\alpha+\beta} \\ a \otimes b &\longmapsto ab. \end{aligned} \tag{1.75}$$

The following theorem gives another characterisation for strongly graded rings.

Theorem 1.5.12 *Let A be a Γ-graded ring. Then A is a strongly graded ring if and only if for any $\gamma \in \Gamma$, the homomorphism*

$$\begin{aligned} \phi_{\gamma,-\gamma} : A_\gamma \otimes_{A_0} A_{-\gamma} &\longrightarrow A_0, \\ a \otimes b &\longmapsto ab \end{aligned}$$

is an isomorphism. In particular, if A is strongly graded, then the homogeneous components A_γ, $\gamma \in \Gamma$, are finitely generated projective A_0-modules.

Proof Suppose that for any $\gamma \in \Gamma$, the map $\phi_{\gamma,-\gamma} : A_\gamma \otimes A_{-\gamma} \to A_0$ is an isomorphism. Thus there are $a_i \in A_\gamma$, $b_i \in A_{-\gamma}$ such that

$$\sum_i a_i b_i = \phi_{\gamma,-\gamma}\Big(\sum_i a_i \otimes b_i\Big) = 1.$$

So $1 \in A_\gamma A_{-\gamma}$. Now by Proposition 1.1.15(1) A is strongly graded.

Conversely, suppose A is a strongly graded ring. We prove that the homomorphism (1.75) is an isomorphism. The definition of strongly graded implies that $\phi_{\alpha,\beta}$ is surjective. Suppose

$$\phi_{\alpha,\beta}\Big(\sum_i a_i \otimes b_i\Big) = \sum_i a_i b_i = 0. \tag{1.76}$$

Using Proposition 1.1.15(1), write $1 = \sum_j x_j y_j$, where $x_j \in A_{-\beta}$ and $y_j \in A_\beta$. Then

$$\begin{aligned}
\sum_i a_i \otimes b_i = \Big(\sum_i a_i \otimes b_i\Big)\Big(\sum_j x_j y_j\Big) &= \sum_i \Big(a_i \otimes \sum_j b_i x_j y_j\Big)\\
= \sum_i \Big(\sum_j (a_i b_i x_j \otimes y_j)\Big) &= \sum_j \sum_i \Big(a_i b_i x_j \otimes y_j\Big)\\
&= \sum_j \Big(\sum_i (a_i b_i) x_j \otimes y_j)\Big) = 0.
\end{aligned}$$

This shows that $\phi_{\alpha,\beta}$ is injective. Now setting $\alpha = \gamma$ and $\beta = -\gamma$ finishes the proof.

Finally, if A is strongly graded, the above argument shows that the homogeneous components A_γ, $\gamma \in \Gamma$, are invertible A_0-modules, which in turn implies that A_α are finitely generated projective A_0-modules. □

1.5.1 Invertible components of strongly graded rings

Let A and B be rings and P be an A–B-bimodule. Then P is called an *invertible* $A-B$-bimodule if there is a $B-A$-bimodule Q such that $P \otimes_B Q \cong A$ as $A-A$-bimodules and $Q \otimes_A P \cong B$ as $B-B$-bimodules and the following diagrams are commutative:

$$\begin{array}{ccc}
P \otimes_B Q \otimes_A P & \longrightarrow & A \otimes_A P \\
\downarrow & & \downarrow \\
P \otimes_B B & \longrightarrow & P
\end{array}
\qquad
\begin{array}{ccc}
Q \otimes_A P \otimes_B Q & \longrightarrow & B \otimes_B Q \\
\downarrow & & \downarrow \\
Q \otimes_A A & \longrightarrow & Q
\end{array}$$

One can prove that P is a finitely generated projective A and B-module.

Now Theorem 1.5.12 shows that for a strongly Γ-graded ring A, the A_0-bimodules A_γ, $\gamma \in \Gamma$, are invertible modules and thus are finitely generated projective A_0-modules. This in return implies that A is a projective A_0-module. Note that, in general, one can easily construct a graded ring A where A is not projective over A_0 (see Example 1.1.5) and A_γ is not a finitely generated A_0-module, such as the $\mathbb{Z}$-graded ring $\mathbb{Z}[x_i \mid i \in \mathbb{N}]$ of Example 1.1.9.

Remark 1.5.13 OTHER TERMINOLOGIES FOR STRONGLY GRADED RINGS

The term "strongly graded" for such rings was coined by E. Dade in [32] and is now commonly in use. Other terms for these rings are *fully graded* and *generalised crossed products*. See [33] for a history of the development of such rings in literature.

1.6 Grading on graph algebras

1.6.1 Grading on free rings

Let X be a nonempty set of symbols and Γ be a group. (As always we assume the groups are abelian, although the entire theory can be written for an arbitrary group.) Let $d : X \to \Gamma$ be a map. One can extend d in a natural way to a map from the set of finite words on X to Γ, which is called d again. For example if $x, y, z \in X$ and xyz is a word, then $d(xyz) = d(x) + d(y) + d(z)$. One can easily see that if w_1, w_2 are two words, then $d(w_1w_2) = d(w_1) + d(w_2)$. If we allow an empty word, which will be the identity element in the free ring, then we assign the identity of Γ to this word.

Let R be a ring and $R(X)$ be the free ring (with or without identity) on a set X with coefficients in R. The elements of $R(X)$ are of the form $\sum_w r_w w$, where $r_w \in R$ and w stands for a word on X. The multiplication is defined by convolution, *i.e.,*

$$\Big(\sum_w r_w w\Big)\Big(\sum_v r_v v\Big) = \sum_z \Big(\sum_{\{w,v | z=wv,\ r_w, r_v \neq 0\}} r_w r_v\Big) z.$$

In order to make $R(X)$ into a graded ring, define

$$R(X)_\gamma = \Big\{ \sum_w r_w w \mid d(w) = \gamma \Big\}.$$

One can check that $R(X) = \bigoplus_{\gamma\in\Gamma} R(X)_\gamma$. Thus $R(X)$ is a Γ-graded ring. Note that if we don't allow the empty word in the construction, then $R(X)$ is a graded ring without identity (see Remark 1.1.14). It is easy to see that $R(X)$ is never a strongly graded ring.

Example 1.6.1 Let R be a ring and $R(X)$ be the free ring on a set X with a graded structure induced by a map $d : X \to \Gamma$. Let Ω be a subgroup of Γ and consider the map

$$\begin{aligned} \overline{d} : X &\longrightarrow \Gamma/\Omega, \\ x &\longmapsto \Omega + d(x). \end{aligned}$$

The map $\overline{d}$ induces a Γ/Ω-graded structure on $R(X)$ which coincides with the general construction of quotient grading given in §1.1.2.

Example 1.6.2 Let $X = \{x\}$ be a set of symbols with one element and $\mathbb{Z}_n$ be the cyclic group with n elements. Assign $1 \in \mathbb{Z}_n$ to x and generate the free ring with identity on X with coefficients in a field F. This ring is the usual polynomial ring $F[x]$ which, by the above construction, is equipped with a $\mathbb{Z}_n$-grading. Namely,

$$F[x] = \bigoplus_{k \in \mathbb{Z}_n} \Big(\sum_{\substack{l \in \mathbb{N}, \\ \overline{l} = k}} F x^l \Big),$$

where $\overline{l}$ is the image of l in the group $\mathbb{Z}_n$. For $a \in F$, since the polynomial $x^n - a$ is a homogeneous element of degree zero, the ideal $\langle x^n - a \rangle$ is a graded ideal and thus the quotient ring $F[x]/\langle x^n - a \rangle$ is also a $\mathbb{Z}_n$-graded ring (see §1.1.5). In particular, if $x^n - a$ is an irreducible polynomial in $F[x]$, then the field $F[x]/\langle x^n - a \rangle$ is a $\mathbb{Z}_n$-graded field as well.

Example 1.6.3 Let $\{x, y\}$ be a set of symbols. Assign $1 \in \mathbb{Z}_2$ to x and y and consider the graded free ring $\mathbb{R}\langle x, y \rangle$. The ideal generated by homogeneous elements $\{x^2 + 1, y^2 + 1, xy + yx\}$ is graded and thus we retrieve the $\mathbb{Z}_2$-graded Hamilton quaternion algebra of Example 1.1.20 as follows:

$$\mathbb{H} \cong \mathbb{R}\langle x, y \rangle / \langle x^2 + 1, y^2 + 1, xy + yx \rangle.$$

Moreover, assigning $(1, 0) \in \mathbb{Z}_2 \times \mathbb{Z}_2$ to x and $(0, 1) \in \mathbb{Z}_2 \times \mathbb{Z}_2$ to y we obtained the $\mathbb{Z}_2 \times \mathbb{Z}_2$-graded quaternion algebra of Example 1.1.20.

Example 1.6.4 THE WEYL ALGEBRA

For a (commutative) ring R, the *Weyl algebra* $R\langle x, y \rangle / \langle xy - yx - 1 \rangle$ can be considered as a $\mathbb{Z}$-graded ring by assigning 1 to x and -1 to y.

Example 1.6.5 THE LEAVITT ALGEBRA $\mathcal{L}(n, k + 1)$

Let K be a field, n and k positive integers and A be the free associative K-algebra with identity generated by symbols $\{x_{ij}, y_{ji} \mid 1 \leq i \leq n + k, 1 \leq j \leq n\}$ subject to relations (coming from)

$$Y \cdot X = I_{n,n} \qquad \text{and} \qquad X \cdot Y = I_{n+k,n+k},$$

where

$$Y = \begin{pmatrix} y_{11} & y_{12} & \dots & y_{1,n+k} \\ y_{21} & y_{22} & \dots & y_{2,n+k} \\ \vdots & \vdots & \ddots & \vdots \\ y_{n,1} & y_{n,2} & \dots & y_{n,n+k} \end{pmatrix}, \quad X = \begin{pmatrix} x_{11} & x_{12} & \dots & x_{1,n} \\ x_{21} & x_{22} & \dots & x_{2,n} \\ \vdots & \vdots & \ddots & \vdots \\ x_{n+k,1} & x_{n+k,2} & \dots & x_{n+k,n} \end{pmatrix}. \tag{1.77}$$

To be concrete, the relations are

$$\sum_{j=1}^{n+k} y_{ij}x_{jl} = \delta_{i,l}, \qquad 1 \le i, l \le n,$$

$$\sum_{j=1}^{n} x_{ij}y_{jl} = \delta_{i,l}, \qquad 1 \le i, l \le n+k.$$

In Example 1.3.19 we studied a special case of this algebra when $n = 1$ and $k = n - 1$. This algebra was studied by Leavitt in relation with its type in [63, p.130] where it is shown that for arbitrary n and k the algebra is of type (n, k) (see §1.7) and when $n \ge 2$ they are domains. We denote this algebra by $\mathcal{L}(n, k+1)$. (Cohn's notation in [28] for this algebra is $V_{n,n+k}$.)

Assigning

$$\deg(y_{ji}) = (0, \dots, 0, 1, 0 \dots, 0),$$
$$\deg(x_{ij}) = (0, \dots, 0, -1, 0 \dots, 0),$$

for $1 \le i \le n+k$, $1 \le j \le n$, in $\bigoplus_n \mathbb{Z}$, where 1 and -1 are in the jth entries respectively, makes the free algebra generated by x_{ij} and y_{ji} a graded ring. Moreover, one can easily observe that the relations coming from (1.77) are all homogeneous with respect to this grading, so that the Leavitt algebra $\mathcal{L}(n, k+1)$ is a $\bigoplus_n \mathbb{Z}$-graded ring. In particular, $\mathcal{L}(1, k)$ is a $\mathbb{Z}$-graded ring (Example 1.3.19).

1.6.2 Corner skew Laurent polynomial rings

Let R be a ring with identity and p an idempotent of R. Let $\phi : R \to pRp$ be a *corner* isomorphism, i.e, a ring isomorphism such that $\phi(1) = p$. A *corner skew Laurent polynomial ring* with coefficients in R, denoted by $R[t_+, t_-, \phi]$, is a unital ring which is constructed as follows: The elements of $R[t_+, t_-, \phi]$ are the formal expressions

$$t_-^j r_{-j} + t_-^{j-1} r_{-j+1} + \dots + t_- r_{-1} + r_0 + r_1 t_+ + \dots + r_i t_+^i,$$

where $r_{-n} \in p_n R$ and $r_n \in Rp_n$, for all $n \geq 0$, where $p_0 = 1$ and $p_n = \phi^n(p_0)$. The addition is component-wise, and the multiplication is determined by the distribution law and the following rules:

$$t_- t_+ = 1, \qquad t_+ t_- = p, \qquad rt_- = t_-\phi(r), \qquad t_+ r = \phi(r)t_+. \tag{1.78}$$

The corner skew Laurent polynomial rings are studied in [6], where their K_1-groups are calculated. This construction is a special case of the so-called fractional skew monoid rings constructed in [7]. Assigning -1 to t_- and 1 to t_+ makes $A := R[t_+, t_-, \phi]$ a $\mathbb{Z}$-graded ring with $A = \bigoplus_{i \in \mathbb{Z}} A_i$, where

$$\begin{aligned} A_i &= Rp_i t_+^i, \text{ for } i > 0, \\ A_i &= t_-^i p_{-i} R, \text{ for } i < 0, \\ A_0 &= R, \end{aligned}$$

(see [7, Proposition 1.6]). Clearly, when $p = 1$ and ϕ is the identity map, then $R[t_+, t_-, \phi]$ reduces to the familiar ring $R[t, t^{-1}]$.

In the next three propositions we will characterise those corner skew Laurent polynomials which are strongly graded rings (§1.1.3), crossed products (§1.1.4) and graded von Neumann regular rings (§1.1.9).

Recall that an idempotent element p of the ring R is called a *full idempotent* if $RpR = R$.

Proposition 1.6.6 *Let R be a ring with identity and $A = R[t_+, t_-, \phi]$ a corner skew Laurent polynomial ring. Then A is strongly graded if and only if $\phi(1)$ is a full idempotent.*

Proof First note that $A_1 = R\phi(1)t_+$ and $A_{-1} = t_-\phi(1)R$. Moreover, since $\phi(1) = p$, we have

$$r_1\phi(1)t_+t_-\phi(1)r_2 = r_1\phi(1)p\phi(1)r_2 = r_1 ppp r_2 = r_1\phi(1)r_2.$$

Suppose A is strongly graded. Then $1 \in A_1 A_{-1}$. That is

$$1 = \sum_i \big(r_i\phi(1)t_+\big)\big(t_-\phi(1)r_i'\big) = \sum_i r_i\phi(1)r_i', \tag{1.79}$$

where $r_i, r_i' \in R$. So $R\phi(1)R = R$, that is $\phi(1)$ is a full idempotent.

On the other hand suppose $\phi(1)$ is a full idempotent. Since $\mathbb{Z}$ is generated by 1, in order to prove that A is strongly graded, it is enough to show that $1 \in A_1 A_{-1}$ and $1 \in A_{-1} A_1$ (see §1.1.3). But

$$t_-\phi(1)\phi(1)t_+ = t_{-1}\phi(1)t_+ = 1t_-t_+ = 1,$$

which shows that $1 \in A_{-1}A_1$. Since $\phi(1)$ is a full idempotent, there are $r_i, r_i' \in R$, $i \in I$ such that $\sum r_i\phi(1)r_i' = 1$. Then Equation (1.79) shows that $1 \in A_1 A_{-1}$. □

Recall that a ring R is called *Dedekind finite* or *directly finite* if any one-sided invertible element is two-sided invertible. That is, if $ab = 1$, then $ba = 1$, where $a, b \in R$. For example, left (right) Noetherian rings are Dedekind finite.

Proposition 1.6.7 *Let R be a ring with identity which is Dedekind finite and $A = R[t_+, t_-, \phi]$ a corner skew Laurent polynomial ring. Then A is crossed product if and only if $\phi(1) = 1$.*

Proof If $\phi(1) = 1$, then from relations (1.78) it follows that $t_-t_+ = t_+t_- = 1$. Therefore all homogeneous components contain invertible elements and thus A is crossed product.

Suppose A is crossed product. Then there are $a, b \in R$ such that $(t_-a)(bt_+) = 1$ and $(bt_+)(t_-a) = 1$. Using relations (1.78), the first equality gives $ab = p$ and the second one gives $bpa = 1$, where $\phi(1) = p$. Now

$$1 = bpa = bppa = babpa = ba.$$

Since R is Dedekind finite, it follows $ab = 1$ and thus $p = \phi(1) = 1$. □

Proposition 1.6.8 *Let R be a ring with identity and $A = R[t_+, t_-, \phi]$ a corner skew Laurent polynomial ring. Then A is a graded von Neumann regular ring if and only if R is a von Neumann regular ring.*

Proof If a graded ring is graded von Neumann regular, then it is easy to see that its zero component ring is von Neumann regular. This proves one direction of the theorem. For the converse, suppose R is regular. Let $x \in A_i$, where $i > 0$. So $x = rp_it_+^i$, for some $r \in R$, where $p_i = \phi^i(1)$. By relations (1.78) and induction, we have $t_+^it_-^i = \phi^i(p_0) = p_i$. Since R is regular, there is an $s \in R$ such that $rp_isrp_i = rp_i$. Then, choosing $y = t_-^ip_is$, we have

$$xyx = (rp_it_+^i)(t_-^ip_is)(rp_it_+^i) = (rp_it_+^it_-^ip_is)(rp_it_+^i) = rp_ip_ip_isrp_it_+^i = rp_it_+^i = x.$$

A similar argument shows that for $x \in A_i$, where $i < 0$, there is a y such that $xyx = x$. This shows that A is a graded von Neumann regular ring. □

Note that in a corner skew Laurent polynomial ring $R[t_+, t_-, \phi]$, t_+ is a left invertible element with a right inverse t_- (see the relations (1.78)). In fact this property characterises such rings. Namely, a graded ring $A = \bigoplus_{i \in \mathbb{Z}} A_i$ such that A_1 has a left invertible element is a corner skew Laurent polynomial ring, as the following theorem shows. The following theorem (first established in [7]) will be used to realise Leavitt path algebras (§1.6.4) as corner skew Laurent polynomial rings (Example 1.6.14).

Theorem 1.6.9 *Let A be a $\mathbb{Z}$-graded ring which has a left invertible element $t_+ \in A_1$. Then t_+ has a right inverse $t_- \in A_{-1}$, and $A = A_0[t_+, t_-, \phi]$, where*

$$\begin{aligned} \phi : A_0 &\longrightarrow t_+t_-A_0t_+t_-, \\ a &\longmapsto t_+at_-. \end{aligned} \tag{1.80}$$

Proof Since t_+ has a right inverse, it follows easily that there is a $t_- \in A_{-1}$ with $t_-t_+ = 1$. Moreover $t_+t_- = t_+t_-t_+t_-$ is a homogeneous idempotent of degree zero. Observe that the map (1.80) is a (unital) ring isomorphism. Consider the corner skew Laurent polynomial ring $\widetilde{A} = A_0[\widetilde{t_+}, \widetilde{t_-}, \phi]$. Since $\phi(a) = t_+at_-$, it follows that $t_-\phi(a) = at_-$ and $\phi(a)t_+ = t_+a$. Thus t_+ and t_- satisfy all the relations in (1.78). Therefore there is a well-defined map $\psi : \widetilde{A} \to A$, such that $\psi(\widetilde{t_\pm}) = t_\pm$ and the restriction of ψ on A_0 is the identity and

$$\psi\Big(\sum_{k=1}^{j} \widetilde{t_-^k} a_{-k} + a_0 + \sum_{k=1}^{i} a_i\widetilde{t_+^i}\Big) = \sum_{k=1}^{j} t_-^k a_{-k} + a_0 + \sum_{k=1}^{i} a_i t_+^i.$$

This also shows that ψ is a graded homomorphism. In order to show that ψ is an isomorphism, it suffices to show that its restriction to each homogeneous component $\psi : \widetilde{A_i} \to A_i$ is a bijection. Suppose $x \in \widetilde{A_i}$, $i > 0$ such that $\psi(x) = 0$. Then $x = d\widetilde{t_+^i}$ for some $d \in A_0p_i$ where $p_i = \phi^i(1)$ and $\psi(x) = dt_+^i$. Note that $\phi^i(1) = t_+^i t_-^i$. Thus $d\phi^i(1) = dt_+^i t_-^i = \psi(x)t_-^i = 0$. It now follows that $x = d\widetilde{t_+^i} = d\phi^i(1)\widetilde{t_+^i} = 0$ in $\widetilde{A_i}$. This shows ψ is injective. Suppose $y \in A_i$. Then $yt_-^i \in A_0$ and $yt_-^i t_+^i t_-^i = yt_-^i\phi^i(1) \in A_0\phi^i(1) = A_0p_i$. This shows $yt_-^i t_+^i t_-^i \widetilde{t_+^i} \in \widetilde{A_i}$. But $\psi(yt_-^i t_+^i t_-^i \widetilde{t_+^i}) = yt_-^i t_+^i t_-^i t_+^i = y$. This shows that $\psi : \widetilde{A_i} \to A_i$, $i > 0$ is a bijection. A similar argument can be written for the case of $i < 0$. The case $i = 0$ is obvious. This completes the proof. □

1.6.3 Graphs

In this subsection we gather some graph-theoretic definitions which are needed for the construction of path algebras in §1.6.4.

A *directed graph* $E = (E^0, E^1, r, s)$ consists of two countable sets E^0, E^1 and maps $r, s : E^1 \to E^0$. The elements of E^0 are called *vertices* and the elements of E^1 *edges*. If $s^{-1}(v)$ is a finite set for every $v \in E^0$, then the graph is called *row-finite*. In this book we will only consider row-finite graphs. In this setting, if the number of vertices, *i.e.*, $|E^0|$, is finite, then the number of edges, *i.e.*, $|E^1|$, is finite as well and we call E a *finite* graph.

For a graph $E = (E^0, E^1, r, s)$, a vertex v for which $s^{-1}(v)$ is empty is called a *sink*, while a vertex w for which $r^{-1}(w)$ is empty is called a *source*. An edge with the same source and range is called a *loop*. A path μ in a graph E is a sequence of edges $\mu = \mu_1 \dots \mu_k$, such that $r(\mu_i) = s(\mu_{i+1}), 1 \leq i \leq k-1$. In this case, $s(\mu) := s(\mu_1)$ is the *source* of μ, $r(\mu) := r(\mu_k)$ is the *range* of μ, and k is the *length* of μ which is denoted by $|\mu|$. We consider a vertex $v \in E^0$ as a *trivial* path of length zero with $s(v) = r(v) = v$. If μ is a nontrivial path in E, and if $v = s(\mu) = r(\mu)$, then μ is called a *closed path based at* v. If $\mu = \mu_1 \dots \mu_k$ is a closed path based at $v = s(\mu)$ and $s(\mu_i) \neq s(\mu_j)$ for every $i \neq j$, then μ is called a *cycle*. Throughout, we denote a cycle of length n by C_n. We call a graph without cycles a *acyclic* graph. A graph consisting of only one cycle and all the paths ending on this cycle is called a *comet* graph. A C_n-comet graph is a comet graph with a cycle of length n. Here are examples of an acyclic and a 2-comet graph.

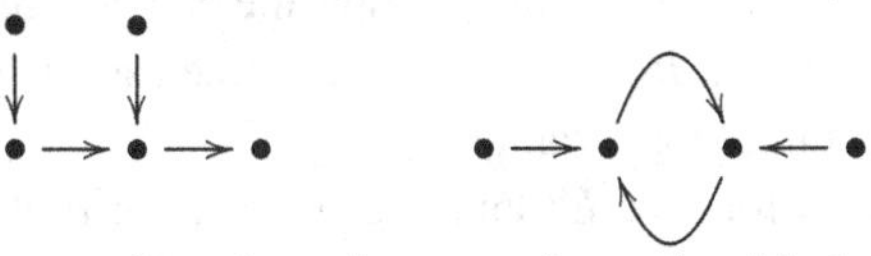

For two vertices v and w, the existence of a path with the source v and the range w is denoted by $v \geq w$. Here we allow paths of length zero. By $v \geq_n w$, we mean there is a path of length n connecting these vertices. Therefore $v \geq_0 v$ represents the vertex v. Also, by $v > w$, we mean a path from v to w where $v \neq w$. In this book, by $v \geq w' \geq w$ it is understood that there is a path connecting v to w and going through w' (*i.e.*, w' is on the path connecting v to w). For $n \geq 2$, we define E^n to be the set of paths of length n and $E^* = \bigcup_{n \geq 0} E^n$, the set of all paths.

For a graph E, let $n_{v,w}$ be the number of edges with the source v and range w. Then the *adjacency matrix* of the graph E is $A_E = (n_{v,w})$. Usually one orders the vertices and then writes A_E based on this ordering. Two different orderings of vertices give different adjacency matrices. However, if A_E and A'_E are two adjacency matrices of E, then there is a permutation matrix P such that $A'_E = PA_EP^{-1}$.

A graph E is called *essential* if E does not have sinks and sources. Moreover, a graph is called *irreducible* if for every ordered pair of vertices v and w there is a path from v to w.

1.6.4 Leavitt path algebras

A path algebra, with coefficients in the field K, is constructed as follows: consider a K-vector space with finite paths as the basis and define the multiplication by concatenation of paths. A path algebra has a natural graded structure by assigning paths as homogeneous elements of degree equal to their lengths. A formal definition of path algebras with coefficients in a ring R is given below.

Definition 1.6.10 For a graph E and a ring R with identity, we define the *path algebra of* E, denoted by $\mathcal{P}_R(E)$, to be the algebra generated by the sets $\{v \mid v \in E^0\}$, $\{\alpha \mid \alpha \in E^1\}$ with coefficients in R, subject to the relations

1 $v_i v_j = \delta_{ij} v_i$ for every $v_i, v_j \in E^0$;
2 $s(\alpha)\alpha = \alpha r(\alpha) = \alpha$ for all $\alpha \in E^1$.

Here the ring R commutes with the generators $\{v, \alpha \mid v \in E^0, \alpha \in E^1\}$. When the coefficient ring R is clear from the context, we simply write $\mathcal{P}(E)$ instead of $\mathcal{P}_R(E)$. When R is not commutative, then we consider $\mathcal{P}_R(E)$ as a left R-module. Using the above two relations, it is easy to see that when the number of vertices is finite, then $\mathcal{P}_R(E)$ is a ring with identity $\sum_{v \in E^0} v$.

When the graph has one vertex and n loops, the path algebra associated with this graph is isomorphic to $R\langle x_1, \dots, x_n \rangle$, *i.e.,* a free associative unital algebra over R with n noncommuting variables.

Setting $\deg(v) = 0$ for $v \in E^0$ and $\deg(\alpha) = 1$ for $\alpha \in E^1$, we obtain a natural $\mathbb{Z}$-grading on the free R-ring generated by $\{v, \alpha \mid v \in E^0, \alpha \in E^1\}$ (§1.6.1). Since the relations in Definition 1.6.10 are all homogeneous, the ideal generated by these relations is homogeneous and thus we have a natural $\mathbb{Z}$-grading on $\mathcal{P}_R(E)$. Note that $\mathcal{P}(E)$ is positively graded, and for any $m, n \in \mathbb{N}$,

$$\mathcal{P}(E)_m \, \mathcal{P}(E)_n = \mathcal{P}(E)_{m+n}.$$

However, by Proposition 1.1.15(2), $\mathcal{P}(E)$ is not a strongly $\mathbb{Z}$-graded ring.

The theory of Leavitt path algebras was introduced in [2, 5] which associate to directed graphs certain types of algebras. These algebras were motivated by Leavitt's construction of universal non-IBN rings [63]. Leavitt path algebras are quotients of path algebras by relations resembling those in the construction of algebras studied by Leavitt (see Example 1.3.19).

Definition 1.6.11 For a row-finite graph E and a ring R with identity, we define the *Leavitt path algebra of* E, denoted by $\mathcal{L}_R(E)$, to be the algebra generated by the sets $\{v \mid v \in E^0\}$, $\{\alpha \mid \alpha \in E^1\}$ and $\{\alpha^* \mid \alpha \in E^1\}$ with the coefficients in R, subject to the relations

1 $v_i v_j = \delta_{ij} v_i$ for every $v_i, v_j \in E^0$;
2 $s(\alpha)\alpha = \alpha r(\alpha) = \alpha$ and $r(\alpha)\alpha^* = \alpha^* s(\alpha) = \alpha^*$ for all $\alpha \in E^1$;
3 $\alpha^* \alpha' = \delta_{\alpha\alpha'} r(\alpha)$, for all $\alpha, \alpha' \in E^1$;
4 $\sum_{\{\alpha \in E^1, s(\alpha) = v\}} \alpha\alpha^* = v$ for every $v \in E^0$ for which $s^{-1}(v)$ is nonempty.

Here the ring R commutes with the generators $\{v, \alpha, \alpha^* \mid v \in E^0, \alpha \in E^1\}$. When the coefficient ring R is clear from the context, we simply write $\mathcal{L}(E)$ instead of $\mathcal{L}_R(E)$. When R is not commutative, then we consider $\mathcal{L}_R(E)$ as a left R-module. The elements α^* for $\alpha \in E^1$ are called *ghost edges*. One can show that $\mathcal{L}_R(E)$ is a ring with identity if and only if the graph E is finite (otherwise, $\mathcal{L}_R(E)$ is a ring with local identities, see [2, Lemma 1.6]).

Setting $\deg(v) = 0$, for $v \in E^0$, $\deg(\alpha) = 1$ and $\deg(\alpha^*) = -1$ for $\alpha \in E^1$, we obtain a natural $\mathbb{Z}$-grading on the free R-ring generated by $\{v, \alpha, \alpha^* \mid v \in E^0, \alpha \in E^1\}$. Since the relations in Definition 1.6.11 are all homogeneous, the ideal generated by these relations is homogeneous and thus we have a natural $\mathbb{Z}$-grading on $\mathcal{L}_R(E)$.

If $\mu = \mu_1 \dots \mu_k$, where $\mu_i \in E^1$, is an element of $\mathcal{L}(E)$, then we denote by μ^* the element $\mu_k^* \dots \mu_1^* \in \mathcal{L}(E)$. Further, we define $v^* = v$ for any $v \in E^0$. Since $\alpha^*\alpha' = \delta_{\alpha\alpha'} r(\alpha)$, for all $\alpha, \alpha' \in E^1$, any word in the generators $\{v, \alpha, \alpha^* \mid v \in E^0, \alpha \in E^1\}$ in $\mathcal{L}(E)$ can be written as $\mu\gamma^*$, where μ and γ are paths in E (recall that vertices were considered paths of length zero). The elements of the form $\mu\gamma^*$ are called *monomials*.

If the graph E is infinite, $\mathcal{L}_R(E)$ is a graded ring without identity (see Remark 1.1.14).

Taking the grading into account, one can write

$$\mathcal{L}_R(E) = \bigoplus_{k \in \mathbb{Z}} \mathcal{L}_R(E)_k,$$

where

$$\boxed{\mathcal{L}_R(E)_k = \Big\{ \sum_i r_i \alpha_i \beta_i^* \mid \alpha_i, \beta_i \text{ are paths, } r_i \in R, \text{ and } |\alpha_i| - |\beta_i| = k \text{ for all } i \Big\}.}$$

For simplicity we denote $\mathcal{L}_R(E)_k$, the homogeneous elements of degree k, by $\mathcal{L}_k$.

Example 1.6.12 A GRADED RING WHOSE MODULES ARE ALL GRADED

Consider the infinite line graph

$$E: \quad \cdots\cdots \dashrightarrow u_{-1} \xrightarrow{e_0} u_0 \xrightarrow{e_1} u_1 \dashrightarrow \cdots\cdots$$

Then the Leavitt path algebra $\mathcal{L}(E)$ is a $\mathbb{Z}$-graded ring. Let X be a right $\mathcal{L}(E)$-module. Set

$$X_i = Xu_i, i \in \mathbb{Z},$$

and observe that $X = \bigoplus_{i \in \mathbb{Z}} X_i$. It is easy to check that X becomes a graded $\mathcal{L}(E)$-module. Moreover, any module homomorphism is a graded homomorphism. Note, however, that the module category Mod-$\mathcal{L}(E)$ is not equivalent to Gr-$\mathcal{L}(E)$. Also notice that although any ideal is a graded module over $\mathcal{L}(E)$, they are not graded ideals of $\mathcal{L}(E)$.

The following theorem was proved in [47] and determines the finite graphs whose associated Leavitt path algebras are strongly graded.

Theorem 1.6.13 *Let E be a finite graph and K a field. Then $\mathcal{L}_K(E)$ is strongly graded if and only if E does not have sinks.*

The proof of this theorem is quite long and does not fit the purpose of this book. However, we can realise the Leavitt path algebras of finite graphs with no source in terms of corner skew Laurent polynomial rings (see §1.6.2). Using this representation, we can provide a short proof for the above theorem when the graph has no sources.

Example 1.6.14 LEAVITT PATH ALGEBRAS AS CORNER SKEW LAURENT RINGS

Let E be a finite graph with no source and $E^0 = \{v_1, \dots, v_n\}$ the set of all vertices of E. For each $1 \leq i \leq n$, we choose an edge e_i such that $r(e_i) = v_i$ and consider $t_+ = e_1 + \cdots + e_n \in \mathcal{L}(E)_1$. Then $t_- = e_1^* + \cdots + e_n^*$ is its right inverse. Thus by Theorem 1.6.9, $\mathcal{L}(E) = \mathcal{L}(E)_0[t_+, t_-, \phi]$, where

$$\begin{aligned} \phi : \mathcal{L}(E)_0 &\longrightarrow t_+t_- \mathcal{L}(E)_0 t_+t_- \\ a &\longmapsto t_+ a t_- \end{aligned}$$

Using this interpretation of Leavitt path algebras we are able to prove the following theorem.

Theorem 1.6.15 *Let E be a finite graph with no source and K a field. Then $\mathcal{L}_K(E)$ is strongly graded if and only if E does not have sinks.*

Proof Write $\mathcal{L}(E) = \mathcal{L}(E)_0[t_+, t_-, \phi]$, where $\phi(1) = t_+t_-$ (see Example 1.6.14). The theorem now follows from an easy to prove observation that t_+t_- is a full idempotent if and only if E does not have sinks, along with Proposition 1.6.6, that $\phi(1)$ is a full idempotent if and only if $\mathcal{L}(E)_0[t_+, t_-, \phi]$ is strongly graded. □

In the following theorem, we use the fact that $\mathcal{L}(E)_0$ is an *ultramatricial algebra, i.e.,* it is isomorphic to the union of an increasing countable chain of a finite product of matrix algebras over a field K (see §3.9.3).

Theorem 1.6.16 *Let E be a finite graph with no source and K a field. Then $\mathcal{L}_K(E)$ is crossed product if and only if E is a cycle.*

Proof Suppose E is a cycle with edges $\{e_1, e_2, \dots, e_n\}$. It is straightforward to check that $e_1 + e_2 + \cdots + e_n$ is an invertible element of degree 1. It then follows that each homogeneous component contains invertible elements and thus $\mathcal{L}(E)$ is crossed product.

Suppose now $\mathcal{L}(E)$ is crossed product. Write $\mathcal{L}(E) = \mathcal{L}(E)_0[t_+, t_-, \phi]$, where $\phi(1) = t_+t_-$ and $t_+ = e_1 + \cdots + e_n \in \mathcal{L}(E)_1$ (see Example 1.6.14). Since $\mathcal{L}(E)_0$ is an ultramatricial algebra, it is Dedekind finite, and thus by Proposition 1.6.7,

$$\phi(1) = e_1e_1^* + e_2e_2^* + \cdots + e_ne_n^* = v_1 + v_2 + \cdots + v_n.$$

From this it follows that (after suitable permutation), $e_ie_i^* = v_i$, for all $1 \leq i \leq n$. This in turn shows that only one edge emits from each vertex, *i.e.*, E is a cycle. □

As a consequence of Theorem 1.6.13, we can show that Leavitt path algebras associated with finite graphs with no sinks are graded regular von Neumann rings (§1.1.9).

Corollary 1.6.17 *Let E be a finite graph with no sinks and K a field. Then $\mathcal{L}_K(E)$ is a graded von Neumann regular ring.*

Proof Since $\mathcal{L}(E)$ is strongly graded (Theorem 1.6.13), by Corollary 1.5.10, $\mathcal{L}(E)$ is von Neumann regular if $\mathcal{L}(E)_0$ is a von Neumann regular ring. But we know that the zero component ring $\mathcal{L}(E)_0$ is an ultramatricial algebra (§3.9.3) which is von Neumann regular (see the proof of [5, Theorem 5.3]). This finishes the proof. □

Example 1.6.18 LEAVITT PATH ALGEBRAS ARE NOT GRADED UNIT REGULAR RINGS

By analogy with the nongraded case, a graded ring is *graded von Neumann unit regular* (or *graded unit regular* for short) if for any homogeneous element x, there is an invertible homogeneous element y such that $xyx = x$. Clearly any graded unit regular ring is von Neumann regular. However, the converse is not the case. For example, Leavitt path algebras are not in general unit regular as the following example shows. Consider the graph:

y_1 y_2

E :

Then it is easy to see that there is no homogeneous invertible element x such that $y_1xy_1 = y_1$ in $\mathcal{L}(E)$.

The following theorem determines the graded structure of Leavitt path algebras associated with acyclic graphs. It turns out that such algebras are natural examples of graded matrix rings (§1.3).

Theorem 1.6.19 *Let K be a field and E a finite acyclic graph with sinks $\{v_1, \ldots, v_t\}$. For any sink v_s, let $R(v_s) = \{p_1^{v_s}, \ldots, p_{n(v_s)}^{v_s}\}$ denote the set of all*

paths ending at v_s. Then there is a $\mathbb{Z}$-graded isomorphism

$$\mathcal{L}_K(E) \cong_{\mathrm{gr}} \bigoplus_{s=1}^{t} \mathbb{M}_{n(v_s)}(K)(|p_1^{v_s}|, \dots, |p_{n(v_s)}^{v_s}|). \tag{1.81}$$

Sketch of proof Fix a sink v_s and denote $R(v_s) = \{p_1, \dots, p_{n(v_s)}\}$. The set

$$I_{v_s} = \Big\{ \sum k p_i p_j^* \mid k \in K, p_i, p_j \in R(v_s) \Big\}$$

is an ideal of $\mathcal{L}_K(E)$, and we have an isomorphism

$$\begin{aligned} \phi : I_{v_s} &\longrightarrow \mathbb{M}_{n(v_s)}(K), \\ k p_i p_j^* &\longmapsto k(\mathbf{e}_{ij}), \end{aligned}$$

where $k \in K$, $p_i, p_j \in R(v_s)$ and $\mathbf{e}_{ij}$ is the standard matrix unit. Now, considering the grading on $\mathbb{M}_{n(v_s)}(K)(|p_1^{v_s}|, \dots, |p_{n(v_s)}^{v_s}|)$, we show that ϕ is a graded isomorphism. Let $p_i p_j^* \in I_{v_s}$. Then

$$\deg(p_i p_j^*) = |p_i| - |p_j| = \deg(\mathbf{e}_{ij}) = \deg(\phi(p_i p_j^*)).$$

So ϕ respects the grading. Hence ϕ is a graded isomorphism. One can check that

$$\mathcal{L}_K(E) = \bigoplus_{s=1}^{t} I_{v_s} \cong_{\mathrm{gr}} \bigoplus_{s=1}^{t} \mathbb{M}_{n(v_s)}(K)(|p_1^{v_s}|, \dots, |p_{n(v_s)}^{v_s}|). \qquad \square$$

Example 1.6.20 Consider the following graphs:

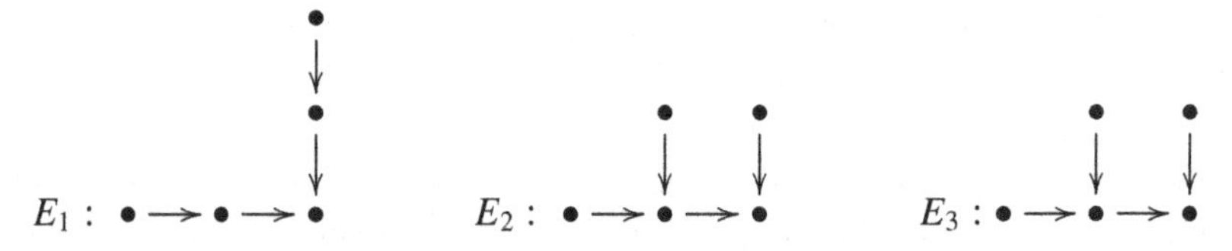

Theorem 1.6.19 shows that the Leavitt path algebras of the graphs E_1 and E_2 with coefficients from the field K are graded isomorphic to $\mathbb{M}_5(K)(0, 1, 1, 2, 2)$ and thus $\mathcal{L}(E_1) \cong_{\mathrm{gr}} \mathcal{L}(E_2)$. However

$$\mathcal{L}(E_3) \cong_{\mathrm{gr}} \mathbb{M}_5(K)(0, 1, 2, 2, 3).$$

Similar to Theorem 1.6.19, we can characterise the graded structure of Leavitt path algebras associated with comet graphs.

Theorem 1.6.21 *Let K be a field and E a C_n-comet with the cycle C of length $n \geq 1$. Let v be a vertex on the cycle C and e be the edge in the cycle with $s(e) = v$. Eliminate the edge e and consider the set $\{p_i \mid 1 \leq i \leq m\}$ of all paths with end in v. Then*

$$\mathcal{L}_K(E) \cong_{\mathrm{gr}} \mathbb{M}_m(K[x^n, x^{-n}])(|p_1|, \dots, |p_m|). \tag{1.82}$$

Sketch of proof One can show that the set of monomials

$$\left\{ p_i C^k p_j^* \mid 1 \le i, j \le m, k \in \mathbb{Z} \right\}$$

is a basis of $\mathcal{L}_K(E)$ as a K-vector space. Define a map

$$\phi : \mathcal{L}_K(E) \longrightarrow \mathbb{M}_m(K[x^n, x^{-n}])(|p_1|, \dots, |p_m|), \text{ by } \quad \phi(p_i C^k p_j^*) = \mathbf{e}_{ij}(x^{kn}),$$

where $\mathbf{e}_{ij}(x^{kn})$ is a matrix with x^{kn} in the ij-position and zero elsewhere. Extend this linearly to $\mathcal{L}_K(E)$. We have

$$\begin{aligned}
\phi((p_i C^k p_j^*)(p_r C^t p_s^*)) &= \phi(\delta_{jr} p_i C^{k+t} p_s^*) \\
&= \delta_{jr} \mathbf{e}_{is}(x^{(k+t)n}) \\
&= (\mathbf{e}_{ij} x^{kn})(\mathbf{e}_{rs} x^{tn}) \\
&= \phi(p_i C^k p_j^*) \phi(p_r C^t p_s^*).
\end{aligned}$$

Thus ϕ is a homomorphism. Also, ϕ sends the basis to the basis, so ϕ is an isomorphism.

We now need to show that ϕ is graded. We have

$$\deg(p_i C^k p_j^*) = |p_i C^k p_j^*| = nk + |p_i| - |p_j|$$

and

$$\deg(\phi(p_i C^k p_j^*)) = \deg(\mathbf{e}_{ij}(x^{kn})) = nk + |p_i| - |p_j|.$$

Therefore ϕ respects the grading. This finishes the proof. □

Example 1.6.22 Consider the Leavitt path algebra $\mathcal{L}_K(E)$, with coefficients in a field K, associated with the following graph:

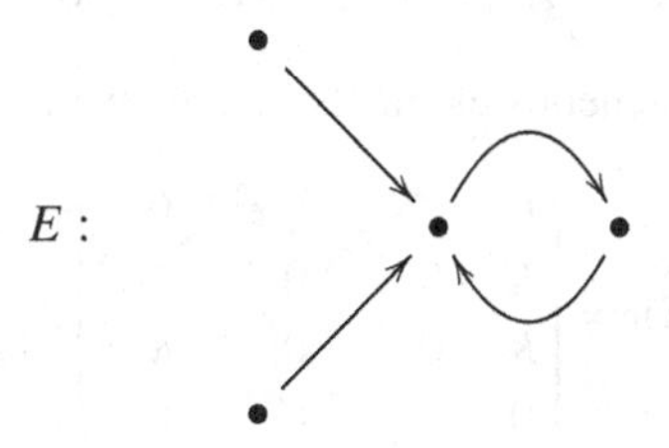

By Theorem 1.6.13, $\mathcal{L}_K(E)$ is strongly graded. Now by Theorem 1.6.21,

$$\mathcal{L}_K(E) \cong_{\mathrm{gr}} \mathbb{M}_4(K[x^2, x^{-2}])(0, 1, 1, 1). \tag{1.83}$$

However, this algebra is not crossed product. Set $B = K[x, x^{-1}]$ with the grading $B = \bigoplus_{n \in \mathbb{Z}} Kx^n$ and consider $A = K[x^2, x^{-2}]$ as a graded subring of B with

$A_n = Kx^n$ if $n \equiv 0 \pmod 2$, and $A_n = 0$ otherwise. Using the graded isomorphism of (1.83), by (1.45) a homogeneous element of degree 1 in $\mathcal{L}_K(E)$ has the form

$$\begin{pmatrix} A_1 & A_2 & A_2 & A_2 \\ A_0 & A_1 & A_1 & A_1 \\ A_0 & A_1 & A_1 & A_1 \\ A_0 & A_1 & A_1 & A_1 \end{pmatrix}.$$

Since $A_1 = 0$, the determinants of these matrices are zero, and thus no homogeneous element of degree 1 is invertible. Thus $\mathcal{L}_K(E)$ is not crossed product (see §1.1.3).

Now consider the following graph:

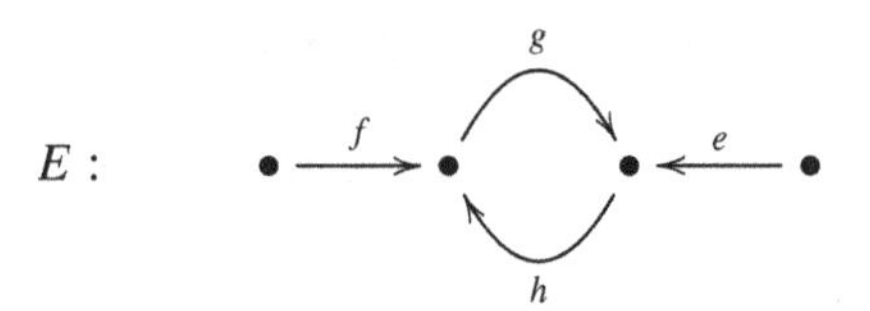

By Theorem 1.6.21,

$$\mathcal{L}_K(E) \cong_{\mathrm{gr}} \mathbb{M}_4(K[x^2, x^{-2}])(0, 1, 1, 2). \tag{1.84}$$

Using the graded isomorphism of (1.84), by (1.47) homogeneous elements of degree 0 in $\mathcal{L}_K(E)$ have the form

$$\mathcal{L}_K(E)_0 = \begin{pmatrix} A_0 & A_1 & A_1 & A_2 \\ A_{-1} & A_0 & A_0 & A_1 \\ A_{-1} & A_0 & A_0 & A_1 \\ A_{-2} & A_{-1} & A_{-1} & A_0 \end{pmatrix} = \begin{pmatrix} K & 0 & 0 & Kx^2 \\ 0 & K & K & 0 \\ 0 & K & K & 0 \\ Kx^{-2} & 0 & 0 & K \end{pmatrix}.$$

In the same manner, homogeneous elements of degree 1 have the form

$$\mathcal{L}_K(E)_1 = \begin{pmatrix} 0 & Kx^2 & Kx^2 & 0 \\ K & 0 & 0 & Kx^2 \\ K & 0 & 0 & Kx^2 \\ 0 & K & K & 0 \end{pmatrix}.$$

Choose

$$u = \begin{pmatrix} 0 & 0 & x^2 & 0 \\ 0 & 0 & 0 & x^2 \\ 1 & 0 & 0 & x^2 \\ 0 & 1 & 0 & 0 \end{pmatrix} \in \mathcal{L}(E)_1$$

and observe that u is invertible; this matrix corresponds to the element

$$g + h + fge^* + ehf^* \in \mathcal{L}_K(E)_1.$$

Thus $\mathcal{L}_K(E)$ is crossed product and therefore a skew group ring as the grading is cyclic (see §1.1.4), *i.e.*,

$$\mathcal{L}_K(E) \cong_{\mathrm{gr}} \bigoplus_{i \in \mathbb{Z}} \mathcal{L}_K(E)_0 u^i$$

and a simple calculation shows that one can describe this algebra as follows:

$$\mathcal{L}_K(E)_0 \cong \mathbb{M}_2(K) \times \mathbb{M}_2(K)$$

and

$$\mathcal{L}_K(E) \cong_{\mathrm{gr}} (\mathbb{M}_2(K) \times \mathbb{M}_2(K)) \star_\tau \mathbb{Z}, \tag{1.85}$$

where

$$\tau\left(\begin{pmatrix} a_{11} & a_{12} \\ a_{21} & a_{22} \end{pmatrix}, \begin{pmatrix} b_{11} & b_{12} \\ b_{21} & b_{22} \end{pmatrix}\right) = \left(\begin{pmatrix} b_{22} & b_{21} \\ b_{12} & b_{11} \end{pmatrix}, \begin{pmatrix} a_{22} & a_{21} \\ a_{12} & a_{11} \end{pmatrix}\right).$$

Remark 1.6.23 NONCANONICAL GRADINGS ON LEAVITT PATH ALGEBRAS

For a graph E, the Leavitt path algebra $\mathcal{L}_K(E)$ has a canonical $\mathbb{Z}$-graded structure. This grading was obtained by assigning 0 to vertices, 1 to edges and -1 to ghost edges. However, one can equip $\mathcal{L}_K(E)$ with other graded structures as well. Let Γ be an arbitrary group with the identity element e. Let $w : E^1 \to \Gamma$ be a *weight* map and further define $w(\alpha^*) = w(\alpha)^{-1}$, for any edge $\alpha \in E^1$ and $w(v) = e$ for $v \in E^0$. The free K-algebra generated by the vertices, edges and ghost edges is a Γ-graded K-algebra (see §1.6.1). Moreover, the Leavitt path algebra is the quotient of this algebra by relations in Definition 1.6.11 which are all homogeneous. Thus $\mathcal{L}_K(E)$ is a Γ-graded K-algebra. One can write Theorems 1.6.19 and 1.6.21 with this general grading.

As an example, consider the graphs

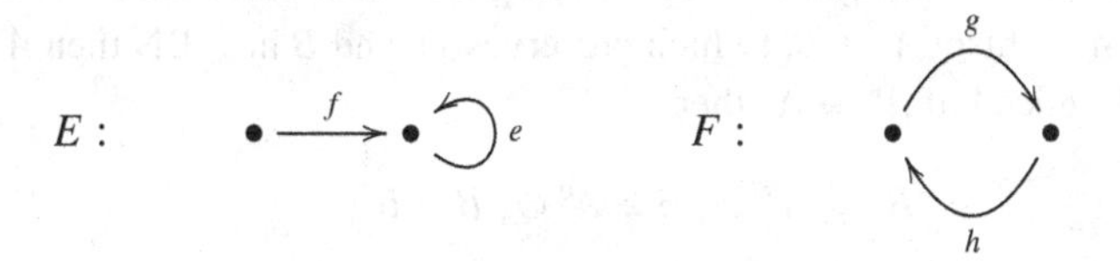

and assign 1 for the degree of f, 2 for the degree of e in E and 1 for the degrees of g and h in F. Then the proof of Theorem 1.6.21 shows that

$$\mathcal{L}_K(E) \cong \mathbb{M}_2(K[x^2, x^{-2}])(0, 1)$$

and

$$\mathcal{L}_K(F) \cong \mathbb{M}_2(K[x^2, x^{-2}])(0, 1)$$

are $\mathbb{Z}$-graded rings. So with these gradings, $\mathcal{L}_K(E) \cong_{\text{gr}} \mathcal{L}_K(F)$.

Example 1.6.24 LEAVITT PATH ALGEBRAS ARE STRONGLY $\mathbb{Z}_2$-GRADED

Let E be a (connected) row-finite graph with at least one edge. By Remark 1.6.23, $A = \mathcal{L}_K(E)$ has a $\mathbb{Z}_2$-grading induced by assigning 0 to vertices and $1 \in \mathbb{Z}_2$ to edges and ghost edges. Since the defining relations of Leavitt path algebras guarantee that for any $v \in E^0$, $v \in A_1A_1$, one can easily check that $\mathcal{L}_K(E)$ is strongly $\mathbb{Z}_2$-graded for any graph (compare this with Theorem 1.6.13). In contrast to the canonical grading, in this case the 0-component ring is not necessarily an ultramatricial ring (see §3.9.3).

1.7 The graded IBN and graded type

A ring A with identity has an *invariant basis number* (IBN) or *invariant basis property* if any two bases of a free (right) A-module have the same cardinality, *i.e.*, if $A^n \cong A^m$ as A-modules, then $n = m$. When A does not have IBN, the *type* of A is defined as a pair of positive integers (n, k) such that $A^n \cong A^{n+k}$ as A-modules and these are the smallest number with this property, that is, (n, k) is the minimum under the usual lexicographic order. This means any two bases of a free A-module have the unique cardinality if one of the bases has the cardinality less than n and, further, if a free module has rank n, then a free module with the smallest cardinality (other than n) isomorphic to this module is of rank $n + k$. Another way to describe a type (n, k) is that $A^n \cong A^{n+k}$ is the first repetition in the list $A, A^2, A^3, \dots$.

It was shown that if A has type (n, k), then $A^m \cong A^{m'}$ if and only if $m = m'$ or $m, m' \geq n$ and $m \equiv m' \pmod{k}$ (see [28, p. 225], [63, Theorem 1]).

One can show that a (right) Noetherian ring has IBN. Moreover, if there is a ring homomorphism $A \to B$, (which preserves 1), and B has IBN then A has IBN as well. Indeed, if $A^m \cong A^n$ then

$$B^m \cong A^m \otimes_A B \cong A^n \otimes_A B \cong B^n, \tag{1.86}$$

so $n = m$. One can describe the type of a ring by using the monoid of isomorphism classes of finitely generated projective modules (see Example 3.1.4). For nice discussions about these rings see [21, 28, 67].

A graded ring A has a *graded invariant basis number* (gr-IBN) if any two homogeneous bases of a graded free (right) A-module have the same cardinality, *i.e.*, if $A^m(\overline{\alpha}) \cong_{\mathrm{gr}} A^n(\overline{\delta})$, where $\overline{\alpha} = (\alpha_1, \dots, \alpha_m)$ and $\overline{\delta} = (\delta_1, \dots, \delta_n)$, then $m = n$. Note that, in contrast to the nongraded case, this does not imply that two graded free modules with bases of the same cardinality are graded isomorphic (see Proposition 1.3.16). A graded ring A has *IBN in* gr-A, if $A^m \cong_{\mathrm{gr}} A^n$ then $m = n$. If A has IBN in gr-A, then A_0 has IBN. Indeed, if $A_0^m \cong A_0^n$ as A_0-modules, then, similarly to (1.86),

$$A^m \cong_{\mathrm{gr}} A_0^m \otimes_{A_0} A \cong A_0^n \otimes_{A_0} A \cong_{\mathrm{gr}} A^n,$$

so $n = m$ (see [75, p. 215]).

When the graded ring A does not have gr-IBN, the *graded type* of A is defined as a pair of positive integers (n, k) such that $A^n(\overline{\delta}) \cong_{\mathrm{gr}} A^{n+k}(\overline{\alpha})$ as A-modules, for some $\overline{\delta} = (\delta_1, \dots, \delta_n)$ and $\overline{\alpha} = (\alpha_1, \dots, \alpha_{n+k})$ and these are the smallest number with this property. In Proposition 1.7.1 we show that the Leavitt algebra $\mathcal{L}(n, k+1)$ (see Example 1.6.5) has graded type (n, k).

Parallel to the nongraded setting, one can show that a graded (right) Noetherian ring has gr-IBN. Moreover, if there is a graded ring homomorphism $A \to B$, (which preserves 1), and B has gr-IBN then A has gr-IBN as well. Indeed, if $A^m(\overline{\alpha}) \cong_{\mathrm{gr}} A^n(\overline{\delta})$, where $\overline{\alpha} = (\alpha_1, \dots, \alpha_m)$ and $\overline{\delta} = (\delta_1, \dots, \delta_n)$, then

$$B^m(\overline{\alpha}) \cong_{\mathrm{gr}} A^m(\overline{\alpha}) \otimes_A B \cong A^n(\overline{\delta}) \otimes_A B \cong_{\mathrm{gr}} B^n(\overline{\delta}),$$

which implies $n = m$. Using this, one can show that any graded commutative ring has gr-IBN. For, there exists a graded maximal ideal and its quotient ring is a graded field which has gr-IBN (see §1.1.5 and Proposition 1.4.2).

Let A be a Γ-graded ring such that $A^m(\overline{\alpha}) \cong_{\mathrm{gr}} A^n(\overline{\delta})$, where $\overline{\alpha} = (\alpha_1, \dots, \alpha_m)$ and $\overline{\delta} = (\delta_1, \dots, \delta_n)$. Then there is a universal Γ-graded ring R such that

$$R^m(\overline{\alpha}) \cong_{\mathrm{gr}} R^n(\overline{\delta})$$

and a graded ring homomorphism $R \to A$ which induces the graded isomorphism

$$A^m(\overline{\alpha}) \cong_{\mathrm{gr}} R^m(\overline{\alpha}) \otimes_R A \cong_{\mathrm{gr}} R^n(\overline{\delta}) \otimes_R A \cong_{\mathrm{gr}} A^n(\overline{\delta}).$$

Indeed, by Proposition 1.3.16, there are matrices $a = (a_{ij}) \in \mathbb{M}_{n\times m}(A)[\overline{\delta}][\overline{\alpha}]$ and $b = (b_{ij}) \in \mathbb{M}_{m\times n}(A)[\overline{\alpha}][\overline{\delta}]$ such that $ab = \mathbb{I}_n$ and $ba = \mathbb{I}_m$. The free ring generated by symbols in place of a_{ij} and b_{ij} subject to relations imposed by $ab = \mathbb{I}_n$ and $ba = \mathbb{I}_m$ is the desired universal graded ring. In detail, let F be a free ring generated by x_{ij}, $1 \leq i \leq n$, $1 \leq j \leq m$ and y_{ij}, $1 \leq i \leq m$, $1 \leq j \leq n$. Assign the degrees $\deg(x_{ij}) = \delta_i - \alpha_j$ and $\deg(y_{ij}) = \alpha_i - \delta_j$ (see §1.6.1). This makes F a Γ-graded ring. Let R be a ring F modulo the relations $\sum_{s=1}^{m} x_{is} y_{sk} = \delta_{ik}$, $1 \leq i, k \leq n$ and $\sum_{t=1}^{n} y_{it} x_{tk} = \delta_{ik}$, $1 \leq i, k \leq m$, where δ_{ik} is the Kronecker delta. Since all the relations are homogeneous, R is a Γ-graded ring. Clearly the map sending x_{ij} to a_{ij} and y_{ij} to b_{ij} induces a graded ring homomorphism $R \to A$. Again Proposition 1.3.16 shows that $R^m(\overline{\alpha}) \cong_{\mathrm{gr}} R^n(\overline{\delta})$.

Proposition 1.7.1 *Let $R = \mathcal{L}(n, k+1)$ be the Leavitt algebra of type (n, k). Then*

(1) *R is a universal $\bigoplus_n \mathbb{Z}$-graded ring which does not have gr-IBN;*
(2) *R has graded type (n, k);*
(3) *for $n = 1$, R has IBN in* gr-R.

Proof (1) Consider the algebra $\mathcal{L}(n, k+1)$ constructed in Example 1.6.5, which is a $\bigoplus_n \mathbb{Z}$-graded ring and is universal. Moreover, (1.77) combined with Proposition 1.3.16(3) shows that $R^n \cong_{\mathrm{gr}} R^{n+k}(\overline{\alpha})$. Here $\overline{\alpha} = (\alpha_1, \dots, \alpha_{n+k})$, where $\alpha_i = (0, \dots, 0, 1, 0 \dots, 0)$ and 1 is in the ith entry. This shows that $R = \mathcal{L}(n, k+1)$ does not have gr-IBN.

(2) By [28, Theorem 6.1], R is of type (n, k). This immediately implies the graded type of R is also (n, k).

(3) Suppose $R^n \cong_{\mathrm{gr}} R^m$ as graded R-modules. Then $R_0^n \cong R_0^m$ as R_0-modules. But R_0 is an ultramatricial algebra, *i.e.,* the direct limit of an increasing chain of a finite product of matrices over a field. Since IBN respects direct limits ([28, Theorem 2.3]), R_0 has IBN. Therefore, $n = m$. □

Remark 1.7.2 Assignment of $\deg(y_{ij}) = 1$ and $\deg(x_{ij}) = -1$, for all i, j, makes $R = \mathcal{L}(n, k+1)$ a $\mathbb{Z}$-graded algebra of graded type (n, k) with $R^n \cong_{\mathrm{gr}} R^{n+k}(1)$.

Remark 1.7.3 Let A be a Γ-graded ring. In [77, Proposition 4.4], it was shown that if Γ is finite then A has gr-IBN if and only if A has IBN.

1.8 The graded stable rank

The notion of stable rank was defined by H. Bass [14] to study the K_1-group of rings that are finitely generated over commutative rings with finite Krull dimension. For a concise introduction to the stable rank, we refer the reader to [59, 60], and for its applications to K-theory to [13, 14]. It seems that the natural notion of graded stable rank in the context of graded ring theory has not yet been investigated in the literature. In this section we propose a definition for the graded stable rank and study the important case of graded rings with graded stable rank 1. This will be used later in Chapter 3 in relation to graded Grothendieck groups.

A row $(a_1, \dots, a_n)$ of homogeneous elements of a Γ-graded ring A is called a *graded left unimodular row* if the graded left ideal generated by a_i, $1 \le i \le n$, is A.

Lemma 1.8.1 *Let $(a_1, \dots, a_n)$ be a row of homogeneous elements of a Γ-graded ring A. The following are equivalent:*

(1) *$(a_1, \dots, a_n)$ is a left unimodular row;*
(2) *$(a_1, \dots, a_n)$ is a graded left unimodular row;*
(3) *the graded homomorphism*

$$\phi_{(a_1,\dots,a_n)} : A^n(\overline{\alpha}) \longrightarrow A,$$

$$(x_1, \dots, x_n) \longmapsto \sum_{i=1}^{n} x_i a_i,$$

where $\overline{\alpha} = (\alpha_1, \dots, \alpha_n)$ and $\alpha_i = -\deg(a_i)$, is surjective.

Proof The proof is straightforward. □

When $n \ge 2$, a graded left unimodular row $(a_1, \dots, a_n)$ is called *stable* if there exist homogeneous elements $b_1, \dots, b_{n-1}$ of A such that the graded left ideal generated by homogeneous elements $a_i + b_i a_n$, $1 \le i \le n-1$, is A.

The *graded left stable rank* of a ring A is defined to be n, denoted $\operatorname{sr}^{\mathrm{gr}}(A) = n$, if any graded unimodular row of length $n+1$ is stable, but there exists an unstable unimodular row of length n. If such an n does not exist (*i.e.*, there are unstable unimodular rows of arbitrary length) we say that the graded stable rank of A is infinite.

In order that this definition is well-defined, one needs to show that if any graded unimodular row of fixed length n is stable, so is any unimodular row of a greater length. This can be proved similarly to the nongraded case and we omit the proof (see for example [59, Proposition 1.3]).

When the grade group Γ is a trivial group, the above definitions reduce to the standard definitions of unimodular rows and stable ranks.

The case of graded stable rank 1 is of special importance. Suppose A is a Γ-graded with $\operatorname{sr}^{\mathrm{gr}}(A) = 1$. Then from the definition it follows that, if $a, b \in A^h$ such that $Aa + Ab = A$, then there is a homogeneous element c such that the homogeneous element $a + cb$ is left invertible. When $\operatorname{sr}^{\mathrm{gr}}(A) = 1$, any left invertible homogeneous element is in fact invertible. For, suppose $c \in A^h$ is left invertible, *i.e.*, there is an $a \in A^h$ such that $ac = 1$. Then the row $(a, 1 - ca)$ is graded left unimodular. Thus, there is an $s \in A^h$ such that $u := a + s(1 - ca)$ is left invertible. But $uc = 1$. Thus u is (left and right) invertible and consequently, c is an invertible homogeneous element.

The graded stable rank 1 is quite a strong condition. In fact if $\operatorname{sr}^{\mathrm{gr}}(A) = 1$ then $\Gamma_A = \Gamma_A^*$. For, if $a \in A_\gamma$ is a nonzero element, then since $(a, 1)$ is unimodular and $\operatorname{sr}^{\mathrm{gr}}(A) = 1$, there is a $c \in A^h$, such that $a + c$ is an invertible homogeneous element, necessarily of degree γ. Thus $\gamma \in \Gamma_A^*$.

Example 1.8.2 GRADED DIVISION RINGS HAVE GRADED STABLE RANK 1

Since any nonzero homogeneous element of a graded division ring is invertible, one shows easily that its graded stable rank is 1. Thus for a field K, $\operatorname{sr}^{\mathrm{gr}}(K[x, x^{-1}]) = 1$, whereas $\operatorname{sr}(K[x, x^{-1}]) = 2$.

Example 1.8.3 FOR A STRONGLY GRADED RING A, $\operatorname{sr}^{\mathrm{gr}}(A) \neq \operatorname{sr}(A_0)$

Let $A = \mathcal{L}(1, 2)$ be the Leavitt algebra generated by x_1, x_2, y_1, y_2 (see Example 1.3.19). Then relations (1.60) show that y_1 is left invertible but it is not invertible. This shows that $\operatorname{sr}^{\mathrm{gr}}(A) \neq 1$. On the other hand, since A_0 is an ultramatricial algebra, $\operatorname{sr}(A_0) = 1$ (see §3.9.3, [59, Corollary 5.5] and [40]).

We have the following theorem which is a graded version of the cancellation theorem with a similar proof (see [60, Theorem 20.11]).

Theorem 1.8.4 (GRADED CANCELLATION THEOREM) *Let A be a Γ-graded ring and let M, N, P be graded right A-modules, with P being finitely generated. If the graded ring* $\operatorname{End}_A(P)$ *has graded left stable rank* 1*, then $P \oplus M \cong_{\mathrm{gr}} P \oplus N$ as A-modules implies $M \cong_{\mathrm{gr}} N$ as A-modules.*

Proof Set $E := \operatorname{End}_A(P)$. Let $h : P \oplus M \to P \oplus N$ be a graded A-module isomorphism. Then the composition of the maps

$$P \xrightarrow{i_1} P \oplus M \xrightarrow{h} P \oplus N \xrightarrow{\pi_1} P,$$

$$M \xrightarrow{i_2} P \oplus M \xrightarrow{h} P \oplus N \xrightarrow{\pi_1} P$$

induces a graded split epimorphism of degree zero, denoted by $(f, g) : P \oplus M \to P$. Here $(f, g)(p, m) = f(p) + g(m)$, where $f = \pi_1 h i_1$ and $g = \pi_1 h i_2$. It is clear that $\ker(f, g) \cong_{\mathrm{gr}} N$. Let $\begin{pmatrix} f' \\ g' \end{pmatrix} : P \to P \oplus M$ be the split homomorphism. Thus

$$1 = (f, g)\begin{pmatrix} f' \\ g' \end{pmatrix} = ff' + gg'.$$

This shows that the left ideal generated by f' and gg' is E. Since E has graded stable rank 1, it follows there is an $e \in E$ of degree 0 such that $u := f' + e(gg')$ is an invertible element of E. Writing $u = (1, eg)\begin{pmatrix} f' \\ g' \end{pmatrix}$ implies that both $\ker(f, g)$

and $\ker(1, eg)$ are graded isomorphic to

$$P \oplus M/\operatorname{Im}\begin{pmatrix} f' \\ g' \end{pmatrix}.$$

Thus $\ker(f, g) \cong_{\mathrm{gr}} \ker(1, eg)$. But $\ker(f, g) \cong_{\mathrm{gr}} N$ and $\ker(1, eg) \cong_{\mathrm{gr}} M$. Thus $M \cong_{\mathrm{gr}} N$. □

The following corollary will be used in Chapter 3 to show that for a graded ring with graded stable rank 1, the monoid of graded finitely generated projective modules injects into the graded Grothendieck group (see Corollary 3.1.8).

Corollary 1.8.5 *Let A be a Γ-graded ring with graded left stable rank* 1 *and M, N, P be graded right A-modules. If P is a graded finitely generated projective A-module, then $P \oplus M \cong_{\mathrm{gr}} P \oplus N$ as A-modules implies $M \cong_{\mathrm{gr}} N$ as A-modules.*

Proof Suppose $P \oplus M \cong_{\mathrm{gr}} P \oplus N$ as A-modules. Since P is a graded finitely generated A-module, there is a graded A-module Q such that $P \oplus Q \cong_{\mathrm{gr}} A^n(\overline{\alpha})$ (see (1.39)). It follows that

$$A^n(\overline{\alpha}) \oplus M \cong_{\mathrm{gr}} A^n(\overline{\alpha}) \oplus N.$$

We prove that if

$$A(\alpha) \oplus M \cong_{\mathrm{gr}} A(\alpha) \oplus N, \tag{1.87}$$

then $M \cong_{\mathrm{gr}} N$. The corollary then follows by an easy induction.

By (1.48) there is a graded ring isomorphism $\operatorname{End}_A(A(\alpha)) \cong_{\mathrm{gr}} A$. Since A has graded stable rank 1, so does $\operatorname{End}_A(A(\alpha))$. Now by Theorem 1.8.4, from (1.87) it follows that $M \cong_{\mathrm{gr}} N$. This finishes the proof. □

The stable rank imposes other finiteness properties on rings such as the IBN property (see [95, Exercise I.1.5(e)]).

Theorem 1.8.6 *Let A be a Γ-graded ring such that $A \cong_{\mathrm{gr}} A^r(\overline{\alpha})$ as left A-modules, for some $\overline{\alpha} = (\alpha_1, \dots, \alpha_r)$, $r > 1$. Then the graded stable rank of A is infinite.*

Proof Suppose the graded stable rank of A is n. Then one can find $\overline{\alpha} = (\alpha_1, \dots, \alpha_r)$, where $r > n$ and $A^r(\overline{\alpha}) \cong_{\mathrm{gr}} A$. Suppose $\phi : A^r(\overline{\alpha}) \to A$ is this given graded isomorphism. Set $a_i = \phi(e_i)$, $1 \leq i \leq r$, where $\{e_i \mid 1 \leq i \leq r\}$ are the standard (homogeneous) basis of A^r. Then for any $x \in A^r(\overline{\alpha})$,

$$\phi(x) = \phi\Big(\sum_{i=1}^{r} x_i e_i\Big) = \sum_{i=1}^{r} x_i a_i = \phi_{(a_1,\dots,a_r)}(x_1, \dots, x_r).$$

Since ϕ is an isomorphism, by Lemma 1.8.1, the row $(a_1, \dots, a_r)$ is graded left unimodular. Since $r > \operatorname{sr}^{\mathrm{gr}}(A)$, there is a homogeneous row $(b_1, \dots, b_{r-1})$ such that $(a_1 + b_1 a_r, \dots, a_{r-1} + b_{r-1} a_r)$ is also left unimodular. Note that $\deg(b_i) = \alpha_r - \alpha_i$. Consider the graded left A-module homomorphism

$$\begin{aligned} \psi : A^{r-1}(\alpha_1, \dots, \alpha_{r-1}) &\longrightarrow A^r(\alpha_1, \dots, \alpha_{r-1}, \alpha_r), \\ (x_1, \dots, x_{r-1}) &\longmapsto (x_1, \dots, x_{r-1}, \sum_{i=1}^{r-1} x_i b_i) \end{aligned}$$

and the commutative diagram

$$\begin{array}{ccc} A^{r-1}(\alpha_1, \dots, \alpha_{r-1}) & \xrightarrow{\psi} & A^r(\alpha_1, \dots, \alpha_{r-1}, \alpha_r) . \\ & \searrow^{\phi_{(a_1+b_1a_r, \dots, a_{r-1}+b_{r-1}a_r)}} \quad {}^{\phi_{(a_1, \dots, a_r)}}\swarrow & \\ & A & \end{array}$$

Since $\phi_{(a_1, \dots, a_r)}$ is an isomorphism and $\phi_{(a_1+b_1a_r, \dots, a_{r-1}+b_{r-1}a_r)}$ is an epimorphism, ψ is also an epimorphism. Thus there is $(x_1, \dots, x_{r-1})$ such that $\psi(x_1, \dots, x_{r-1}) = (0, \dots, 0, 1)$ which immediately gives a contradiction. □

Corollary 1.8.7 *The graded stable rank of the Leavitt algebra $\mathcal{L}(1, n)$ is infinite.*

Proof This follows from Proposition 1.7.1 and Theorem 1.8.6. □

Example 1.8.8 GRADED VON NEUMANN REGULAR RINGS WITH STABLE RANK 1

One can prove, similarly to the nongraded case [40, Proposition 4.12], that a graded von Neumann regular ring has a stable rank 1 if and only if it is a graded von Neumann unit regular.

1.9 Graded rings with involution

Let A be a ring with an involution denoted by *, *i.e.*, $^* : A \to A$, $a \mapsto a^*$, is an anti-automorphism of order two. Throughout this book we call A also a $*$-ring. If M is a right A-module, then M can be given a left A-module structure by defining

$$am := ma^*. \tag{1.88}$$

This gives an equivalent

$$\text{Mod-}A \approx A\text{-Mod}, \tag{1.89}$$

where Mod-A is the category of right A-modules and A-Mod is the category of left A-modules.

Now let $A = \bigoplus_{\gamma \in \Gamma} A_\gamma$ be a Γ-graded ring. We call A a *graded $*$-ring* if there is an involution on A such that for $a \in A_\gamma$, $a^* \in A_{-\gamma}$, where $\gamma \in \Gamma$. It follows that $A_\gamma^* = A_{-\gamma}$, for any $\gamma \in \Gamma$.

Remark 1.9.1 Depending on the circumstances, one can also set another definition that $A_\gamma^* = A_\gamma$, where $\gamma \in \Gamma$.

If A is a graded $*$-ring, and M is a graded right A-module, then the multiplication in (1.88) makes M a graded left $A^{(-1)}$-module and makes $M^{(-1)}$ a graded left A-module, where $A^{(-1)}$ and $M^{(-1)}$ are Veronese rings and modules (see Examples 1.1.19 and 1.2.7). These give graded equivalences

$$\begin{aligned} \mathfrak{I} : \text{Gr-}A &\longrightarrow A^{(-1)}\text{-Gr}, \\ M &\longmapsto M \end{aligned} \tag{1.90}$$

and

$$\begin{aligned} \mathfrak{J} : \text{Gr-}A &\longrightarrow A\text{-Gr}, \\ M &\longmapsto M^{(-1)}. \end{aligned} \tag{1.91}$$

Here for $\alpha \in \Gamma$, $\mathfrak{I}(M(\alpha)) = \mathfrak{I}(M)(\alpha)$ (*i.e.*, $\mathfrak{I}$ is a graded functor, see Definition 2.3.3), whereas $\mathfrak{J}(M(\alpha)) = \mathfrak{J}(M)(-\alpha)$.

Clearly, if the grade group Γ is trivial, the equivalences (1.90) and (1.91) both reduce to (1.89).

Let A be a graded $*$-field (*i.e.*, a graded field with $*$-involution) and R a graded A-algebra with involution denoted by * again. Then R is a graded $*$-A-algebra if $(ar)^* = a^*r^*$ (*i.e.*, the graded homomorphism $A \to R$ is a $*$-homomorphism).

Example 1.9.2 GROUP RINGS

For a group Γ (denoted multiplicatively here), the group ring $\mathbb{Z}[\Gamma]$ with a natural Γ-grading

$$\mathbb{Z}[\Gamma] = \bigoplus_{\gamma \in \Gamma} \mathbb{Z}[\Gamma]_\gamma, \text{ where } \mathbb{Z}[\Gamma]_\gamma = \mathbb{Z}\gamma,$$

and the natural involution $* : \mathbb{Z}[\Gamma] \to \mathbb{Z}[\Gamma], \gamma \mapsto \gamma^{-1}$ is a graded $*$-ring.

Example 1.9.3 HERMITIAN TRANSPOSE

If A is a graded $*$-ring, then for $a = (a_{ij}) \in \mathbb{M}_n(A)(\delta_1, \dots, \delta_n)$, the *Hermitian transpose* $a^* = (a_{ji}^*)$, makes $\mathbb{M}_n(A)(\delta_1, \dots, \delta_n)$ a graded $*$-ring (see 1.45).

Example 1.9.4 LEAVITT PATH ALGEBRAS ARE GRADED $*$-ALGEBRAS

A Leavitt path algebra has a natural $*$-involution. Let K be a $*$-field (*i.e.,* a field with $*$-involution). Define a homomorphism from the free K-algebra generated by the vertices, edges and ghost edges of the graph E to $\mathcal{L}(E)^{\mathrm{op}}$, by $k \mapsto k^*$, $v \mapsto v$, $\alpha \mapsto \alpha^*$ and $\alpha^* \mapsto \alpha$, where $k \in K$, $v \in E^0$, $\alpha \in E^1$ and α^* is the ghost edge. The relations in the definition of a Leavitt path algebra, Definition 1.6.11, show that this homomorphism induces an isomorphism from $\mathcal{L}(E)$ to $\mathcal{L}(E)^{\mathrm{op}}$. This makes $\mathcal{L}(E)$ a $*$-algebra. Moreover, considering the grading, it is easy to see that in fact, $\mathcal{L}(E)$ is a graded $*$-algebra.

Example 1.9.5 CORNER SKEW LAURENT RINGS AS GRADED $*$-ALGEBRAS

Recall the corner skew Laurent polynomial ring $A = R[t_+, t_-, \phi]$, where R is a ring with identity and $\phi : R \to pRp$ a corner isomorphism (see §1.6.2). Let R be a $*$-ring, p a *projection* (*i.e.,* $p = p^* = p^2$), and ϕ a $*$-isomorphism. Then A has a $*$-involution defined on generators by $(t_-^j r_{-j})^* = r_{-j}^* t_+^j$ and $(r_i t_+^i)^* = t_-^i r_i^*$. With this involution A becomes a graded $*$-ring. A $*$-ring is called *$*$-proper*, if $xx^* = 0$ implies $x = 0$. It is called *positive-definite* if $\sum_{i=1}^n x_i x_i{}^* = 0$, $n \in \mathbb{N}$, implies $x_i = 0$, $1 \le i \le n$. A graded $*$-ring is called *graded $*$-proper*, if $x \in A^h$ and $xx^* = 0$ then $x = 0$. The following lemma is easy to prove and we leave part of it to the reader.

Lemma 1.9.6 *Let R be a $*$-ring and $A = R[t_+, t_-, \phi]$ a $*$-corner skew Laurent polynomial ring. We have*

(1) *R is positive-definite if and only if A is positive-definite.*
(2) *R is $*$-proper if and only if A is graded $*$-proper.*

Proof (1) Since R is a $*$-subring of A, if A is positive-definite, then so is R.

For the converse, suppose R is positive-definite and

$$\sum_{k=1}^{l} x_k x_k^* = 0, \tag{1.92}$$

where $x_k \in A$. Write

$$x_k = t_-^{j_k} r_{-j_k}^k + t_-^{j_k - 1} r_{-j_k+1}^k + \cdots + t_- r_{-1}^k + r_0^k + r_1^k t_+ + \cdots + r_{i_k}^k t_+^{i_k}.$$

It is easy to observe that the constant term of $x_k x_k^*$ is

$$\phi^{-j_k}(r_{-j_k}^k r^{k*}_{-j_k}) + \cdots + \phi^{-1}(r_{-1}^k r^{k*}_{-1}) + r_0^k r^{k*}_0 + r_1^k r^{k*}_1 + \cdots + r_{i_k}^k r^{k*}_{i_k}.$$

Now Equation 1.92 implies that the sum of these constant terms are zero. Since

R is positive-definite and ϕ is an $*$-isomorphism, it follows that all $r^k_{-j_k}$ and $r^k_{i_k}$ for $1 \leq k \leq l$ are zero and thus $x_k = 0$. This finishes the proof of (1).

(2) The proof is similar to (1) and is left to the reader. □

2
Graded Morita theory

Starting from a right A-module P one can construct a 6-tuple

$$(A, P, P^*, B; \phi, \psi),$$

where

$$\begin{aligned} P^* &= \operatorname{Hom}_A(P, A), \\ B &= \operatorname{Hom}_A(P, P), \\ \phi &: P^* \otimes_B P \to A, \\ \psi &: P \otimes_A P^* \to B, \end{aligned}$$

which have appropriate bimodule structures. This is called the *Morita context* associated with P. When P is a progenerator (*i.e.,* finitely generated projective and a generator), then one can show that Mod-$A \approx$ Mod-B, *i.e.,* the category of (right) A-modules is equivalent to the category of (right) B-modules. Conversely, if for two rings A and B, Mod-$A \approx$ Mod-B, then one can construct an A-progenerator P such that $B \cong \operatorname{End}_A(P)$. The feature of this theory, called the Morita theory, is that the whole process involves working with Hom and tensor functors, both of which respect the grading. Thus, starting from a graded ring A and a graded progenerator P, and carrying out the Morita theory, one can naturally extend the equivalence from Gr-$A \approx$ Gr-B to Mod-$A \approx$ Mod-B and vice versa.

Extending the equivalence from the (sub)categories of graded modules to the categories of modules is not at first glance obvious. Recall that two categories $\mathcal{C}$ and $\mathcal{D}$ are equivalent if and only if there is a functor $\phi : \mathcal{C} \to \mathcal{D}$ which is fully faithful and for any $D \in \mathcal{D}$, there is a $C \in \mathcal{C}$ such that $\phi(C) = D$.

Suppose ϕ : Gr-$A \to$ Gr-B is a graded equivalence (see Definition 2.3.3). Then for two graded A-modules M and N (where M is finitely generated) we have

$$\begin{aligned}\operatorname{Hom}_A(M,N)_0 &= \operatorname{Hom}_{\text{Gr-}A}(M,N)\\ &\cong \operatorname{Hom}_{\text{Gr-}B}(\phi(M),\phi(N)) = \operatorname{Hom}_B(\phi(M),\phi(N))_0.\end{aligned}$$

The fact that this can be extended to

$$\operatorname{Hom}_A(M,N) \cong \operatorname{Hom}_B(\phi(M),\phi(N))$$

is not immediate. We will show that this is indeed the case.

In this chapter we study the equivalences between the categories of graded modules and their relations with the categories of modules. The main theorem of this chapter shows that for two graded rings A and B, if Gr-$A \approx_{\text{gr}}$ Gr-B then Mod-$A \approx$ Mod-B (Theorem 2.3.7). This was first studied by Gordon and Green [42] in the setting of $\mathbb{Z}$-graded rings.

Throughout this chapter, A is a Γ-graded ring unless otherwise stated. Moreover, all functors are *additive* functors. For the theory of (nongraded) Morita theory we refer the reader to [61] and [4]

2.1 First instance of the graded Morita equivalence

Before describing the general graded Morita equivalence, we treat a special case of matrix algebras, which, in the word of T. Y. Lam, is the "first instance" of equivalence between module categories [61, §17B]. This special case is quite explicit and will help us in calculating the graded Grothendieck group of matrix algebras. Recall from §1.3 that for a Γ-graded ring A, and $\overline{\delta} = (\delta_1, \dots, \delta_n)$

$$A^n(\overline{\delta}) = A(\delta_1) \oplus \cdots \oplus A(\delta_n) \tag{2.1}$$

is a graded free A-bimodule. Moreover, $A^n(\overline{\delta})$ is a graded right $\mathbb{M}_n(A)(\overline{\delta})$-module and $A^n(-\overline{\delta})$ is a graded left $\mathbb{M}_n(A)(\overline{\delta})$-module. Here $-\overline{\delta} = (-\delta_1, \dots, -\delta_n)$.

Proposition 2.1.1 *Let A be a Γ-graded ring and let $\overline{\delta} = (\delta_1, \dots, \delta_n)$, where $\delta_i \in \Gamma$, $1 \leq i \leq n$. Then the functors*

$$\begin{aligned}\psi : \text{Gr-}\,\mathbb{M}_n(A)(\overline{\delta}) &\longrightarrow \text{Gr-}A,\\ P &\longmapsto P \otimes_{\mathbb{M}_n(A)(\overline{\delta})} A^n(-\overline{\delta})\end{aligned}$$

and

$$\varphi : \text{Gr-}A \longrightarrow \text{Gr-}\,\mathbb{M}_n(A)(\overline{\delta}),$$
$$Q \longmapsto Q \otimes_A A^n(\overline{\delta})$$

form equivalences of categories and commute with suspensions, i.e, $\psi\mathcal{T}_\alpha = \mathcal{T}_\alpha\psi$, $\alpha \in \Gamma$.

Proof One can check that there is a Γ-graded A-bimodule isomorphism

$$f : A^n(\overline{\delta}) \otimes_{\mathbb{M}_n(A)(\overline{\delta})} A^n(-\overline{\delta}) \longrightarrow A, \tag{2.2}$$
$$(a_1, \dots, a_n) \otimes \begin{pmatrix} b_1 \\ \vdots \\ b_n \end{pmatrix} \longmapsto a_1 b_1 + \cdots + a_n b_n$$

with

$$f^{-1} : A \longrightarrow A^n(\overline{\delta}) \otimes_{\mathbb{M}_n(A)(\overline{\delta})} A^n(-\overline{\delta}),$$
$$a \longmapsto (a, 0, \dots, 0) \otimes \begin{pmatrix} 1 \\ 0 \\ \vdots \\ 0 \end{pmatrix}.$$

Moreover, there is a Γ-graded $\mathbb{M}_n(A)(\overline{\delta})$-bimodule isomorphism

$$g : A^n(-\overline{\delta}) \otimes_A A^n(\overline{\delta}) \longrightarrow \mathbb{M}_n(A)(\overline{\delta}), \tag{2.3}$$
$$\begin{pmatrix} a_1 \\ \vdots \\ a_n \end{pmatrix} \otimes (b_1, \dots, b_n) \longmapsto \begin{pmatrix} a_1 b_1 & \cdots & a_1 b_n \\ \vdots & & \vdots \\ a_n b_1 & \cdots & a_n b_n \end{pmatrix}$$

with

$$g^{-1} : \mathbb{M}_n(A)(\overline{\delta}) \longrightarrow A^n(-\overline{\delta}) \otimes_A A^n(\overline{\delta}),$$
$$(a_{i,j}) \longmapsto \begin{pmatrix} a_{1,1} \\ a_{2,1} \\ \vdots \\ a_{n,1} \end{pmatrix} \otimes (1, 0, \dots, 0) + \cdots + \begin{pmatrix} a_{1,n} \\ \vdots \\ a_{n-1,n} \\ a_{n,n} \end{pmatrix} \otimes (0, 0, \dots, 1).$$

Now using (2.2) and (2.3), it follows easily that $\varphi\psi$ and $\psi\varphi$ are equivalent to identity functors. The general fact that for $\alpha \in \Gamma$,

$$(P \otimes Q)(\alpha) = P(\alpha) \otimes Q = P \otimes Q(\alpha),$$

(see §1.2.6) shows that the suspension functor commutes with ψ and ϕ. □

Since the functors ϕ and ψ, being tensor functors, preserve the projectivity, and send finitely generated modules to finitely generated modules, we immediately get the following corollary which we will use to compute the graded K_0 of matrix algebras (see Example 3.7.8).

Corollary 2.1.2 *Let A be a Γ-graded ring and let $\overline{\delta} = (\delta_1, \dots, \delta_n)$, where $\delta_i \in \Gamma$, $1 \leq i \leq n$. Then the functors*

$$\psi : \text{Pgr-}\mathbb{M}_n(A)(\overline{\delta}) \longrightarrow \text{Pgr-}A,$$
$$P \longmapsto P \otimes_{\mathbb{M}_n(A)(\overline{\delta})} A^n(-\overline{\delta})$$

and

$$\varphi : \text{Pgr-}A \longrightarrow \text{Pgr-}\mathbb{M}_n(A)(\overline{\delta}),$$
$$Q \longmapsto Q \otimes_A A^n(\overline{\delta})$$

form equivalences of categories and commute with suspensions, i.e., $\psi\mathcal{T}_\alpha = \mathcal{T}_\alpha\psi$, $\alpha \in \Gamma$.

Example 2.1.3 Let $\mathbf{e}_{ii}$, $1 \leq i \leq n$, be the matrix unit in $\mathbb{M}_n(A)(\overline{\delta})$. Then $\mathbf{e}_{ii}\,\mathbb{M}_n(A)(\overline{\delta})$ is the graded finitely generated projective right $\mathbb{M}_n(A)(\overline{\delta})$-module. It is easy to see that there is a graded $\mathbb{M}_n(A)(\overline{\delta})$-module isomorphism

$$\mathbf{e}_{ii}\,\mathbb{M}_n(A)(\overline{\delta}) \longrightarrow A(\delta_1 - \delta_i) \oplus A(\delta_2 - \delta_i) \oplus \cdots \oplus A(\delta_n - \delta_i),$$
$$\mathbf{e}_{ii}X \longmapsto (x_{i1}, x_{i2}, \dots, x_{in}).$$

Thus, by Proposition 2.1.1, the module $\mathbf{e}_{ii}\,\mathbb{M}_n(A)(\overline{\delta})$ in Gr-$\mathbb{M}_n(A)(\overline{\delta})$ corresponds to $A(-\delta_i)$ in Gr-A as follows:

$$\begin{aligned}
&\mathbf{e}_{ii}\,\mathbb{M}_n(A)(\overline{\delta}) \otimes_{\mathbb{M}_n(A)(\overline{\delta})} A^n(-\overline{\delta}) \\
&\quad \cong_{\text{gr}} A(\delta_1 - \delta_i) \oplus A(\delta_2 - \delta_i) \oplus \cdots \oplus A(\delta_n - \delta_i) \otimes_{\mathbb{M}_n(A)(\overline{\delta})} A^n(-\overline{\delta}) \\
&\quad \cong_{\text{gr}} A^n(\overline{\delta})(-\delta_i) \otimes_{\mathbb{M}_n(A)(\overline{\delta})} A^n(-\overline{\delta}) \cong_{\text{gr}} (A^n(\overline{\delta}) \otimes_{\mathbb{M}_n(A)(\overline{\delta})} A^n(-\overline{\delta}))(-\delta_i) \\
&\quad \cong_{\text{gr}} A(-\delta_i).
\end{aligned}$$

Example 2.1.4 STRONGLY GRADED IS NOT A MORITA INVARIANT PROPERTY

Let A be a strongly Γ-graded ring. One can see that $\mathbb{M}_n(A)(\alpha_1, \dots, \alpha_n)$, $\alpha_i \in \Gamma$, is also strongly graded. However, the converse is not true. For example, let K be a field and consider the $\mathbb{Z}$-graded ring $K[x^2, x^{-2}]$ which has the support $2\mathbb{Z}$. By Proposition 2.1.1, $A = \mathbb{M}_2(K[x^2, x^{-2}])(0, 1)$ is graded Morita equivalent to $B = K[x^2, x^{-2}]$. However, one can easily see that A is strongly graded whereas B is not.

One can also observe that although A and B are graded Morita equivalent, the support set of A is $\mathbb{Z}$, whereas the support of B is $2\mathbb{Z}$ (see (1.49)).

Remark 2.1.5 RESTRICTING THE EQUIVALENCE TO A SUBGROUP OF GRADE GROUP

Let A be a Γ-graded ring and let $\overline{\delta} = (\delta_1, \dots, \delta_n)$, where $\delta_i \in \Gamma$, $1 \leq i \leq n$. Suppose Ω is a subgroup of Γ such that $\Gamma_A \subseteq \Omega$ and $\{\delta_1, \dots, \delta_n\} \subseteq \Omega$. Since the support of $\mathbb{M}_n(A)(\overline{\delta})$ is a subset of Ω (see (1.49)), we can naturally consider A and $\mathbb{M}_n(A)(\overline{\delta})$ as Ω-graded rings (see Remark 1.2.5). Since the Γ-isomorphisms (2.2) and (2.3) in the proof of Proposition 2.1.1 can be considered as Ω-isomorphisms, we can easily see that Gr^{Ω}- $\mathbb{M}_n(A)(\overline{\delta})$ is equivalent to Gr^{Ω}-A as well. The converse of this statement is valid for the general graded Morita theory, see Remark 2.3.11.

Remark 2.1.6 THE MORITA EQUIVALENCE FOR NONABELIAN GRADE GROUP

As mentioned in the Introduction, most of the results of this book can be carried over to the setting of nonabelian grade groups as well. In some cases this will be done by adjusting the arrangements and shifts. Below, we adjust the arrangements so that Proposition 2.1.1 holds for the nonabelian grade group as well.

Let Γ be a (nonabelian) group and A be a Γ-graded ring. Further, let M and N be Γ-graded left and right A-modules, respectively. Define the δ-shifted graded left and right A-modules $M(\delta)$ and $(\delta)N$, respectively, as

$$M(\delta) = \bigoplus_{\gamma \in \Gamma} M(\delta)_\gamma, \text{ where } M(\delta)_\gamma = M_{\gamma\delta},$$

$$(\delta)N = \bigoplus_{\gamma \in \Gamma} (\delta)N_\gamma, \text{ where } (\delta)N_\gamma = N_{\delta\gamma}.$$

One can easily check that $M(\delta)$ and $(\delta)N$ are left and right graded A-modules, respectively, $M(\delta)(\gamma) = M(\gamma\delta)$ and $(\gamma)(\delta)N = (\delta\gamma)N$. Note that with this arrangement, for $\overline{\delta} = (\delta_1, \dots, \delta_n)$ and $\overline{\delta}^{-1} = (\delta_1^{-1}, \dots, \delta_n^{-1})$, $(\overline{\delta})A^n$ is an $A - \mathbb{M}_n(A)(\overline{\delta})$-bimodule and $A^n(\overline{\delta}^{-1})$ is an $\mathbb{M}_n(A)(\overline{\delta}) - A$-bimodule, where $\mathbb{M}_n(A)(\overline{\delta})$ is as in Remark 1.3.2. Moreover, the isomorphisms (2.2) and (2.3) take the form

$$f : A^n(\overline{\delta}^{-1}) \otimes_{\mathbb{M}_n(A)(\overline{\delta})} (\overline{\delta})A^n \longrightarrow A,$$

$$(a_1, \dots, a_n) \otimes \begin{pmatrix} b_1 \\ \vdots \\ b_n \end{pmatrix} \longmapsto a_1 b_1 + \dots + a_n b_n$$

and

$$g : (\overline{\delta})A^n \otimes_A A^n(\overline{\delta}^{-1}) \longrightarrow \mathbb{M}_n(A)(\overline{\delta}),$$

$$\begin{pmatrix} a_1 \\ \vdots \\ a_n \end{pmatrix} \otimes (b_1, \dots, b_n) \longmapsto \begin{pmatrix} a_1 b_1 & \cdots & a_1 b_n \\ \vdots & & \vdots \\ a_n b_1 & \cdots & a_n b_n \end{pmatrix}.$$

Thus we can write Proposition 2.1.1 in the nonabelian grade group setting.

2.2 Graded generators

We start with a categorical definition of graded generators. In this section, as usual, the modules are (graded) right A-modules.

Definition 2.2.1 Let A be a Γ-graded ring. A graded A-module P is called a *graded generator* if whenever $f : M \to N$ is a nonzero graded A-module homomorphism, then there exists $\alpha \in \Gamma$ and a graded homomorphism $g : P(\alpha) \to M$ such that $fg : P(\alpha) \to N$ is a nonzero map.

Let A be a Γ-graded ring and P be a graded right A-module. Define the *graded trace ideal* of P as follows:

$$\boxed{\operatorname{Tr}^{\mathrm{gr}}(P) := \Big\{ \sum_i f_i(p_i) \mid f_i \in \operatorname{Hom}(P, A)_\alpha, \alpha \in \Gamma,\ p_i \in P^h \Big\}.}$$

One can check that $\operatorname{Tr}^{\mathrm{gr}}(P)$ is a graded two-sided ideal of A. The following theorem provides a set theoretic way to define a graded generator.

Theorem 2.2.2 *For any graded A-module P, the following are equivalent:*

(1) *P is a graded generator;*
(2) $\operatorname{Tr}^{\mathrm{gr}}(P) = A$;
(3) *A is a direct summand of a finite direct sum $\bigoplus_i P(\alpha_i)$, where $\alpha_i \in \Gamma$;*
(4) *A is a direct summand of a direct sum $\bigoplus_i P(\alpha_i)$, where $\alpha_i \in \Gamma$;*
(5) *every graded A-module M is a homomorphic image of $\bigoplus_i P(\alpha_i)$, where $\alpha_i \in \Gamma$.*

Proof (1) $\Rightarrow$ (2) First note that

$$\operatorname{Tr}^{\mathrm{gr}}(P) = \Big\{ \sum_i f_i(P(\alpha)) \mid f_i \in \operatorname{Hom}_{A\text{-}\mathrm{Gr}}(P(\alpha), A), \alpha \in \Gamma \Big\}.$$

Suppose that $\operatorname{Tr}^{\mathrm{gr}}(P) \neq A$. Then the graded canonical projection

$$f : A \to A/\operatorname{Tr}^{\mathrm{gr}}(P)$$

is not a zero map. Since P is a graded generator, there is a $g \in \operatorname{Hom}_{A\text{-Gr}}(P(\alpha), A)$ such that fg is not zero. But this implies $g(P(\alpha)) \not\subseteq \operatorname{Tr}^{\mathrm{gr}}(P)$ which is a contradiction.

(2) $\Rightarrow$ (3) Since $\operatorname{Tr}^{\mathrm{gr}}(P) = A$, one can find $g_i \in \operatorname{Hom}_{A\text{-Gr}}(P(\alpha_i), A)$, $\alpha_i \in \Gamma$, $1 \leq i \leq n$ such that $\sum_{i=1}^n g_i(P(\alpha_i)) = A$. Consider the graded A-module epimorphism

$$\bigoplus_{i=1}^{n} P(\alpha_i) \longrightarrow A,$$

$$(p_1, \dots, p_n) \longmapsto \sum_{i=1}^{n} g_i(p_i).$$

Since A is graded projective, this map splits. Thus

$$\bigoplus_{i=1}^{n} P(\alpha_i) \cong_{\mathrm{gr}} A \oplus Q,$$

for some graded A-module Q. This gives (3).

(3) $\Rightarrow$ (4) This is immediate.

(4) $\Rightarrow$ (5) Since any module is a homomorphic image of a graded free A-module, there is a graded epimorphism $\bigoplus_j A(\alpha'_j) \to M$. By (3) there is a graded epimorphism $\bigoplus_i P(\alpha_i) \to A$, so an epimorphism

$$\bigoplus_{i} P(\alpha_i + \alpha'_j) \to A(\alpha'_j).$$

Therefore

$$\bigoplus_{j} \bigoplus_{i} P(\alpha_i + \alpha'_j) \to \oplus_j A(\alpha'_j) \to M$$

is a graded epimorphism.

(5) $\Rightarrow$ (1) Let $f : M \to N$ be a nonzero graded homomorphism. By (4) there is an epimorphism $\bigoplus_i P(\alpha_i) \to M$. So the composition $\bigoplus_i P(\alpha_i) \to M \to N$ is not zero. This immediately implies there is an i such that the composition $P(\alpha_i) \to M \to N$ is not zero. This gives (1). □

Example 2.2.3 Let A be a graded simple ring and P be a graded projective A-module. Since $\operatorname{Tr}^{\mathrm{gr}}(P)$ is a graded two-sided ideal, one can easily show, using Proposition 2.2.2(2), that P is a graded generator.

Recall that in the category of right A-modules, Mod-A, a generator is defined as in Definition 2.2.1 by dropping all the graded adjectives. Moreover, a similar theorem as in Theorem 2.2.2 can be written in the nongraded case, by considering Γ to be the trivial group (see [61, Theorem 18.8]). In particular, P is a generator if and only if $\mathrm{Tr}(P) = A$, where

$$\mathrm{Tr}(P) := \Big\{ \sum_i f_i(p_i) \mid f_i \in \mathrm{Hom}_A(P, A),\ p_i \in P \Big\}.$$

Remark 2.2.4 Considering Gr-A as a subcategory of Mod-A (which is clearly not a full subcategory), one can define generators for Gr-A. In this case one can easily see that $\bigoplus_{\gamma\in\Gamma} A(\gamma)$ is a generator for the category Gr-A. Recall that, in comparison, A is a generator for Mod-A and A is a graded generator for Gr-A.

An A-module P is called a *progenerator* if it is both a finitely generated projective and a generator, *i.e.*, there is an $n \in \mathbb{N}$ such that $A^n \cong P \oplus K$ and $P^n \cong A \oplus L$, where K and L are A-modules. Similarly, a graded A-module is called a *graded progenerator* if it is both a graded finitely generated projective and a graded generator.

Reminiscent of the case of graded projective modules (Proposition 1.2.15), we have the following relation between generators and graded generators.

Theorem 2.2.5 *Let P be a finitely generated A-module. Then P is a graded generator if and only if P is graded and is a generator.*

Proof Suppose P is a graded generator. By Theorem 2.2.2, $\mathrm{Tr}^{\mathrm{gr}}(P) = A$ which implies that there are graded homomorphisms f_i (of possibly different degrees) in $\mathrm{Hom}_{A\text{-Gr}}(P(\alpha_i), A)$ and $p_i \in P^h$, such that $\sum_i f_i(p_i) = 1$. This immediately implies $\mathrm{Tr}(P)$, being an ideal, is A. Thus P is a generator.

Conversely, suppose P is graded and is a generator. Then there are homomorphisms f_i in $\mathrm{Hom}(P, A)$ and $p_i \in P$ such that $\sum_i f_i(p_i) = 1$. Since P is finitely generated, by Theorem 1.2.6, f_i can be written as a sum of graded homomorphisms, and p_i as sum of homogeneous elements in P. This shows $1 \in \mathrm{Tr}^{\mathrm{gr}}(P)$. Since $\mathrm{Tr}^{\mathrm{gr}}(P)$ is an ideal, $\mathrm{Tr}^{\mathrm{gr}}(P) = A$ and so P is a graded generator by Theorem 2.2.2. □

2.3 General graded Morita equivalence

Let P be a right A-module. Consider the ring $B = \mathrm{Hom}_A(P, P)$. Then P has a natural $B{-}A$-bimodule structure. The actions of A and B on P are defined by $p.a = pa$ and $g.p = g(p)$, respectively, where $g \in B$, $p \in P$ and $a \in A$. Consider the dual $P^* = \mathrm{Hom}_A(P, A)$. Then P^* has a natural $A - B$-bimodule structure. The actions of A and B on P^* defined by $(a.q)(p) = aq(p)$ and $q.g = q \circ g$, respectively, where $g \in B$, $q \in P^*$, $p \in P$ and $a \in A$. Moreover, one defines

$$\begin{aligned} \phi : P^* \otimes_B P &\longrightarrow A, \\ q \otimes p &\longmapsto q(p) \end{aligned} \tag{2.4}$$

and

$$\begin{aligned} \psi : P \otimes_A P^* &\longrightarrow B, \\ p \otimes q &\longmapsto pq, \end{aligned} \tag{2.5}$$

where $pq(p') = p(q(p'))$. One can check that ϕ is a $A-A$-bimodule homomorphism and ψ is a $B-B$-bimodule homomorphism. We leave it to the reader to check these and that with the actions introduced above, P has a B–A-bimodule structure and P^* has a $A-B$-bimodule structure (see [61, §18C]). These compatibility conditions amount to the fact that the *Morita ring*

$$M = \begin{pmatrix} A & P^* \\ P & B \end{pmatrix}, \tag{2.6}$$

with matrix multiplication has the associativity property. In fact M is a formal matrix ring as defined in Example 1.1.4.

As part of the Morita theory, one proves that when P is a generator, then ϕ is an isomorphism. Similarly, if P is finitely generated and projective, then ψ is an isomorphism (see [61, §18C]). Putting these facts together, it is an easy observation that

$$- \otimes_A P^* : \text{Mod-}A \to \text{Mod-}B \quad \text{and} \quad - \otimes_B P : \text{Mod-}B \to \text{Mod-}A \tag{2.7}$$

are inverses of each other and so these two categories are (Morita) equivalent.

If P is a graded finitely generated right A-module, then by Theorem 1.2.6, $B = \operatorname{End}_A(P, P)$ is also a graded ring and P^* a graded left A-module. In fact, one can easily check that with the actions defined above, P is a graded $B-A$-bimodule, P^* is a graded $A-B$-module and similarly ϕ and ψ are graded $A-A$ and $B-B$-module homomorphisms, respectively. The Morita ring M of (2.6) is a graded formal matrix ring (see Example 1.2.9), with

$$M_\alpha = \begin{pmatrix} A_\alpha & P^*_\alpha \\ P_\alpha & B_\alpha \end{pmatrix} = \begin{pmatrix} A_\alpha & \operatorname{Hom}_{\text{Gr-}A}(P, A(\alpha)) \\ P_\alpha & \operatorname{Hom}_{\text{Gr-}A}(P, P(\alpha)) \end{pmatrix}. \tag{2.8}$$

We demonstrate here that P^* is a graded A–B-bimodule and leave the others which are similar and routine to the reader. Recall that

$$P^* = \bigoplus_{\alpha \in \Gamma} \operatorname{Hom}_A(P, A)_\alpha.$$

Let $a \in A_\alpha$ and $q \in P^*_\beta = \operatorname{Hom}_A(P, A)_\beta$. Then $a.q \in P^*$, where $(a.q)(p) = aq(p)$.

If $p \in P_\gamma$, then one easily sees that $(a.q)(p) \in A_{\alpha+\beta+\gamma}$. This shows that $aq \in \operatorname{Hom}_A(P, A)_{\alpha+\beta} = P^*_{\alpha+\beta}$. On the other hand, if $q \in P^*_\alpha$ and $g \in \operatorname{Hom}(P, P)_\beta$, then $q.g \in P^*$, where $q.g(p) = qg(p)$. So if $p \in P_\gamma$ then $(q.g)(p) = qg(p) \in A_{\alpha+\beta+\gamma}$. So $q.g \in P^*_{\alpha+\beta}$ as well.

Remark 2.3.1 A NATURAL GROUPOID GRADED RING

Consider the Morita ring M of (2.6) with homogeneous components (2.8). Further, consider the following homogeneous elements:

$$m_\gamma = \begin{pmatrix} a_\gamma & 0 \\ 0 & 0 \end{pmatrix} \in M_\gamma, \quad m_\delta = \begin{pmatrix} 0 & 0 \\ p_\delta & 0 \end{pmatrix} \in M_\delta.$$

By definition $m_\gamma m_\delta \in M_{\gamma+\delta}$. However, $m_\gamma m_\delta = 0$. It would be more informative to recognise this from the outset. Thus we introduce a groupoid grading on the Morita ring M in (2.6) with a groupoid graded structure which gives a "finer" grading than one defined in (2.8) (see Remark 1.1.12 for the groupoid graded rings). Consider the groupoid $2 \times \Gamma \times 2$ and for $\gamma \in \Gamma$, set

$$M_{(1,\gamma,1)} = \begin{pmatrix} A_\gamma & 0 \\ 0 & 0 \end{pmatrix}, M_{(1,\gamma,2)} = \begin{pmatrix} 0 & P^*_\gamma \\ 0 & 0 \end{pmatrix}, M_{(2,\gamma,1)} = \begin{pmatrix} 0 & 0 \\ P_\gamma & 0 \end{pmatrix}, M_{(2,\gamma,2)} = \begin{pmatrix} 0 & 0 \\ 0 & B_\gamma \end{pmatrix}.$$

Clearly, for $\gamma \in \Gamma$

$$M_\gamma = \bigoplus_{1 \leq i,j \leq 2} M_{(i,\gamma,j)}$$

and

$$M = \bigoplus_{g \in 2 \times \Gamma \times 2} M_g.$$

This makes M a $2 \times \Gamma \times 2$-groupoid graded ring.

By Theorem 2.2.5 and Proposition 1.2.15, if P is a graded finitely generated projective and graded generator, then P is finitely generated projective and a generator. Since the grading is preserved under tensor products, the restriction of the functors $- \otimes_A P^*$ and $- \otimes_B P$ of (2.7) induces an equivalence

$$- \otimes_A P^* : \text{Gr-}A \longrightarrow \text{Gr-}B \quad \text{and} \quad - \otimes_B P : \text{Gr-}B \longrightarrow \text{Gr-}A. \tag{2.9}$$

Moreover, these functors commute with suspensions. Thus we get a commutative diagram where the vertical maps are forgetful functors (see §1.2.7),

$$\begin{array}{ccc} \text{Gr-}A & \xrightarrow{-\otimes_A P^*} & \text{Gr-}B \\ \downarrow U & & \downarrow U \\ \text{Mod-}A & \xrightarrow{-\otimes_A P^*} & \text{Mod-}B. \end{array} \tag{2.10}$$

We call $- \otimes_A P^*$ a graded equivalence functor (see Definition 2.3.3).

Example 2.3.2 Let A be a graded ring and e be a *full homogeneous idempotent* of A, *i.e.*, e is a homogeneous element, $e^2 = e$ and $AeA = A$. Clearly e has degree zero. Consider $P = eA$. One can readily see that P is a graded right progenerator. Then $P^* = \operatorname{Hom}_A(eA, A) \cong_{\mathrm{gr}} Ae$ as graded left A-modules and $B = \operatorname{End}_A(eA) \cong_{\mathrm{gr}} eAe$ as graded rings (see §1.3.2). The maps ϕ and ψ, described above as a part of the Morita context ((2.4) and (2.5)), take the form $\phi : Ae \otimes_{eAe} eA \to A$ and $\psi : eA \otimes_A Ae \to eAe$ which are graded isomorphisms. Thus we get a (graded) equivalence between Gr-A and Gr-eAe which lifts to a (graded) equivalence between Mod-A and Mod-eAe, as is shown in the diagram below.

$$\begin{array}{ccc} \text{Gr-}A & \xrightarrow{-\otimes_A Ae} & \text{Gr-}eAe \\ \downarrow U & & \downarrow U \\ \text{Mod-}A & \xrightarrow{-\otimes_A Ae} & \text{Mod-}eAe. \end{array} \tag{2.11}$$

Before stating the general graded Morita equivalence, we need to make some definitions. Recall from §1.1 that for $\alpha \in \Gamma$, the α-suspension functor $\mathcal{T}_\alpha$: Gr-$A \to$ Gr-A, $M \mapsto M(\alpha)$ is an isomorphism with the property $\mathcal{T}_\alpha \mathcal{T}_\beta = \mathcal{T}_{\alpha+\beta}$, $\alpha, \beta \in \Gamma$.

Definition 2.3.3 Let A and B be Γ-graded rings.

1 A functor ϕ : Gr-$A \to$ Gr-B is called a *graded functor* if $\phi \mathcal{T}_\alpha = \mathcal{T}_\alpha \phi$.

2 A graded functor ϕ : Gr-$A \to$ Gr-B is called a *graded equivalence* if there is a graded functor ψ : Gr-$B \to$ Gr-A such that $\psi\phi \cong 1_{\text{Gr-}A}$ and $\phi\psi \cong 1_{\text{Gr-}B}$.

3 If there is a graded equivalence between Gr-A and Gr-B, we say A and B are *graded equivalent* or *graded Morita equivalent* and we write Gr-$A \approx_{\mathrm{gr}}$ Gr-B, or Gr^Γ-$A \approx_{\mathrm{gr}} \text{Gr}^\Gamma$-$B$ to emphasise that the categories are Γ-graded.

4 A functor ϕ' : Mod-$A \to$ Mod-B is called a *graded functor* if there is a graded functor ϕ : Gr-$A \to$ Gr-B such that the following diagram, where the vertical functors are forgetful functors (see §1.2.7), is commutative.

$$\begin{array}{ccc} \text{Gr-}A & \xrightarrow{\phi} & \text{Gr-}B \\ \downarrow U & & \downarrow U \\ \text{Mod-}A & \xrightarrow{\phi'} & \text{Mod-}B. \end{array} \tag{2.12}$$

The functor ϕ is called an *associated graded functor* of ϕ'.

5 A functor ϕ : Mod-$A \to$ Mod-B is called a *graded equivalence* if it is graded and an equivalence.

Definition 2.3.3 of graded functors is formulated for the category of (graded) right modules. A similar definition can be written for the category of (graded) left modules. We will see that the notion of graded equivalence is left–right symmetric (see Remark 2.3.10).

Remark 2.3.4 Note that although we require the graded functors to commute with the suspensions, we don't require the natural transformations between these functors to have these properties.

Example 2.3.5 The equivalence between Mod-A and Mod-eAe in Example 2.3.2 is a graded equivalence as is demonstrated in Diagram (2.11).

If Q is an object in Gr-A, then we denote $U(Q) \in$ Mod-A also by Q, forgetting its graded structure (see §1.2.7). Also, when working with graded matrix rings, say $\mathbb{M}_n(A)(\overline{\delta})$, we write simply Mod-$\mathbb{M}_n(A)$ when considering the category of (nongraded) modules over this matrix algebra. These should not cause any confusion in the text.

Example 2.3.6 Proposition 2.1.1 can be written for the category of A-modules, which gives us a nongraded version of Morita equivalence. We have the commutative diagram

$$\begin{array}{ccc} \text{Gr-}A & \xrightarrow{-\otimes_A A^n(\overline{\delta})} & \text{Gr-}\,\mathbb{M}_n(A)(\overline{\delta}) \\ \Big\downarrow U & & \Big\downarrow U \\ \text{Mod-}A & \xrightarrow{-\otimes_A A^n} & \text{Mod-}\mathbb{M}_n(A), \end{array}$$

which shows that the functor

$$-\otimes_A A^n : \text{Mod-}A \to \text{Mod-}\mathbb{M}_n(A)$$

is a graded equivalence.

We are in a position to state the main theorem of this chapter.

Theorem 2.3.7 *Let A and B be two Γ-graded rings. Let ϕ : Gr-$A \to$ Gr-B be a graded equivalence. Then there is a graded equivalence ϕ' : Mod-$A \to$ Mod-B with an associated graded functor isomorphic to ϕ. Indeed, there is a graded A–B-bimodule Q such that $\phi \cong -\otimes_A Q$ and consequently the following diagram* *commutes:*

$$\begin{array}{ccc} \text{Gr-}A & \xrightarrow{-\otimes_A Q} & \text{Gr-}B \\ {\scriptstyle U}\downarrow & & \downarrow{\scriptstyle U} \\ \text{Mod-}A & \xrightarrow{-\otimes_A Q} & \text{Mod-}B. \end{array}$$

Proof Suppose the graded functor $\psi : \text{Gr-}B \to \text{Gr-}A$ is an inverse of the functor ϕ with

$$f : \psi\phi \cong 1_{\text{Gr-}A}.$$

Since B is a graded finitely generated projective and a graded generator in Gr-B, it follows that $P = \psi(B)$ is a graded finitely generated projective and a graded generator in Gr-A. Thus by Theorem 2.2.5 and Proposition 1.2.15, P is a finitely generated projective and a generator in Mod-A as well. This shows that $\operatorname{Hom}_A(P, -) : \text{Mod-}A \to \text{Mod-}B$ is a graded equivalence. We will show that $\phi \cong \operatorname{Hom}_A(P, -)$ on the category of Gr-A.

By Theorem 1.2.6, $\operatorname{Hom}_B(B, B) = \bigoplus_{\alpha\in\Gamma} \operatorname{Hom}_B(B, B)_\alpha$, and by (1.15) we can write

$$\operatorname{Hom}_B(B, B)_\alpha = \operatorname{Hom}_{\text{Gr-}B}(B, B(\alpha)).$$

Applying ψ to each of these components, since ψ is a graded functor, we get a group homomorphism

$$\operatorname{Hom}_B(B, B) = \bigoplus_{\alpha\in\Gamma} \operatorname{Hom}_B(B, B)_\alpha \xrightarrow{\psi} \bigoplus_{\alpha\in\Gamma} \operatorname{Hom}_A(P, P)_\alpha = \operatorname{Hom}_A(P, P). \tag{2.13}$$

One can immediately see that this in fact gives a graded isomorphism of rings between $\operatorname{Hom}_B(B, B)$ and $\operatorname{Hom}_A(P, P)$.

For any $b \in B$ consider the right B-module homomorphism

$$\begin{aligned} \eta_b : B &\longrightarrow B \\ x &\longmapsto bx. \end{aligned}$$

Then the regular representation map $\eta : B \to \operatorname{Hom}_B(B, B)$, $\eta(b) = \eta_b$, is a graded isomorphism of rings. Thus we have graded isomorphisms of rings

$$B \xrightarrow{\eta} \operatorname{Hom}_B(B, B) \xrightarrow{\psi} \operatorname{Hom}_A(P, P), \tag{2.14}$$

where P is a graded A-progenerator.

Now since for any graded A-module X,

$$\operatorname{Hom}_A(P, X)$$

is a graded right $\operatorname{Hom}_A(P, P)$-module, the isomorphisms (2.14) induce a graded

B-module structure on $\operatorname{Hom}_A(P,X)$. Namely, for homogeneous elements $b \in B$ and $t \in \operatorname{Hom}_A(P,X)$

$$t.b = t\psi(\eta_b) \tag{2.15}$$

which extends linearly to all elements.

We show that $\phi \cong \operatorname{Hom}_A(P,-)$. Let X be a graded A-module. Then

$$\begin{aligned}\phi(X) \cong_{\rm gr} \operatorname{Hom}_B(B,\phi(X)) = \bigoplus_{\alpha\in\Gamma} \operatorname{Hom}_B(B,\phi(X))_\alpha &\overset{\psi}{\longrightarrow} \bigoplus_{\alpha\in\Gamma} \operatorname{Hom}_A(P,\psi\phi(X))_\alpha \\ &\overset{\operatorname{Hom}(1,f)}{\longrightarrow} \bigoplus_{\alpha\in\Gamma} \operatorname{Hom}_A(P,X)_\alpha = \operatorname{Hom}_A(P,X).\end{aligned} \tag{2.16}$$

This shows that $\phi(X)$ is isomorphic to $\operatorname{Hom}_A(P,X)$ as a graded abelian group. We need to show that this isomorphism, call it Θ, is a B-module isomorphism. Let $z \in \phi(X)$ and $b \in B$ be homogeneous elements. Since

$$\phi(X) \cong_{\rm gr} \operatorname{Hom}_B(B,\phi(X)),$$

for $z \in \phi(X)$, we denote the corresponding homomorphism in $\operatorname{Hom}_B(B,\phi(X))$ also by z. Then

$$\Theta(z.b) = \Theta(z\,\eta_b) = \operatorname{Hom}(1,f)\psi(z\,\eta_b) = f\psi(z)\psi(\eta_b).$$

But

$$\Theta(z).b = \big(\operatorname{Hom}(1,f)\psi(z)\big).b = f\psi(z)\psi(\eta_b) \qquad \text{(by (2.15))}.$$

Thus $\phi \cong \operatorname{Hom}_A(P,-)$ on Gr-A. Now consider

$$\phi' = \operatorname{Hom}_A(P,-) : \text{Mod-}A \to \text{Mod-}B.$$

This gives the first part of the theorem.

Setting $P^* = \operatorname{Hom}_A(P,A)$, one can easily check that for any graded right A-module M, the map

$$\begin{aligned} M \otimes_A P^* &\longrightarrow \operatorname{Hom}_A(P,M),\\ m \otimes q &\longmapsto (m\otimes q)(p) = mq(p),\end{aligned}$$

where $m \in M$, $q \in P^*$ and $p \in P$, is a graded B-homomorphism (recall that by (2.14), $B \cong_{\rm gr} \operatorname{End}_A(P)$). Since P is a graded progenerator, by Theorem 2.2.5, it is a progenerator, which in return gives that the above homomorphism is in fact an isomorphism (see [61, Remark 18.25]). Thus $\phi \cong \operatorname{Hom}_A(P,-) \cong - \otimes_A P^*$. This gives the second part of the theorem. □

Theorem 2.3.8 *Let A and B be two Γ-graded rings. The following are equivalent:*

(1) Mod-A *is graded equivalent to* Mod-B*;*
(2) Gr-A *is graded equivalent to* Gr-B*;*
(3) $B \cong_{\mathrm{gr}} \mathrm{End}_A(P)$ *for a graded A-progenerator P;*
(4) $B \cong_{\mathrm{gr}} e\,\mathbb{M}_n(A)(\overline{\delta})e$ *for a full homogeneous idempotent* $e \in \mathbb{M}_n(A)(\overline{\delta})$*, where* $\overline{\delta} = (\delta_1, \dots, \delta_n)$*,* $\delta_i \in \Gamma$.

Proof (1) $\Rightarrow$ (2) Let ϕ : Mod-$A \rightarrow$ Mod-B be a graded equivalence. Using (2.12), it follows that $\phi(A) = P$ is a graded right B-module. Also a similar argument as in (2.14) shows that P is a graded left A-module. Since ϕ is an equivalence, from the (nongraded) Morita theory it follows that $\phi \cong - \otimes_A P$ with an inverse $-\otimes_B P^*$. Since the tensor product respects the grading, the same functor ϕ induces a graded equivalence between Gr-A and Gr-B.

(2) $\Rightarrow$ (3) This is (2.14) in the proof of Theorem 2.3.7.

(3) $\Rightarrow$ (4) Since P is a graded finitely generated projective A-module,

$$P \oplus Q \cong_{\mathrm{gr}} A^n(-\overline{\delta}),$$

where $\overline{\delta} = (\delta_1, \dots, \delta_n)$ and $n \in \mathbb{N}$. Let

$$e \in \mathrm{End}_A(A^n(-\overline{\delta})) \cong_{\mathrm{gr}} \mathbb{M}_n(A)(\overline{\delta})$$

be the graded homomorphism which sends Q to zero and acts as identity on P. Thus

$$e \in \mathrm{End}_A(A^n(-\overline{\delta}))_0 \cong \mathbb{M}_n(A)(\overline{\delta})_0$$

and $P = eA^n(-\overline{\delta})$. Define the map

$$\theta : \mathrm{End}_A(P) \longrightarrow e\,\mathrm{End}_A(A^n(-\overline{\delta}))e$$

by $\theta(f)|_P = f$ and $\theta(f)|_Q = 0$. Since e is homogeneous of degree zero, it is straightforward to see that this is a graded isomorphism of rings (which preserves the identity). Thus

$$B \cong_{\mathrm{gr}} \mathrm{End}_A(P) \cong_{\mathrm{gr}} e\,\mathrm{End}_A(A^n(-\overline{\delta}))e \cong_{\mathrm{gr}} e\,\mathbb{M}_n(A)(\overline{\delta})e.$$

We are left to show that e is full. By [61, Exercise 2.8],

$$\mathbb{M}_n(A)e\,\mathbb{M}_n(A) = \mathbb{M}_n(\mathrm{Tr}(P)),$$

where $P = eA^n$. Since P is graded progenerator, it is a finitely generated projective A-module and a generator (see Theorem 2.2.5), thus $\mathrm{Tr}(P) = A$ and therefore e is a (homogeneous) full idempotent.

(4) $\Rightarrow$ (1) Example 2.3.2 shows that there is a graded equivalence between Mod-$\mathbb{M}_n(A)$ and Mod-$e\,\mathbb{M}_n(A)e$. On the other hand, Example 2.3.6 shows that there is a graded equivalence between Mod-$\mathbb{M}_n(A)$ and Mod-A. This finishes the proof. □

Example 2.3.9 Gr-$A \cong$ Gr-B DOES NOT IMPLY Gr-$A \cong_{\text{gr}}$ Gr-B

One can easily construct examples of two Γ-graded rings A and B such that the categories Gr-A and Gr-B are equivalent, but not graded equivalent, *i.e.*, the equivalent functors do not commute with suspensions (see Definition 2.3.3). Let A be a Γ-graded ring and $\phi : \Gamma \to \operatorname{Aut}(A)$ be a group homomorphism. Consider the group ring $A[\Gamma]$ and the skew group ring $A\star_\phi\Gamma$ (see §1.1.4). These rings are strongly Γ-graded, and thus by Dade's Theorem 1.5.1, Gr-$A[\Gamma]$ and Gr-$A \star_\phi \Gamma$ are equivalent to Mod-A and thus Gr-$A[\Gamma] \cong$ Gr-$A \star_\phi \Gamma$. However, one can easily show that these two graded rings are not necessarily graded equivalent.

Remark 2.3.10

1 The graded Morita theory has a left–right symmetry property. Indeed, starting from the category of graded left modules, one can prove a similar statement as in Proposition 2.3.8, in which in turn, part (4) is independent of the left–right assumption. This shows that Gr-A is graded equivalent to Gr-B if and only if A-Gr is graded equivalent to B-Gr
2 Proposition 2.3.8(3) shows that if all graded finitely generated projective A-modules are graded free, then Mod-A is graded equivalence to Mod-B if and only if $B \cong_{\text{gr}} \mathbb{M}_n(A)(\overline{\delta})$ for some $n \in \mathbb{N}$ and $\overline{\delta} = (\delta_1, \dots, \delta_n)$, $\delta_i \in \Gamma$.

Remark 2.3.11 $\text{Gr}^{\Omega}\text{-}A \approx_{\text{gr}} \text{Gr}^{\Omega}\text{-}B$ IMPLIES $\text{Gr}^{\Gamma}\text{-}A \approx_{\text{gr}} \text{Gr}^{\Gamma}\text{-}B$

Recall from Definition 2.3.3 that we write $\text{Gr}^{\Gamma}\text{-}A \approx_{\text{gr}} \text{Gr}^{\Gamma}\text{-}B$, if there is a graded equivalence between the categories of Γ-graded A-modules $\text{Gr}^{\Gamma}\text{-}A$ and Γ-graded B-modules $\text{Gr}^{\Gamma}\text{-}B$.

Let A and B be Γ-graded rings and Ω a subgroup of Γ such that $\Gamma_A, \Gamma_B \subseteq \Omega \subseteq \Gamma$. Then A and B can be naturally considered as Ω-graded rings. If $\text{Gr}^{\Omega}\text{-}A \approx_{\text{gr}} \text{Gr}^{\Omega}\text{-}B$, then by Theorem 2.3.8 there is an Ω-isomorphism ϕ : $B \cong_{\text{gr}} e\,\mathbb{M}_n(A)(\overline{\delta})e$ for a full homogeneous idempotent $e \in \mathbb{M}_n(A)(\overline{\delta})$, where $\overline{\delta} = (\delta_1, \dots, \delta_n)$, $\delta_i \in \Omega$. Since $\Omega \subseteq \Gamma$, another application of Theorem 2.3.8 shows that $\text{Gr}^{\Gamma}\text{-}A \approx_{\text{gr}} \text{Gr}^{\Gamma}\text{-}B$. One can also use Theorem 1.2.11 to obtain this statement.

Remark 2.3.12 $\text{Gr}^{\Gamma}\text{-}A \approx_{\text{gr}} \text{Gr}^{\Gamma}\text{-}B$ IMPLIES $\text{Gr}^{\Gamma/\Omega}\text{-}A \approx_{\text{gr}} \text{Gr}^{\Gamma/\Omega}\text{-}B$

Let A and B be two Γ-graded rings. Theorem 2.3.8 shows the equivalence Gr-$A \approx_{\text{gr}}$ Gr-B induces an equivalence Mod-$A \approx$ Mod-B. Haefner in [44] observed that Gr-$A \approx_{\text{gr}}$ Gr-B induces other equivalences between different "layers" of grading. We briefly recount this result here.

Let A be a Γ-graded ring and let Ω be subgroup of Γ. Recall from §1.1.2

and §1.2.8 that A can be considered a Γ/Ω-graded ring. Recall also that the category of Γ/Ω-graded A-modules, denoted by $\text{Gr}^{\Gamma/\Omega}\text{-}A$, consists of the Γ/Ω-graded A-modules as objects and A-module homomorphisms $\phi : M \to N$ which are grade-preserving in the sense that $\phi(M_{\Omega+\alpha}) \subseteq N_{\Omega+\alpha}$ for all $\Omega + \alpha \in \Gamma/\Omega$ as morphisms of the category. In the two extreme cases $\Omega = 0$ and $\Omega = \Gamma$, we have $\text{Gr}^{\Gamma/\Omega}\text{-}A = \text{Gr-}A$ and $\text{Gr}^{\Gamma/\Omega}\text{-}A = \text{Mod-}A$, respectively.

In [44] Haefner shows that, for any two Γ-graded equivalent rings A and B and for any subgroup Ω of Γ, there are equivalences between the categories $\text{Gr}^{\Gamma/\Omega}\text{-}A$ and $\text{Gr}^{\Gamma/\Omega}\text{-}B$. In fact, Haefner works with an arbitrary (nonabelian) group Γ and any subgroup Ω. In this case one needs to adjust the definitions as follows.

Let Γ/Ω denote a set of (right) cosets of Ω (we use the multiplication notation here). A Γ/Ω-graded right A-module M is defined as a right A-module M with an internal direct sum decomposition $M = \bigoplus_{\Omega\alpha\in\Gamma/\Omega} M_{\Omega\alpha}$, where each $M_{\Omega\alpha}$ is an additive subgroup of M such that $M_{\Omega\alpha}A_\beta \subseteq M_{\Omega\alpha\beta}$ for all $\Omega\alpha \in \Gamma/\Omega$ and $\beta \in \Gamma$. The decomposition is called a Γ/Ω-grading of M. With an abuse of notation we denote the category of Γ/Ω-graded right A-modules with $\text{Gr}^{\Gamma/\Omega}\text{-}A$. Then in [44] it was shown that $\text{Gr}^{\Gamma}\text{-}A \approx_{\text{gr}} \text{Gr}^{\Gamma}\text{-}B$ implies $\text{Gr}^{\Gamma/\Omega}\text{-}A \approx_{\text{gr}} \text{Gr}^{\Gamma/\Omega}\text{-}B$.

Consider the truncation of A at Ω, *i.e.*, $A_\Omega = \oplus_{\gamma\in\Omega}A_\gamma$ (see §1.1.2). As usual, let $\text{Mod-}A_\Omega$ denote the category of right A_Ω-modules. If A is strongly Γ-graded, one can show that the categories $\text{Gr}^{\Gamma/\Omega}\text{-}A$ and $\text{Mod-}A_\Omega$ are equivalent via the functors truncation $(-)_\Omega : \text{Gr}^{\Gamma/\Omega}\text{-}A \to \text{Mod-}A_\Omega$ and induction $- \otimes_{A_\Omega} A : \text{Mod-}A_\Omega \to \text{Gr}^{\Gamma/\Omega}\text{-}A$ (see [44, Lemma 7.3]). Again, in the case that the subgroup Ω is trivial, this gives the equivalences of Gr-A and Mod-A_0 (see Theorem 1.5.1).

Remark 2.3.13 THE CASE OF $\text{Gr-}A \approx \text{Mod-}R$

If a Γ-graded ring A is strongly graded, then by Theorem 1.5.1, Gr-A is equivalent to Mod-A_0. In [70], the conditions under which Gr-A is equivalent to Mod-R, for some ring R with identity, are given. It was shown that, among other things, $\text{Gr-}A \approx \text{Mod-}R$, for some ring R with identity, if and only if $\text{Gr-}A \approx_{\text{gr}} \text{Gr-}B$, for a strongly Γ-graded ring B, if and only if there is a finite subset Ω of Γ such that for any $\tau \in \Gamma$,

$$A_0 = \sum_{\gamma\in\Omega} A_{\tau-\gamma}A_{\gamma-\tau}. \tag{2.17}$$

Moreover, it was shown that if $\text{Gr-}A \approx \text{Mod-}R$ then $A\text{-Gr} \approx R\text{-Mod}$ and $A_0 \cong e\,\mathbb{M}_n(R)e$ for some $n \in \mathbb{N}$, and some idempotent $e \in \mathbb{M}_n(R)$.

If the grade group Γ is finite, then (2.17) shows that there is a ring R such that Gr-A is equivalent to Mod-R. In fact, Cohen and Montgomery in [29] construct a so called *smash product* $A\#\mathbb{Z}[\Gamma]^*$ and show that Gr-A is equivalent to Mod-$A\#\mathbb{Z}[\Gamma]^*$. In [76] it was shown that $A\#\mathbb{Z}[\Gamma]^*$ is isomorphic to the ring $\operatorname{End}_{\text{Gr-}A}\left(\bigoplus_{\alpha\in\Gamma} A(\alpha)\right)$ (see Example 1.3.7).

Remark 2.3.14 THE CASE OF $\mathrm{Gr}^{\Gamma}\text{-}A \approx \mathrm{Gr}^{\Lambda}\text{-}B$

One can consider several variations under which two categories are equivalent. For example, for a Γ-graded ring A and Λ-graded ring B, the situation when the categories Gr^{Γ}-A and Gr^{Λ}-B are equivalent is investigated in [82, 83] (see also Remark 1.1.26).

Moreover, for two Γ-graded rings A and B, when Gr-A is equivalent to Gr-B (not necessarily respecting the shift) has been studied in [86, 99].

3
Graded Grothendieck groups

Starting from a ring, the isomorphism classes of finitely generated (projective) modules with direct sum form an abelian monoid. The "enveloping group" of this monoid is called the Grothendieck group of the ring. If the ring comes equipped with an extra structure, this structure should pass to its modules and thus should be reflected in the Grothendieck group. For example, if the ring has an involution, then the Grothendieck group has a $\mathbb{Z}_2$-module structure. If the ring has a coring structure, or it is a Hopf algebra, then the Grothendieck group becomes a ring with involution thanks to the co-multiplication and antipode of the Hopf algebra.

For a Γ-graded ring A, one of the main aims of this book is to study the Grothendieck group constructed from the graded projective modules of A, called the graded Grothendieck group and denoted by $K_0^{\mathrm{gr}}(A)$. In fact, $K_0^{\mathrm{gr}}(A)$ is not just an abelian group but it also has an extra $\mathbb{Z}[\Gamma]$-module structure. As we will see throughout this book, this extra structure carries substantial information about the graded ring A. In §3.1 we construct in detail the graded Grothendieck groups using the concept of group completions. Here we give a brief overview of different equivalent constructions.

For an abelian monoid V, we denote by V^+ the group completion of V. This gives a left adjoint functor to the forgetful functor from the category of abelian groups to abelian monoids. When the monoid V has a Γ-module structure, where Γ is a group, then V^+ inherits a natural Γ-module structure, or equivalently, $\mathbb{Z}[\Gamma]$-module structure. We study this construction in §3.1.

There is also a more direct way to construct V^+ which we recall here. Consider the set of symbols $\{ [m] \mid m \in V \}$ and let V^+ be the free abelian group generated by this set modulo the relations $[m] + [n] - [m+n]$, $m, n \in V$. There is a natural (monoid) homomorphism $V \longrightarrow V^+$, $m \mapsto [m]$, which is universal. Using the universality, one can show that the group V^+ obtained here coincides with the one constructed above using the group completion.

Now let Γ be a group which acts on a monoid V. Then Γ acts in a natural way on the free abelian group generated by symbols $\{[m] \mid m \in V\}$. Moreover the subgroup generated by relations $[m]+[n]-[m+n]$, $m, n \in V$ is closed under this action. Thus V^+ has a natural structure of a Γ-module, or equivalently, V^+ is a $\mathbb{Z}[\Gamma]$-module.

For a ring A with identity and a finitely generated projective (right) A-module P, let $[P]$ denote the class of A-modules isomorphic to P. Then the set

$$\mathcal{V}(A) = \{\, [P] \mid P \text{ is a finitely generated projective A-module} \,\} \tag{3.1}$$

with addition $[P]+[Q] = [P \bigoplus Q]$ forms an abelian monoid. The *Grothendieck group* of A, denoted by $K_0(A)$, is by definition $\mathcal{V}(A)^+$.

The graded Grothendieck group of a graded ring is constructed similarly, by using graded finitely generated projective modules everywhere in the above process. Namely, for a Γ-graded ring A with identity and a graded finitely generated projective (right) A-module P, let $[P]$ denote the class of graded A-modules graded isomorphic to P. Then the monoid

$$\mathcal{V}^{\text{gr}}(A) = \{\, [P] \mid P \text{ is graded finitely generated projective A-module} \,\} \tag{3.2}$$

has a Γ-module structure defined as follows: for $\gamma \in \Gamma$ and $[P] \in \mathcal{V}^{\text{gr}}(A)$, $\gamma.[P] = [P(\gamma)]$. The group $\mathcal{V}^{\text{gr}}(A)^+$ is called the *graded Grothendieck group* and is denoted by $K_0^{\text{gr}}(A)$, which as the above discussion shows is a $\mathbb{Z}[\Gamma]$-module.

The above construction of the graded Grothendieck groups carries over to the category of graded right A-modules. Since for a graded finitely generated right A-module P, the dual module $P^* = \operatorname{Hom}_A(P, A)$ is a graded left A-module, and furthermore, if P is projective, then also $P^{**} \cong_{\text{gr}} P$, passing to the dual defines an equivalence between the category of graded finitely generated projective right A-modules and the category of graded finitely generated projective left A-modules, *i.e.*,

$$A\text{-Pgr} \approx_{\text{gr}} \text{Pgr-}A.$$

This implies that constructing a graded Grothendieck group, using the graded left-modules, call it $K_0^{\text{gr}}(A)_l$, is isomorphic $K_0^{\text{gr}}(A)_r$, the group constructed using graded right A-modules. Moreover, defining the action of Γ on generators of $K_0^{\text{gr}}(A)_l$, by $\alpha[P] = [P(-\alpha)]$ and extending it to the whole group, and defining the action of Γ on $K_0^{\text{gr}}(A)_r$, in the usual way, by $\alpha[P] = [P(\alpha)]$, then (1.19) shows that these two groups are $\mathbb{Z}[\Gamma]$-module isomorphic.

As emphasised, one of the main differences between a graded Grothendieck group and the nongraded version is that the former has an extra module structure. For instance, in Example 3.7.3, we will see that for the $\mathbb{Z}$-graded ring $A = K[x^n, x^{-n}]$, where K is a field and $n \in \mathbb{N}$,

$$K_0^{\mathrm{gr}}(A) \cong \bigoplus_n \mathbb{Z},$$

which is a $\mathbb{Z}[x, x^{-1}]$-module, with the action of x on $(a_1, \dots, a_n) \in \bigoplus_n \mathbb{Z}$ is as follows:

$$x(a_1, \dots, a_n) = (a_n, a_1, \dots, a_{n-1}).$$

In this chapter we study graded Grothendieck groups in detail. We will calculate these groups (or rather modules) for certain types of graded rings, including the graded division rings (§3.7), graded local rings (§3.8) and path algebras (§3.9). In Chapter 5 we use the graded Grothendieck groups to classify the so-called graded ultramatricial algebras. We will show that certain information encoded in the graded Grothendieck group can be used to give a complete invariant for such algebras.

In general, the category of graded finitely generated projective modules is an exact category with the usual notion of a (split) short exact sequence. The Quillen K-groups of this category (see [81, §2] for the construction of K-groups of an exact category) is denoted by $K_i^{\mathrm{gr}}(A)$, $i \geq 0$. The group Γ acts on this category via $(\alpha, P) \mapsto P(\alpha)$. By functoriality of K-groups this equips $K_i^{\mathrm{gr}}(A)$ with the structure of a $\mathbb{Z}[\Gamma]$-module. When $i = 0$, Quillen's construction coincides with the above construction via the group completion. In Chapter 6 we compare the graded versus nongraded Grothendieck groups as well as the higher K-groups.

For a comprehensive study of nongraded Grothendieck groups see [58, 67, 84].

3.1 The graded Grothendieck group K_0^{gr}

3.1.1 Group completions

Let V be a monoid and Γ be a group which acts on V. The *group completion* of V (this is also called the *Grothendieck group* of V) has a natural Γ-module structure. We recall the construction here. Define a relation $\sim$ on $V \times V$ as follows: $(x_1, y_1) \sim (x_2, y_2)$ if there is a $z \in V$ such that

$$x_1 + y_2 + z = y_1 + x_2 + z. \tag{3.3}$$

This is an equivalence relation. We denote the equivalence classes of $V \times V/\sim$ by $V^+ := \{\, [(x, y)] \mid (x, y) \in V \times V \,\}$. Define

$$[(x_1, y_1)] + [(x_2, y_2)] = [(x_1 + x_2, y_1 + y_2)],$$

$$\alpha[(x, y)] = [(\alpha x, \alpha y)].$$

One can easily check that these operations are well-defined, V^+ is an abelian group and further it is a Γ-module. The map

$$\begin{aligned} \phi : V &\longrightarrow V^+ \\ x &\longmapsto [(x, 0)] \end{aligned} \tag{3.4}$$

is a Γ-module homomorphism and ϕ is *universal, i.e.,* if there is a Γ-module G and a Γ-module homomorphism $\psi : V \to G$, then there exists a unique Γ-module homomorphism $f : V^+ \to G$, defined by $f([x, y]) = \psi(x) - \psi(y)$, such that $f\phi = \psi$.

We record the following properties of V^+ whose proofs are easy and are left to the reader.

Lemma 3.1.1 *Let V be a monoid, Γ a group which acts on V, and let*

$$\begin{aligned} \phi : V &\longrightarrow V^+, \\ x &\longmapsto [(x, 0)] \end{aligned}$$

be the universal homomorphism.

(1) *For $x, y \in V$, if $\phi(x) = \phi(y)$ then there is a $z \in V$ such that $x + z = y + z$ in V.*
(2) *Each element of V^+ is of the form $\phi(x) - \phi(y)$ for some $x, y \in V$.*
(3) *V^+ is generated by the image of V under ϕ as a group.*

Example 3.1.2 Let G be a group and $\mathbb{N}$ be the monoid of nonnegative integers. Then $\mathbb{N}[G]$ is a monoid equipped by the natural action of G. Its group completion is $\mathbb{Z}[G]$ which has a natural G-module structure. This will be used in Proposition 3.7.1 to calculate the graded Grothendieck group of graded fields.

Example 3.1.3 <u>A NONZERO MONOID WHOSE GROUP COMPLETION IS ZERO</u>

Let V be a monoid with the trivial module structure (*i.e.,* Γ is trivial). Suppose that $V\backslash\{0\}$ is an abelian group. Then one can check that $V^+ \cong V\backslash\{0\}$.

Now let $V = \{0, v\}$, with $v + v = v$ and 0 as the trivial element. One can check that with this operation V is a monoid and $V\backslash\{0\}$ is an abelian group which is a trivial group. In general, the group completion of any monoid with a zero element, *i.e.*, an element z such that $x + z = z$, where x is any element of the monoid, is a trivial group.

Example 3.1.4 THE TYPE, IBN AND THE MONOID OF PROJECTIVE MODULES

Let A be a ring (with trivial grading) and $\mathcal{V}(A)$ the monoid of the isomorphism classes of finitely generated projective A-modules. Then one can observe that A has IBN if and only if the submonoid generated by $[A]$ in $\mathcal{V}(A)$ is isomorphic to $\mathbb{N}$. Moreover, a ring A has type (n, k) if and only if the submonoid generated by $[A]$ is isomorphic to a free monoid generated by a symbol v subject to $nv = (n + k)v$ (see §1.7).

3.1.2 K_0^{gr}-groups

Let A be a Γ-graded ring (with identity as usual) and let $\mathcal{V}^{\text{gr}}(A)$ denote the monoid of graded isomorphism classes of graded finitely generated projective modules over A with the direct sum as the addition operation. For a graded finitely generated projective A-module P, we denote the graded isomorphism class of P by $[P]$, which is an element of $\mathcal{V}^{\text{gr}}(A)$ (see (3.2)). Thus for $[P], [Q] \in \mathcal{V}^{\text{gr}}(A)$, we have $[P] + [Q] = [P \oplus Q]$. Note that for $\alpha \in \Gamma$, the α-suspension functor $\mathcal{T}_\alpha : \text{Gr-}A \to \text{Gr-}A$, $M \mapsto M(\alpha)$ is an isomorphism with the property $\mathcal{T}_\alpha \mathcal{T}_\beta = \mathcal{T}_{\alpha+\beta}$, $\alpha, \beta \in \Gamma$. Moreover, $\mathcal{T}_\alpha$ restricts to Pgr-A, the category of graded finitely generated projective A-modules. Thus the abelian group Γ acts on $\mathcal{V}^{\text{gr}}(A)$ via

$$(\alpha, [P]) \mapsto [P(\alpha)]. \tag{3.5}$$

The *graded Grothendieck group*, $K_0^{\text{gr}}(A)$, is defined as the group completion of the monoid $\mathcal{V}^{\text{gr}}(A)$ which naturally inherits the Γ-module structure via (3.5). This makes $K_0^{\text{gr}}(A)$ a $\mathbb{Z}[\Gamma]$-module. In particular if A is a $\mathbb{Z}$-graded then $K_0^{\text{gr}}(A)$ is a $\mathbb{Z}[x, x^{-1}]$-module. This extra structure plays a crucial role in the applications of graded Grothendieck groups.

For a graded finitely generated projective A-module P, we denote the image of $[P] \in \mathcal{V}^{\text{gr}}(A)$ under the natural homomorphism $\mathcal{V}^{\text{gr}}(A) \to K_0^{\text{gr}}(A)$ by $[P]$ again (see (3.4)). When the ring has graded stable rank 1, this map is injective (see Corollary 3.1.8).

Example 3.1.5 In the following "trivial" cases one can determine the graded Grothendieck group based on the Grothendieck group of the ring.

Trivial group grading Let A be a ring and Γ be a trivial group. Then A is a Γ-graded ring in an obvious way, Gr-A = Mod-A and $K_0^{\text{gr}}(A) = K_0(A)$ as a $\mathbb{Z}[\Gamma]$-module. However, here $\mathbb{Z}[\Gamma] \cong \mathbb{Z}$ and $K_0^{\text{gr}}(A)$ does not have any extra module structure. This shows that the statements we prove in the graded setting will cover the results in the classical nongraded setting, by considering rings with trivial gradings from the outset.

Trivial grading Let A be a ring and Γ be a group. Consider A with the trivial Γ-grading, *i.e.*, A is concentrated in degree zero (see §1.1.1). Then by Corollary 1.2.12 and Remark 1.2.16, $\text{Pgr-}A = \bigoplus_\Gamma \text{Pr-}A$. Consequently, $K_0^{\mathrm{gr}}(A) \cong \bigoplus_\Gamma K_0(A)$. Considering the shift which induces a $\mathbb{Z}[\Gamma]$-module structure (see (1.35)), we have

$$K_0^{\mathrm{gr}}(A) \cong K_0(A)[\Gamma]$$

as a $\mathbb{Z}[\Gamma]$-module.

Remark 3.1.6 The group completion of all projective modules is trivial

One reason for restricting ourselves to the finitely generated projective modules is that the group completion of the monoid of the isomorphism classes of (all) projective modules gives a trivial group. To see this, first observe that for a monoid V, V^+ is the trivial group if and only if for any $x, y \in V$ there is a $z \in V$ such that $x + z = y + z$. Now consider graded projective modules P and Q. Then for $M = \bigoplus_\infty (P \oplus Q)$, we have $P \oplus M \cong Q \oplus M$. Indeed,

$$\begin{aligned} Q \oplus M &\cong Q \oplus (P \oplus Q) \oplus (P \oplus Q) \oplus \cdots \\ &\cong (Q \oplus P) \oplus (Q \oplus P) \oplus \cdots \\ &\cong (P \oplus Q) \oplus (P \oplus Q) \oplus \cdots \\ &\cong P \oplus (Q \oplus P) \oplus (Q \oplus P) \oplus \cdots \\ &\cong P \oplus M. \end{aligned}$$

Lemma 3.1.7 *Let A be a Γ-graded ring.*

(1) *Each element of $K_0^{\mathrm{gr}}(A)$ is of the form $[P] - [Q]$ for some graded finitely generated projective A-modules P and Q.*
(2) *Each element of $K_0^{\mathrm{gr}}(A)$ is of the form $[P] - [A^n(\overline{\alpha})]$ for some graded finitely generated projective A-module P and some $\overline{\alpha} = (\alpha_1, \ldots, \alpha_n)$.*
(3) *Let P, Q be graded finitely generated projective A-modules. Then $[P] = [Q]$ in $K_0^{\mathrm{gr}}(A)$ if and only if $P \oplus A^n(\overline{\alpha}) \cong_{\mathrm{gr}} Q \oplus A^n(\overline{\alpha})$, for some $\overline{\alpha} = (\alpha_1, \ldots, \alpha_n)$.*

Proof (1) This follows immediately from Lemma 3.1.1(2) by considering $\mathcal{V}^{\mathrm{gr}}(A)$ as the monoid and the fact that $[P]$ also represents the image of $[P] \in \mathcal{V}^{\mathrm{gr}}(A)$ in $K_0^{\mathrm{gr}}(A)$.

(2) This follows from (1) and the fact that for a graded finitely generated projective A-module Q, there is a graded finitely generated projective module Q'

such that $Q \oplus Q' \cong_{\text{gr}} A^n(\overline{\alpha})$, for some $\overline{\alpha} = (\alpha_1, \dots, \alpha_n)$ (see Proposition 1.2.15 and (1.39)).

(3) Suppose $[P] = [Q]$ in $K_0^{\text{gr}}(A)$. Then by Lemma 3.1.1(1) (for $V = \mathcal{V}^{\text{gr}}(A)$) there is a graded finitely generated projective A-module T such that $P \oplus T \cong_{\text{gr}} Q \oplus T$. Since T is graded finitely generated projective, there is an S such that $T \oplus S \cong_{\text{gr}} A^n(\overline{\alpha})$, for some $\overline{\alpha} = (\alpha_1, \dots, \alpha_n)$ (see (1.39)). Thus

$$P \oplus A^n(\overline{\alpha}) \cong_{\text{gr}} P \oplus T \oplus S \cong_{\text{gr}} Q \oplus T \oplus S \cong_{\text{gr}} Q \oplus A^n(\overline{\alpha}).$$

Since $K_0^{\text{gr}}(A)$ is a group, the converse is immediate. □

For graded finitely generated projective A-modules P and Q, we say P and Q are *graded stably isomorphic* if $[P] = [Q]$ in $K_0^{\text{gr}}(A)$ or equivalently by Lemma 3.1.7(3), if $P \oplus A^n(\overline{\alpha}) \cong_{\text{gr}} Q \oplus A^n(\overline{\alpha})$, for some $\overline{\alpha} = (\alpha_1, \dots, \alpha_n)$.

Corollary 3.1.8 *Let A be a Γ-graded ring with the graded stable rank* 1. *Then the natural map $\mathcal{V}^{\text{gr}}(A) \to K_0^{\text{gr}}(A)$ is injective.*

Proof Let P and Q be graded finitely generated projective A-modules such that $[P] = [Q]$ in $K_0^{\text{gr}}(A)$. Then by Lemma 3.1.7(3), $P \oplus A^n(\overline{\alpha}) \cong_{\text{gr}} Q \oplus A^n(\overline{\alpha})$, for some $\overline{\alpha} = (\alpha_1, \dots, \alpha_n)$. Now by Corollary 1.8.5, $P \cong_{\text{gr}} Q$. Thus $[P] = [Q]$ in $\mathcal{V}^{\text{gr}}(A)$. □

3.1.3 K_0^{gr} of strongly graded rings

Let A be a strongly Γ-graded ring. By Dade's Theorem 1.5.1 and Remark 1.5.4, the functor $(-)_0 : \text{Pgr-}A \to \text{Pr-}A_0$, $M \mapsto M_0$, is an additive functor with an inverse $- \otimes_{A_0} A : \text{Pr-}A_0 \to \text{Pgr-}A$ so that it induces an equivalence between the category of graded finitely generated projective A-modules and the category of finitely generated projective A_0-modules. This implies that

$$\boxed{K_0^{\text{gr}}(A) \cong K_0(A_0).} \tag{3.6}$$

(In fact, this implies that $K_i^{\text{gr}}(A) \cong K_i(A_0)$, for all $i \geq 0$, where $K_i^{\text{gr}}(A)$ and $K_i(A_0)$ are Quillen's K-groups. These groups will be discussed in Chapter 6.) Moreover, since $A_\alpha \otimes_{A_0} A_\beta \cong A_{\alpha+\beta}$ as an A_0-bimodule, the functor $\mathcal{T}_\alpha$ on gr-A induces a functor on the level of mod-A_0, $\mathcal{T}_\alpha : \text{mod-}A_0 \to \text{mod-}A_0$, $M \mapsto M \otimes_{A_0} A_\alpha$ such that $\mathcal{T}_\alpha \mathcal{T}_\beta \cong \mathcal{T}_{\alpha+\beta}$, $\alpha, \beta \in \Gamma$, so that the following diagram is commutative up to isomorphism.

$$\begin{array}{ccc} \text{Pgr-}A & \xrightarrow{\mathcal{T}_\alpha} & \text{Pgr-}A \\ \downarrow{\scriptstyle (-)_0} & & \downarrow{\scriptstyle (-)_0} \\ \text{Pr-}A_0 & \xrightarrow{\mathcal{T}_\alpha} & \text{Pr-}A_0 \end{array} \tag{3.7}$$

Therefore $K_i(A_0)$ is also a $\mathbb{Z}[\Gamma]$-module and

$$K_i^{\mathrm{gr}}(A) \cong K_i(A_0), \tag{3.8}$$

as $\mathbb{Z}[\Gamma]$-modules.

Also note that if A is a graded commutative ring then the isomorphism (3.8), for $i = 0$, is a ring isomorphism.

Example 3.1.9 K_0^{gr} OF CROSSED PRODUCTS

Let A be a Γ-graded crossed product ring. Thus $A = A_0{}_{\psi}^{\phi}[\Gamma]$ (see §1.1.4). By Proposition 1.1.15(3), A is strongly graded, and so $\text{Gr-}A \approx \text{Mod-}A_0$ and $K_0^{\mathrm{gr}}(A) \cong K_0(A_0)$ (see (3.6)).

On the other hand, by Corollary 1.3.18(4), the restriction of the shift functor on $\text{Pgr-}A$, $\mathcal{T}_\alpha : \text{Pgr-}A \to \text{Pgr-}A$, is isomorphic to the trivial functor. This shows that the action of Γ on $K_0^{\mathrm{gr}}(A) \cong K_0(A_0)$ (and indeed on all K-groups $K_i^{\mathrm{gr}}(A)$) is trivial.

Example 3.1.10 K_0^{Γ} VERSES $K_0^{\Gamma/\Omega}$

Let A be a Γ-graded ring and Ω be a subgroup of Γ. Recall the construction of Γ/Ω-graded ring A from §1.1.2. The canonical forgetful functor

$$U : \mathrm{Gr}^{\Gamma}\text{-}A \to \mathrm{Gr}^{\Gamma/\Omega}\text{-}A$$

is an exact functor (see §1.2.8). Proposition 1.2.15 guarantees that U restricts to the categories of graded finitely generated projective modules, *i.e.*,

$$U : \mathrm{Pgr}^{\Gamma}\text{-}A \to \mathrm{Pgr}^{\Gamma/\Omega}\text{-}A.$$

This induces a group homomorphism

$$\theta : K_0^{\Gamma}(A) \longrightarrow K_0^{\Gamma/\Omega}(A),$$

such that for $\alpha \in \Gamma$ and $a \in K_0^{\Gamma}(A)$, we have $\theta(\alpha a) = (\Omega + \alpha)\theta(a)$. Here, to distinguish the grade groups, we denote by K_0^{Γ} and $K_0^{\Gamma/\Omega}$ the graded Grothendieck groups of A as Γ and Γ/Ω-graded rings, respectively (we will use this notation again in Chapter 6).

Now let A be a strongly Γ-graded ring. Then A is also a strongly Γ/Ω-graded (Example 1.1.17). Using (3.6), we have

$$K_0^{\Gamma}(A) \cong K_0(A_0),$$
$$K_0^{\Gamma/\Omega}(A) \cong K_0(A_\Omega),$$

where $A_\Omega = \bigoplus_{\gamma \in \Omega} A_\gamma$.

Example 3.1.11 K_0^{Γ} VERSES K_0^{Ω}

Let A be a Γ-graded ring and Ω be a subgroup of Γ such that $\Gamma_A \subseteq \Omega$. By Theorem 1.2.11 the category Gr^{Γ}-A is equivalent to $\bigoplus_{\Gamma/\Omega} \mathrm{Gr}^{\Omega}$-$A_{\Omega}$. Since this equivalence preserves the projective modules (Remark 1.2.16), we have

$$K_0^{\Gamma}(A) \cong \bigoplus_{\Gamma/\Omega} K_0^{\Omega}(A).$$

In fact, the same argument shows that for any (higher) K-groups (see Chapter 6), if A is a Γ-graded ring such that for $\Omega := \Gamma_A$, A_{Ω} is strongly Ω-graded, then we have

$$\boxed{K_i^{\Gamma}(A) \cong \bigoplus_{\Gamma/\Omega} K_i^{\Omega}(A) \cong \bigoplus_{\Gamma/\Omega} K_i(A_0).} \tag{3.9}$$

Moreover, if A_{Ω} is a crossed product, we can determine the action of Γ on K_i^{gr} concretely. First, using Corollary 1.3.18, in (1.30) for any $\omega \in \Omega$, $M(w)_{\Omega+\alpha_i} \cong M_{\Omega+\alpha_i}$ as a graded A_{Ω}-module. Thus the functor $\overline{\rho}_{\beta}$ in (1.30) reduces to permutations of categories. Representing $\bigoplus_{\Gamma/\Omega} K_i(A_0)$ by the additive group of the group ring $K_i(A_0)[\Gamma/\Omega]$, by Theorem 1.2.11, the $\mathbb{Z}[\Gamma]$-module structure of $K_i^{\Gamma}(A)$ can be descried as $K_i(A_0)[\Gamma/\Omega]$ with the natural $\mathbb{Z}[\Gamma]$-module structure.

This can be used to calculate the (lower) graded K-theory of graded division algebras (see Proposition 3.7.1, Remark 3.7.2 and Example 3.12.3).

3.1.4 The reduced graded Grothendieck group $\widetilde{K_0^{\mathrm{gr}}}$

For a ring A, the element $[A] \in K_0(A)$ generates the "obvious" part of the Grothendieck group, *i.e.*, all the elements presented by $\pm[A^n]$ for some $n \in \mathbb{N}$. The *reduced Grothendieck group*, denoted by $\widetilde{K_0}(A)$, is the quotient of $K_0(A)$ by this cyclic subgroup. There is a unique ring homomorphism $\mathbb{Z} \to A$, which induces the group homomorphism $K_0(\mathbb{Z}) \to K_0(A)$. Since the ring of integers $\mathbb{Z}$ is a PID and finitely generated projective modules over a PID ring are free of unique rank (see [84, Theorem 1.3.1]), we have

$$\boxed{\widetilde{K_0}(A) = \operatorname{coker}(K_0(\mathbb{Z}) \to K_0(A)).} \tag{3.10}$$

This group appears quite naturally. For example when A is a Dedekind domain, $\widetilde{K_0}(A)$ coincides with $C(A)$, the class group of A (see [84, Chapter 1,§4]).

We can thus write an exact sequence,

$$0 \longrightarrow \ker\theta \longrightarrow K_0(\mathbb{Z}) \xrightarrow{\theta} K_0(A) \longrightarrow \widetilde{K_0}(A) \longrightarrow 0.$$

For a commutative ring A, one can show that $\ker\theta = 0$ and the exact sequence is split, *i.e.*,

$$K_0(A) \cong \mathbb{Z} \bigoplus \widetilde{K_0}(A). \tag{3.11}$$

In fact, $\ker\theta$ determines if a ring has IBN (see §1.7). Namely, $\ker\theta = 0$ if and only if A has IBN. For, suppose $\ker\theta = 0$. If $A^n \cong A^m$ as A-modules, for $n, m \in \mathbb{N}$, then $(n-m)[A] = 0$. It follows that $n = m$. Thus A has IBN. On the other hand, suppose A has IBN. If $n \in \ker\theta$, then we can consider $n \geq 0$. Then $[A^n] = 0$ implies $A^{n+k} \cong A^k$, for some $k \in \mathbb{N}$. Thus $n = 0$ and so $\ker\theta = 0$.

Here we develop the graded version of reduced Grothendieck groups. Let A be a Γ-graded ring. In this setting the obvious part of $K_0^{\mathrm{gr}}(A)$ is the subgroup generated not only by $[A]$ but by all the shifts of $[A]$, *i.e.*, the $\mathbb{Z}[\Gamma]$-module generated by $[A]$. The *reduced graded Grothendieck group* of the Γ-graded ring A, denoted by $\widetilde{K_0^{\mathrm{gr}}}(A)$, is the $\mathbb{Z}[\Gamma]$-module defined by the quotient of $K_0^{\mathrm{gr}}(A)$ by the submodule generated by $[A]$. Similarly to the nongraded case, considering $\mathbb{Z}$ as a Γ-graded ring concentrated in degree zero, the unique graded ring homomorphism $\mathbb{Z} \to A$, indices a $\mathbb{Z}[\Gamma]$-module homomorphism $K_0^{\mathrm{gr}}(\mathbb{Z}) \to K_0^{\mathrm{gr}}(A)$ (see Example 3.1.5(2)). Then

$$\widetilde{K_0^{\mathrm{gr}}}(A) = \operatorname{coker}\big(K_0^{\mathrm{gr}}(\mathbb{Z}) \to K_0^{\mathrm{gr}}(A)\big). \tag{3.12}$$

In order to obtain a graded version of the splitting formula (3.11), we need to take out a part of Γ for which its action on $[A]$ via a shift would not change $[A]$ in $K_0^{\mathrm{gr}}(A)$.

First note that the $\mathbb{Z}[\Gamma]$-module homomorphism $K_0^{\mathrm{gr}}(\mathbb{Z}) \to K_0^{\mathrm{gr}}(A)$ can be written as

$$\begin{aligned} \phi : \mathbb{Z}[\Gamma] &\longrightarrow K_0^{\mathrm{gr}}(A), \\ \sum_\alpha n_\alpha \alpha &\longmapsto \sum_\alpha n_\alpha [A(\alpha)]. \end{aligned} \tag{3.13}$$

Moreover, since by Corollary 1.3.17, $A(\alpha) \cong_{\mathrm{gr}} A(\beta)$ as right A-modules if and only if $\alpha - \beta \in \Gamma_A^*$, the above map induces

$$\begin{aligned} \psi : \mathbb{Z}[\Gamma/\Gamma_A^*] &\longrightarrow K_0^{\mathrm{gr}}(A), \\ \sum_\alpha n_\alpha (\Gamma_A^* + \alpha) &\longmapsto \sum_\alpha n_\alpha [A(\alpha)], \end{aligned}$$

so that the following diagram is naturally commutative.

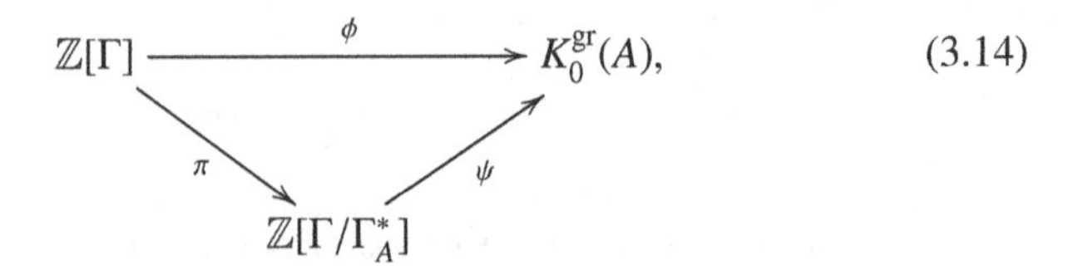

$$\mathbb{Z}[\Gamma] \xrightarrow{\phi} K_0^{\mathrm{gr}}(A), \quad \mathbb{Z}[\Gamma] \xrightarrow{\pi} \mathbb{Z}[\Gamma/\Gamma_A^*] \xrightarrow{\psi} K_0^{\mathrm{gr}}(A) \tag{3.14}$$

Since π is surjective, $\widetilde{K_0^{\mathrm{gr}}}(A)$ is also the cokernel of the map ψ.

In Example 3.7.8 we calculate the reduced Grothendieck group of the graded division algebras. In Example 3.9.6 we show that for a strongly graded ring, this group does not necessarily coincide with the reduced Grothendieck group of its 0-component ring.

3.1.5 K_0^{gr} as a $\mathbb{Z}[\Gamma]$-algebra

Let A be a Γ-graded commutative ring. Then, as in the nongraded case, $K_0^{\mathrm{gr}}(A)$ forms a commutative ring, with multiplication defined on generators by tensor products, *i.e.*, $[P].[Q] = [P \otimes_A Q]$ (see §1.2.6). Consider $\mathbb{Z}[\Gamma]$ as a group ring. To get the natural group ring operations, here we use the multiplicative notation for the group Γ. With this convention, since for any $\alpha, \beta \in \Gamma$, by (1.23), $A(\alpha) \otimes_A A(\beta) \cong_{\mathrm{gr}} A(\alpha\beta)$, the map

$$\begin{aligned} \mathbb{Z}[\Gamma] &\longrightarrow K_0^{\mathrm{gr}}(A), \\ \sum_\alpha n_\alpha \alpha &\longmapsto \sum_\alpha n_\alpha [A(\alpha)] \end{aligned}$$

induces a ring homomorphism. This makes $K_0^{\mathrm{gr}}(A)$ a $\mathbb{Z}[\Gamma]$-algebra. In fact the argument before Diagram (3.14) shows that $K_0^{\mathrm{gr}}(A)$ can be considered as a $\mathbb{Z}[\Gamma/\Gamma_A^*]$-algebra.

In Examples 3.7.10 and 3.8.4 we calculate these algebras for graded fields and graded local rings.

3.2 K_0^{gr} from idempotents

One can describe the K_0-group of a ring A with identity in terms of idempotent matrices of A. This description is quite helpful as we can work with the conjugate classes of idempotent matrices instead of isomorphism classes of finitely generated projective modules. For example, we will use this description to show that the graded K_0 is a continuous map (Theorem 3.2.4). Also, this description allows us to define the monoid $\mathcal{V}$ for rings without identity in a natural way (see §3.5). We briefly recall the construction of K_0 from idempotents here.

Let A be a ring with identity. In the following, we can always enlarge matrices of different sizes over A by adding zeros in the lower right hand corner, so that they can be considered in a ring $\mathbb{M}_k(A)$ for a suitable $k \in \mathbb{N}$. This means that we work in the matrix ring $\mathbb{M}_\infty(A) = \varinjlim_n \mathbb{M}_n(A)$, where the connecting maps are the nonunital ring homomorphism

$$\begin{aligned} \mathbb{M}_n(A) &\longrightarrow \mathbb{M}_{n+1}(A), \\ p &\longmapsto \begin{pmatrix} p & 0 \\ 0 & 0 \end{pmatrix}. \end{aligned} \tag{3.15}$$

Any idempotent matrix $p \in \mathbb{M}_n(A)$ (*i.e.*, $p^2 = p$) gives rise to the finitely generated projective right A-module pA^n. On the other hand, any finitely generated projective right module P gives rise to an idempotent matrix p such that $pA^n \cong P$. We say two idempotent matrices p and q are equivalent if (after suitably enlarging them) there are matrices x and y such that $xy = p$ and $yx = q$. One can show that p and q are equivalent if and only if they are conjugate if and only if the corresponding finitely generated projective modules are isomorphic. Therefore $K_0(A)$ can be defined as the group completion of the monoid of equivalence classes of idempotent matrices with addition defined by $[p] + [q] = \left[\begin{pmatrix} p & 0 \\ 0 & q \end{pmatrix}\right]$. In fact, this is the definition that one adapts for $\mathcal{V}(A)$ when the ring A does not have identity (see for example [30, Chapter 1] or [69, p.296]).

A similar construction can be given in the setting of graded rings. This does not seem to be documented in the literature and we provide the details here.

Let A be a Γ-graded ring with identity and let $\overline{\alpha} = (\alpha_1, \dots, \alpha_n)$, where $\alpha_i \in \Gamma$. Recall that $\mathbb{M}_n(A)(\overline{\alpha})$ is a graded ring (see §1.3). In the following, if we need to enlarge a homogeneous matrix $p \in \mathbb{M}_n(A)(\overline{\alpha})$, by adding zeros in the lower right hand corner (and call it p), then we add zeros in the right hand side of $\overline{\alpha} = (\alpha_1, \dots, \alpha_n)$ as well, accordingly (and call it $\overline{\alpha}$ again) so that p is a homogeneous matrix in $\mathbb{M}_k(A)(\overline{\alpha})$, where $k \geq n$. Recall also the definition of $\mathbb{M}_k(A)[\overline{\alpha}][\overline{\delta}]$ from §1.3.4 and note that if $x \in \mathbb{M}_k(A)[\overline{\alpha}][\overline{\delta}]$ and $y \in \mathbb{M}_k(A)[\overline{\delta}][\overline{\alpha}]$, then by Lemma 1.3.15 and (1.59), $xy \in \mathbb{M}_k(-\overline{\alpha})_0$ and $yx \in \mathbb{M}_k(-\overline{\delta})_0$.

Definition 3.2.1 Let A be a Γ-graded ring and let $\overline{\alpha} = (\alpha_1, \dots, \alpha_n)$ and $\overline{\delta} = (\delta_1, \dots, \delta_m)$, where $\alpha_i, \delta_j \in \Gamma$. Let $p \in \mathbb{M}_n(A)(\overline{\alpha})_0$ and $q \in \mathbb{M}_m(A)(\overline{\delta})_0$ be idempotent matrices. Then p and q are *graded equivalent*, denoted by $p \sim$

q, if (after suitably enlarging them) there are $x \in \mathbb{M}_k(A)[-\overline{\alpha}][-\overline{\delta}]$ and $y \in \mathbb{M}_k(A)[-\overline{\delta}][-\overline{\alpha}]$ such that $xy = p$ and $yx = q$. Moreover, we say p and q are *graded conjugate* if there is an invertible matrix $g \in \mathbb{M}_k(A)[-\overline{\delta}][-\overline{\alpha}]$ with $g^{-1} \in \mathbb{M}_k(A)[-\overline{\alpha}][-\overline{\delta}]$ such that $gpg^{-1} = q$.

The relation $\sim$ defined above is an equivalence relation. Indeed, if $p \in \mathbb{M}_n(A)(\overline{\alpha})_0$ is a homogeneous idempotent, then considering

$$x = p \in \mathbb{M}_n(A)[-\overline{\alpha}][-\overline{\alpha}]$$

(see (1.59)) and

$$y = 1 \in \mathbb{M}_n(A)[-\overline{\alpha}][-\overline{\alpha}],$$

Definition 3.2.1 shows that $p \sim p$. Clearly $\sim$ is reflexive. The following trick shows that $\sim$ is also transitive. Suppose $p \sim q$ and $q \sim r$. Then $p = xy$, $q = yx$ and $q = vw$ and $r = wv$. Thus $p = p^2 = xyxy = xqy = (xv)(wy)$ and $r = r^2 = wvwv = wqv = (wy)(xv)$. This shows that $p \sim r$.

Let $p \in \mathbb{M}_n(A)(\overline{\alpha})_0$ be a homogeneous idempotent. Then one can easily see that for any $\beta = (\beta_1, \dots, \beta_m)$, the idempotent $\begin{pmatrix} p & 0 \\ 0 & 0 \end{pmatrix} \in \mathbb{M}_{n+m}(A)(\overline{\alpha}, \overline{\beta})_0$ is graded equivalent to p. We call this element an *enlargement* of p.

Lemma 3.2.2 *Let $p \in \mathbb{M}_n(A)(\overline{\alpha})_0$ and $q \in \mathbb{M}_m(A)(\overline{\delta})_0$ be idempotent matrices. Then p and q are graded equivalent if and only if some enlargements of p and q are conjugate.*

Proof Let p and q be graded equivalent idempotent matrices. By Definition 3.2.1, there are $x' \in \mathbb{M}_k(A)[-\overline{\alpha}][-\overline{\delta}]$ and $y' \in \mathbb{M}_k(A)[-\overline{\delta}][-\overline{\alpha}]$ such that $x'y' = p$ and $y'x' = q$. Let $x = px'q$ and $y = qy'p$. Then

$$xy = px'qy'p = p(x'y')^2p = p$$

and similarly $yx = q$. Moreover, $x = px = xq$ and $y = yp = qy$. Using Lemma 1.3.15, one can check that $x \in \mathbb{M}_k(A)[-\overline{\alpha}][-\overline{\delta}]$ and $y \in \mathbb{M}_k(A)[-\overline{\delta}][-\overline{\alpha}]$. We now use the standard argument, taking into account the shift of the matrices. Consider the matrix

$$\begin{pmatrix} 1-p & x \\ y & 1-q \end{pmatrix} \in \mathbb{M}_{2k}(A)[-\overline{\alpha}, -\overline{\delta}][-\overline{\alpha}, -\overline{\delta}].$$

This matrix has order two and thus is its own inverse. Then

$$\begin{pmatrix} 1-p & x \\ y & 1-q \end{pmatrix} \begin{pmatrix} p & 0 \\ 0 & 0 \end{pmatrix} \begin{pmatrix} 1-p & x \\ y & 1-q \end{pmatrix} = \begin{pmatrix} 1-p & x \\ y & 1-q \end{pmatrix} \begin{pmatrix} 0 & x \\ 0 & 0 \end{pmatrix} = \begin{pmatrix} 0 & 0 \\ 0 & q \end{pmatrix}.$$

Moreover,

$$\begin{pmatrix} 0 & 1 \\ 1 & 0 \end{pmatrix}\begin{pmatrix} 0 & 0 \\ 0 & q \end{pmatrix}\begin{pmatrix} 0 & 1 \\ 1 & 0 \end{pmatrix} = \begin{pmatrix} q & 0 \\ 0 & 0 \end{pmatrix}.$$

Now considering the enlargements of p and q

$$\begin{pmatrix} p & 0 \\ 0 & 0 \end{pmatrix} \in \mathbb{M}_{2k}(A)(\overline{\alpha}, \overline{\delta})_0,$$

$$\begin{pmatrix} q & 0 \\ 0 & 0 \end{pmatrix} \in \mathbb{M}_{2k}(A)(\overline{\delta}, \overline{\alpha})_0,$$

we have $g\begin{pmatrix} p & 0 \\ 0 & 0 \end{pmatrix}g^{-1} = \begin{pmatrix} q & 0 \\ 0 & 0 \end{pmatrix}$, where

$$g = \begin{pmatrix} 0 & 1 \\ 1 & 0 \end{pmatrix}\begin{pmatrix} 1-p & x \\ y & 1-q \end{pmatrix} \in \mathbb{M}_{2k}(A)[-\overline{\delta}, -\overline{\alpha}][-\overline{\alpha}, -\overline{\delta}].$$

This shows that some enlargements of p and q are conjugates. Conversely, suppose some enlargements of p and q are conjugates. Thus there is a

$$g \in \mathbb{M}_{2k}(A)[-\overline{\delta}, -\overline{\mu}][-\overline{\alpha}, -\overline{\beta}]$$

such that $g\begin{pmatrix} p & 0 \\ 0 & 0 \end{pmatrix}g^{-1} = \begin{pmatrix} q & 0 \\ 0 & 0 \end{pmatrix}$, where $\begin{pmatrix} p & 0 \\ 0 & 0 \end{pmatrix} \in \mathbb{M}_k(A)(\alpha, \beta)_0$ and $\begin{pmatrix} q & 0 \\ 0 & 0 \end{pmatrix} \in \mathbb{M}_k(A)(\delta, \mu)_0$. Now setting $x = g\begin{pmatrix} p & 0 \\ 0 & 0 \end{pmatrix}$ and $y = g^{-1}$ gives that the enlargements of p and q and consequently p and q are graded equivalent. □

The following lemma relates graded finitely generated projective modules to homogeneous idempotent matrices which eventually leads to an equivalent definition of the graded Grothendieck group by idempotents.

Lemma 3.2.3 *Let A be a Γ-graded ring.*

(1) *Any graded finitely generated projective A-module P gives rise to a homogeneous idempotent matrix $p \in \mathbb{M}_n(A)(\overline{\alpha})_0$, for some $n \in \mathbb{N}$ and $\overline{\alpha} = (\alpha_1, \dots, \alpha_n)$, such that $P \cong_{\mathrm{gr}} pA^n(-\overline{\alpha})$.*

(2) *Any homogeneous idempotent matrix $p \in \mathbb{M}_n(A)(\overline{\alpha})_0$ gives rise to a graded finitely generated projective A-module $pA^n(-\overline{\alpha})$.*

(3) *Two homogeneous idempotent matrices are graded equivalent if and only if the corresponding graded finitely generated projective A-modules are graded isomorphic.*

Proof (1) Let P be a graded finitely generated projective (right) A-module. Then there is a graded module Q such that $P \oplus Q \cong_{\rm gr} A^n(-\overline{\alpha})$ for some $n \in \mathbb{N}$ and $\overline{\alpha} = (\alpha_1, \dots, \alpha_n)$, where $\alpha_i \in \Gamma$ (see (1.39)). Identify $A^n(-\overline{\alpha})$ with $P \oplus Q$ and define the homomorphism $p \in \operatorname{End}_A(A^n(-\overline{\alpha}))$ which sends Q to zero and acts as identity on P. Clearly, p is an idempotent and graded homomorphism of degree 0. By (1.17) $p \in \operatorname{End}_A(A^n(-\overline{\alpha}))_0$. By (1.45), the homomorphism p can be represented by a matrix $p \in \mathbb{M}_n(A)(\overline{\alpha})_0$ acting from the left, so

$$P \cong_{\rm gr} pA^n(-\overline{\alpha}).$$

(2) Let $p \in \mathbb{M}_n(A)(\overline{\alpha})_0$, $\overline{\alpha} = (\alpha_1, \dots, \alpha_n)$, where $\alpha_i \in \Gamma$. Since for any $\gamma \in \Gamma$, $pA^n(-\overline{\alpha})_\gamma \subseteq A^n(-\overline{\alpha})_\gamma$, we have

$$pA^n(-\overline{\alpha}) = \bigoplus_{\gamma \in \Gamma} pA^n(-\overline{\alpha})_\gamma.$$

This shows that $pA^n(-\overline{\alpha})$ is a graded finitely generated A-module. Moreover, $1 - p \in \mathbb{M}_n(A)(\overline{\alpha})_0$ and

$$A^n(-\overline{\alpha}) = pA^n(-\overline{\alpha}) \bigoplus (1-p)A^n(-\overline{\alpha}).$$

Thus $pA^n(-\overline{\alpha})$ is a graded finitely generated projective A-module.

(3) Let $p \in \mathbb{M}_n(A)(\overline{\alpha})_0$ and $q \in \mathbb{M}_m(A)(\overline{\delta})_0$ be graded equivalent idempotent matrices. The first part is similar to the proof of Lemma 3.2.2. By Definition 3.2.1, there are $x' \in \mathbb{M}_k(A)[-\overline{\alpha}][-\overline{\delta}]$ and $y' \in \mathbb{M}_k(A)[-\overline{\delta}][-\overline{\alpha}]$ such that $x'y' = p$ and $y'x' = q$. Let $x = px'q$ and $y = qy'p$. Then $xy = px'qy'p = p(x'y')^2p = p$ and similarly $yx = q$. Moreover, $x = px = xq$ and $y = yp = qy$. Now the left multiplications by x and y induce graded right A-homomorphisms $qA^k(-\overline{\delta}) \to pA^k(-\overline{\alpha})$ and $pA^k(-\overline{\alpha}) \to qA^k(-\overline{\delta})$, respectively, which are inverses of each other. Therefore $pA^k(-\overline{\alpha}) \cong_{\rm gr} qA^k(-\overline{\delta})$.

On the other hand, if $f : pA^k(-\overline{\alpha}) \cong_{\rm gr} qA^k(-\overline{\delta})$, then extend f to $A^k(-\overline{\alpha})$ by sending $(1-p)A^k(-\overline{\alpha})$ to zero and thus define a map

$$\theta : A^k(-\overline{\alpha}) = pA^k(-\overline{\alpha}) \oplus (1-p)A^k(-\overline{\alpha}) \longrightarrow qA^k(-\overline{\delta}) \oplus (1-q)A^k(-\overline{\delta}) = A^k(-\overline{\delta}).$$

Similarly, extending f^{-1} to $A^k(-\overline{\delta})$, we get a map

$$\phi : A^k(-\overline{\delta}) = qA^k(-\overline{\delta}) \oplus (1-q)A^k(-\overline{\delta}) \longrightarrow pA^k(-\overline{\alpha}) \oplus (1-p)A^k(-\overline{\alpha}) = A^k(-\overline{\alpha})$$

such that $\phi\theta = p$ and $\theta\phi = q$. It follows that $\theta \in \mathbb{M}_k(A)[-\overline{\alpha}][-\overline{\delta}]$, whereas $\phi \in \mathbb{M}_k(A)[-\overline{\delta}][-\overline{\alpha}]$ (see §1.3.4). This gives that p and q are graded equivalent. □

For a homogeneous idempotent matrix p of degree zero, we denote the graded equivalence class of p, by $[p]$ (see Definition 3.2.1) and we define

$[p] + [q] = \left[\begin{pmatrix} p & 0 \\ 0 & q \end{pmatrix}\right]$. This makes the set of equivalence classes of homogeneous idempotent matrices of degree zero a monoid. Lemma 3.2.3 shows that this monoid is isomorphic to $\mathcal{V}^{\mathrm{gr}}(A)$, via $[p] \mapsto [pA^n(-\overline{\alpha})]$, where $p \in \mathbb{M}_n(A)(\overline{\alpha})_0$ is a homogeneous matrix. Thus $K_0^{\mathrm{gr}}(A)$ can be defined as the group completion of this monoid. In fact, this is the definition we adopt for $\mathcal{V}^{\mathrm{gr}}$ when the graded ring A does not have identity (see §3.5).

3.2.1 Stability of idempotents

Recall that two graded finitely generated projective A-modules P and Q are stably isomorphic if $[P] = [Q] \in K_0^{\mathrm{gr}}(A)$ (see Lemma 3.1.7(3)). Here we describe this stability when the elements of K_0^{gr} are represented by idempotents.

Suppose $p \in \mathbb{M}_n(A)(\overline{\alpha})_0$ and $q \in \mathbb{M}_m(A)(\overline{\delta})_0$ are homogeneous idempotent matrices of degree zero such that $[p] = [q]$ in $K_0^{\mathrm{gr}}(A)$. As in Definition 3.2.1, we add zeros to the lower right hand corner of p and q so that p, q can be considered as matrices in $\mathbb{M}_k(A)$ for some $k \in \mathbb{N}$. Now $[p]$ represents the isomorphism class $[pA^k(-\overline{\alpha})]$ and $[q]$ represents $[qA^k(-\overline{\delta})]$ in $K_0^{\mathrm{gr}}(A)$ (see Lemma 3.2.3). Since $[p] = [q]$, $[pA^k(-\overline{\alpha})] = [qA^k(-\overline{\delta})]$ which by Lemma 3.1.7(3), implies

$$pA^k(-\overline{\alpha}) \oplus A^n(\overline{\beta}) \cong_{\mathrm{gr}} qA^k(-\overline{\delta}) \oplus A^n(\overline{\beta}),$$

for some $n \in \mathbb{N}$. Again by Lemma 3.2.3 this implies $p \oplus I_n = \begin{pmatrix} p & 0 \\ 0 & I_n \end{pmatrix}$ is graded equivalent to $q \oplus I_n = \begin{pmatrix} q & 0 \\ 0 & I_n \end{pmatrix}$, where I_n is the identity element of the ring $\mathbb{M}_n(A)$. (This can also be seen directly from Definition 3.2.1.) This observation will be used in Theorem 3.2.4 and later in Lemma 5.1.5.

3.2.2 Action of Γ on idempotents

We define the action of Γ on the idempotent matrices as follows. For $\gamma \in \Gamma$ and $p \in \mathbb{M}_n(A)(\overline{\alpha})_0$, γp is represented by the same matrix as p but considered in $\mathbb{M}_n(A)(\overline{\alpha} - \gamma)_0$, where $\overline{\alpha} - \gamma = (\alpha_1 - \gamma, \dots, \alpha_n - \gamma)$. Note that if $p \in \mathbb{M}_n(A)(\overline{\alpha})_0$ and $q \in \mathbb{M}_m(A)(\overline{\delta})_0$ are equivalent, then there are $x \in \mathbb{M}_k(A)[-\overline{\alpha}][-\overline{\delta}]$ and $y \in \mathbb{M}_k(A)[-\overline{\delta}][-\overline{\alpha}]$ such that $xy = p$ and $yx = q$ (Definition 3.2.1). Since $x \in \mathbb{M}_k(A)[\gamma - \overline{\alpha}][\gamma - \overline{\delta}]$ and $y \in \mathbb{M}_k(A)[\gamma - \overline{\delta}][\gamma - \overline{\alpha}]$, it follows that γp is equivalent to γq. Thus $K_0^{\mathrm{gr}}(A)$ becomes a $\mathbb{Z}[\Gamma]$-module with this definition.

Now a quick inspection of the proof of Lemma 3.2.3 shows that the action of Γ is compatible in both definitions of K_0^{gr} in §3.1.2 and §3.2.

Let A and B be Γ-graded rings and $\phi : A \to B$ be a Γ-graded homomorphism.

Using the graded homomorphism ϕ, one can consider B as a graded $A-B$-bimodule in a natural way (§1.2.5). Moreover, if P is a graded right A-module, then $P \otimes_A B$ is a graded B-module (§1.2.6). Moreover, if P is finitely generated projective, so is $P \otimes_A B$. Thus one can define a group homomorphism $\overline{\phi}$: $K_0^{\mathrm{gr}}(A) \to K_0^{\mathrm{gr}}(B)$, where $[P] \mapsto [P \otimes_A B]$ and extended to all $K_0^{\mathrm{gr}}(A)$. On the other hand, if $p \in \mathbb{M}_n(A)(\overline{\alpha})_0$ is an idempotent matrix over A then $\phi(p) \in \mathbb{M}_n(B)(\overline{\alpha})_0$ is an idempotent matrix over B obtained by applying ϕ to each entry of p. This also induces a homomorphism on the level of K_0^{gr} using the idempotent presentations. Since

$$pA^n(-\overline{\alpha}) \otimes_A B \cong_{\mathrm{gr}} \phi(p)B^n(-\overline{\alpha}),$$

we have the following commutative diagram:

$$\begin{array}{ccc} [p] & \longrightarrow & [\phi(p)] \\ \downarrow & & \downarrow \\ [pA^n(-\overline{\alpha})] & \longrightarrow & [pA^n(-\overline{\alpha}) \otimes_A B]. \end{array} \tag{3.16}$$

This shows that the homomorphisms induced on K_0^{gr} by ϕ are compatible, whether using the idempotent presentation or the module presentation for the graded Grothendieck groups.

3.2.3 K_0^{gr} is a continuous functor

Recall the construction of the direct limit of graded rings from Example 1.1.9. We are in a position to determine their graded Grothendieck groups.

Theorem 3.2.4 *Let A_i, $i \in I$, be a direct system of Γ-graded rings and $A = \varinjlim A_i$ be a Γ-graded ring. Then $K_0^{\mathrm{gr}}(A) \cong \varinjlim K_0^{\mathrm{gr}}(A_i)$ as $\mathbb{Z}[\Gamma]$-modules.*

Proof First note that $K_0^{\mathrm{gr}}(A_i)$, $i \in I$, is a direct system of abelian groups so $\varinjlim K_0^{\mathrm{gr}}(A_i)$ exists with $\mathbb{Z}[\Gamma]$-module homomorphisms

$$\phi_i : K_0^{\mathrm{gr}}(A_i) \to \varinjlim K_0^{\mathrm{gr}}(A_i).$$

On the other hand, for any $i \in I$, there is a map $\psi_i : A_i \to A$ which induces $\overline{\psi}_i : K_0^{\mathrm{gr}}(A_i) \to K_0^{\mathrm{gr}}(A)$. Due to the universality of the direct limit, we have a $\mathbb{Z}[\Gamma]$-module homomorphism $\phi : \varinjlim K_0^{\mathrm{gr}}(A_i) \to K_0^{\mathrm{gr}}(A)$ such that for any $i \in I$, the diagram

$$\begin{array}{ccc} \varinjlim K_0^{\mathrm{gr}}(A_i) & \xrightarrow{\phi} & K_0^{\mathrm{gr}}(A) \\ \uparrow{\scriptstyle \phi_i} & \nearrow{\scriptstyle \overline{\psi}_i} & \\ K_0^{\mathrm{gr}}(A_i) & & \end{array} \tag{3.17}$$

is commutative. We show that ϕ is in fact an isomorphism. We use the idempotent description of the K_0^{gr} group to show this. Note that if p is an idempotent matrix over A_i for some $i \in I$, which gives the element $[p] \in K_0^{\mathrm{gr}}(A_i)$, then $\overline{\psi}_i([p]) = [\psi_i(p)]$, where $\psi_i(p)$ is an idempotent matrix over A obtained by applying ψ_i to each entry of p.

Let p be an idempotent matrix over $A = \varinjlim A_i$. Then p has a finite number of entires and each is coming from some A_i, $i \in I$. Since I is directed, there is a $j \in I$, such that p is the image of an idempotent matrix in A_j. Thus the class $[p]$ in $K_0^{\mathrm{gr}}(A)$ is the image of an element of $K_0^{\mathrm{gr}}(A_j)$. Since Diagram (3.17) is commutative, there is an element in $\varinjlim K_0(A_i)$ which maps to [p] in $K_0^{\mathrm{gr}}(A)$. Since $K_0^{\mathrm{gr}}(A)$ is generated by elements $[p]$, this shows that ϕ is surjective. We are left to show that ϕ is injective. Suppose $x \in \varinjlim K_0^{\mathrm{gr}}(A_i)$ such that $\phi(x) = 0$. Since there is $j \in I$ such that x is the image of an element of $K_0^{\mathrm{gr}}(A_j)$, we have $[p] - [q] \in K_0^{\mathrm{gr}}(A_j)$ such that $\phi_j([p] - [q]) = x$, where p and q are idempotent matrices over A_j. Again, since Diagram (3.17) is commutative, we have $\overline{\psi}_j([p] - [q]) = 0$. Thus

$$[\psi_j(p)] = \overline{\psi}_j([p]) = \overline{\psi}_j([q]) = [\psi_j(q)] \in K_0^{\mathrm{gr}}(A).$$

This shows that $a = \begin{pmatrix} \psi_j(p) & 0 \\ 0 & I_n \end{pmatrix}$ is equivalent to $b = \begin{pmatrix} \psi_j(q) & 0 \\ 0 & I_n \end{pmatrix}$ in R (see §3.2.3). Thus there are matrices x and y over R such that $xy = a$ and $yx = b$. Since the entires of x and y are finite, one can find $k \geq j$ such that x and y are images of matrices from R_k. Thus $\begin{pmatrix} \psi_{jk}(p) & 0 \\ 0 & I_n \end{pmatrix}$ is equivalent to $\begin{pmatrix} \psi_{jk}(q) & 0 \\ 0 & I_n \end{pmatrix}$ in R_k. This shows that $\overline{\psi}_{jk}([p]) = \overline{\psi}_{jk}([q])$, *i.e.*, the image of $[p] - [q]$ in $K_0^{\mathrm{gr}}(A_j)$ is zero. Thus x, being the image of this element, is also zero. This finishes the proof. □

3.2.4 The Hattori–Stallings (Chern) trace map

Recall that, for a ring A, one can relate $K_0(A)$ to the Hochschild homology $HH_0(A) = A/[A, A]$, where $[A, A]$ is the subgroup generated by additive commutators $ab - ba$, $a, b \in A$. (Or, more generally, if A is a k-algebra, where k is a commutative ring, then $HH_0(A)$ is a k-module.) The construction is as follows.

Let P be a finitely generated projective right A-module. Then, as in the introduction of §3.2, there is an idempotent matrix $p \in \mathbb{M}_n(A)$ such that $P \cong pA^n$. Define

$$\begin{aligned} T : \mathcal{V}(A) &\longrightarrow A/[A,A], \\ [P] &\longmapsto [A,A] + \operatorname{Tr}(p), \end{aligned}$$

where Tr is the trace map of the matrices. If P is isomorphic to Q then the idempotent matrices associated with them, call them p and q, are equivalent, *i.e.*, $p = xy$ and $q = yx$. This shows that $\operatorname{Tr}(p) - \operatorname{Tr}(q) \in [A,A]$, so the map T is a well-defined homomorphism of groups. Since K_0 is the universal group completion (§3.1.1), the map T induces a map on the level of K_0, which is called T again, *i.e.*,

$$T : K_0(A) \to A/[A,A].$$

This map is called the *Hattori–Stallings trace map* or the *Chern map*.

We carry out a similar construction in the graded setting. Let A be a Γ-graded ring and P be a graded finitely generated projective A-module. Then there is a homogeneous idempotent $p \in \mathbb{M}_n(A)(\overline{\alpha})_0$, where $\alpha = (\alpha_1, \dots, \alpha_n)$, such that $P \cong_{\mathrm{gr}} pA^n(-\overline{\alpha})$. Note that by (1.47), $\operatorname{Tr}(p) \in A_0$. Set $[A,A]_0 := A_0 \cap [A,A]$ and define

$$\begin{aligned} T : \mathcal{V}^{\mathrm{gr}}(A) &\longrightarrow A_0/[A,A]_0, \\ [P] &\longmapsto [A,A]_0 + \operatorname{Tr}(p). \end{aligned}$$

Similarly to the nongraded case, Lemma 3.2.3 applies to show that T is a well-defined homomorphism. Note that the description of action of Γ on idempotents (§3.2.2) shows that $T([P]) = T([P(\alpha)])$ for any $\alpha \in \Gamma$. Again, this map induces a group homomorphism

$$T : K_0^{\mathrm{gr}}(A) \to A_0/[A,A]_0.$$

Further, the forgetful functor $U : \text{Gr-}A \to \text{Mod-}A$ (§1.2.7) induces the right hand side of the following commutative diagram, whereas the left hand side is the natural map induced by inclusion, $A_0 \subseteq A$,

$$\begin{array}{ccc} K_0^{\mathrm{gr}}(A) & \xrightarrow{T} & A_0/[A,A]_0 \\ {\scriptstyle U}\downarrow & & \downarrow \\ K_0(A) & \xrightarrow{T} & A/[A,A]. \end{array}$$

3.3 K_0^{gr} of graded $*$-rings

Let A be a graded $*$-ring, *i.e.*, A is a Γ-graded ring with an involution * such that for $a \in A_\gamma$, $a^* \in A_{-\gamma}$, where $\gamma \in \Gamma$ (§1.9).

Recall that for a graded finitely generated projective right A-module P, the dual module $P^* = \mathrm{Hom}_A(P, A)$ is a graded finitely generated projective left A-module. Thus $P^{*(-1)}$ is a graded right A-module (see (1.91)). Since $P^{**} \cong_{\mathrm{gr}} P$ as graded right A-modules, the map

$$[P] \mapsto [P^{*(-1)}] \tag{3.18}$$

induces a $\mathbb{Z}_2$-action on $\mathcal{V}^{\mathrm{gr}}(A)$.

Since $P(\alpha)^* = P^*(-\alpha)$ (see (1.19)), one can easily check that

$$P(\alpha)^{*(-1)} = P^{*(-1)}(\alpha).$$

This shows that the actions of Γ and $\mathbb{Z}_2$ on $\mathcal{V}^{\mathrm{gr}}(A)$ commute. This makes $\mathcal{V}^{\mathrm{gr}}(A)$ a $\Gamma{-}\mathbb{Z}_2$-bimodule or equivalently a $\Gamma \times \mathbb{Z}_2$-module.

The graded Grothendieck group $K_0^{\mathrm{gr}}(A)$, being the group completion of the monoid $\mathcal{V}^{\mathrm{gr}}(A)$, naturally inherits the Γ-module structure via (3.5) and the $\mathbb{Z}_2$-module structure via (3.18). This makes $K_0^{\mathrm{gr}}(A)$ a $\mathbb{Z}[\Gamma]$-module and a $\mathbb{Z}[\mathbb{Z}_2]$-module. In particular, if A is a $\mathbb{Z}$-graded $*$-ring then $K_0^{\mathrm{gr}}(A)$ is a $\mathbb{Z}[x, x^{-1}]$-module and a $\mathbb{Z}[x]/(x^2 - 1)$-module.

Example 3.3.1 <u>Action of $\mathbb{Z}_2$ on K_0</u>

1. If $A = F \times F$, where F is a field, with $^* : (a, b) \mapsto (b, a)$ then the involution induces a homomorphism on the level of K_0 as follows:

$$\bar{*} : \mathbb{Z} \times \mathbb{Z} \longrightarrow \mathbb{Z} \times \mathbb{Z},$$
$$(n, m) \longmapsto (m, n).$$

2. Let R be a $*$-ring and consider $\mathbb{M}_n(R)$ as a $*$-ring with the Hermitian transpose. Then there is a $\mathbb{Z}_2$-module isomorphism $K_0(R) \cong K_0(\mathbb{M}_n(R))$.
3. Let R be a commutative von Neumann regular ring. It is known that any idempotent of $\mathbb{M}_n(R)$ is conjugate to the diagonal matrix with nonzero elements all 1 [12]. Thus if R is a $*$-ring, then the action of $\mathbb{Z}_2$ induced on $K_0(R)$ is trivial. In particular, if F is a $*$-field, then the action of $\mathbb{Z}_2$ on $K_0(\mathbb{M}_n(F))$ is trivial (see (2)).

Example 3.3.2 <u>Action of $\mathbb{Z}_2$ on K_0 of Leavitt path algebras is trivial</u>

Let E be a finite graph and $A = \mathcal{L}_K(E)$ be a Leavitt path algebra associated with E. Recall from Example 1.9.4 that A is equipped with a graded

$*$-involution. We know that the finitely generated projected A-modules are generated by modules of the form uA, where $u \in E^0$. To show that $\mathbb{Z}_2$ acts trivially on $\mathcal{V}(A)$ (and thus on $K_0(A)$), it is enough to show that $uA^* = \operatorname{Hom}_A(uA, A)$ as a right A-module is isomorphic to uA. Consider the map

$$\begin{aligned}\operatorname{Hom}_A(uA, A) &\longrightarrow uA,\\ f &\longmapsto uf(u)^*.\end{aligned}$$

It is easy to see that this map is a right A-module isomorphism.

3.4 Relative K_0^{gr}-groups

Let A be a ring and I be a two-sided ideal of A. The canonical epimorphism $f : A \to A/I$ induces a homomorphism on the level of K-groups. For example, on the level of Grothendieck groups, we have

$$\begin{aligned}\overline{f} : K_0(A) &\longrightarrow K_0(A/I),\\ [P] - [Q] &\longmapsto [P/PI] - [Q/QI].\end{aligned}$$

In order to complete this into a long exact sequence, one needs to introduce the relative K-groups, $K_i(A, I)$, $i \geq 0$. This is done for lower K-groups by Bass and Milnor and the following long exact sequence has been established (see [84, Theorem 4.3.1]),

$$\begin{aligned}K_2(A) \longrightarrow K_2(A/I) \longrightarrow K_1(A, I) \longrightarrow K_1(A) \longrightarrow K_1(A/I)&\\ \longrightarrow K_0(A, I) \longrightarrow K_0(A) \longrightarrow K_0(A/I)&.\end{aligned}$$

In this section, we define the relative K_0^{gr}-group and establish the above sequence only on the level of graded Grothendieck groups. As we will see, this requires a careful arrangement of the degrees of the matrices.

Let A be a Γ-graded ring and I be a graded two-sided ideal of A. Define the *graded double ring* of A along I

$$D^{\Gamma}(A, I) = \{ (a, b) \in A \times A \mid a - b \in I \}.$$

One can check that $D^{\Gamma}(A, I)$ is a Γ-graded ring with

$$D^{\Gamma}(A, I)_{\gamma} = \{ (a, b) \in A_{\gamma} \times A_{\gamma} \mid a - b \in I_{\gamma} \},$$

for any $\gamma \in \Gamma$.

Let

$$\pi_1, \pi_2 : D^{\Gamma}(A, I) \longrightarrow A$$

be the projections of the first and second components of $D^\Gamma(A,I)$ to A, respectively. The maps π_i, $i = 1,2$, are Γ-graded ring homomorphisms and induce $\mathbb{Z}[\Gamma]$-module homomorphisms

$$\overline{\pi}_1, \overline{\pi}_2 : K_0^{\mathrm{gr}}(D^\Gamma(A,I)) \longrightarrow K_0^{\mathrm{gr}}(A).$$

Define the *relative graded Grothendieck group* of the ring A with respect to I as

$$\boxed{K_0^{\mathrm{gr}}(A,I) := \ker\big(K_0^{\mathrm{gr}}(D^\Gamma(A,I)) \xrightarrow{\overline{\pi}_1} K_0^{\mathrm{gr}}(A)\big).} \tag{3.19}$$

The restriction of $\overline{\pi}_2$ to $K_0^{\mathrm{gr}}(A,I)$ gives a $\mathbb{Z}[\Gamma]$-module homomorphism

$$\overline{\pi}_2 : K_0^{\mathrm{gr}}(A,I) \longrightarrow K_0^{\mathrm{gr}}(A).$$

The following theorem relates these groups.

Theorem 3.4.1 *Let A be a Γ-graded ring and I be a graded two-sided ideal of A. Then there is an exact sequence of $\mathbb{Z}[\Gamma]$-modules*

$$K_0^{\mathrm{gr}}(A,I) \xrightarrow{\overline{\pi}_2} K_0^{\mathrm{gr}}(A) \xrightarrow{\overline{f}} K_0^{\mathrm{gr}}(A/I).$$

Proof We will use the presentation of K_0^{gr} by idempotents (§3.2) to prove the theorem. If p is a matrix over the ring A, we will denote by $\overline{p}$ the image of p under the canonical graded homomorphism $f : A \to A/I$. Let $[p] - [q] \in K_0^{\mathrm{gr}}(A,I)$. By the construction of $D^\Gamma(A,I)$, $p = (p_1, p_2)$, where p_1, p_2 are homogeneous idempotent matrices of A such that $p_1 - p_2$ is a matrix over I (in fact over I_0), namely $\overline{p}_1 = \overline{p}_2$. Similarly $q = (q_1, q_2)$, where q_1, q_2 are homogeneous idempotent matrices and $\overline{q}_1 = \overline{q}_2$. Moreover, since

$$[(p_1, p_2)] - [(q_1, q_2)] \in K_0^{\mathrm{gr}}(A,I),$$

by the definition of the relative K-group, $[p_1] - [q_1] = 0$. So

$$\overline{f}([p_1] - [q_1]) = [\overline{p}_1] - [\overline{q}_1] = 0. \tag{3.20}$$

Taking into account that $\overline{p}_1 = \overline{p}_2$ and $\overline{q}_1 = \overline{q}_2$, we have

$$\overline{f}\overline{\pi}_2([p] - [q]) = \overline{f}\overline{\pi}_2([(p_1,p_2)] - [(q_1,q_2)]) = [\overline{p}_2] - [\overline{q}_2] = [\overline{p}_1] - [\overline{q}_1] = 0.$$

This shows that $\operatorname{Im}(\overline{\pi}_2) \subseteq \ker \overline{f}$.

Next we show that $\ker \overline{f} \subseteq \operatorname{Im}(\overline{\pi}_2)$. Let $p \in \mathbb{M}_n(A)(\overline{\alpha})_0$ and $q \in \mathbb{M}_m(A)(\overline{\delta})_0$ be idempotent matrices with $\overline{\alpha} = (\alpha_1, \dots, \alpha_n)$ and $\overline{\delta} = (\delta_1, \dots, \delta_m)$, where $\alpha_i, \delta_j \in \Gamma$. Suppose $x = [p] - [q] \in K_0^{\mathrm{gr}}(A)$ and

$$\overline{f}(x) = \overline{f}([p] - [q]) = [\overline{p}] - [\overline{q}] = 0.$$

Since $[\overline{p}] = [\overline{q}]$ in $K_0^{\mathrm{gr}}(A/I)$, there is an $l \in \mathbb{N}$, such that $\overline{p} \oplus 1_l$ is graded equivalent to $\overline{q} \oplus 1_l$ in A/I (see §3.2.1). We can replace p by $p \oplus 1_l$ and q by $q \oplus 1_l$ without changing x, and consequently we get that $\overline{p}$ is graded equivalent to $\overline{q}$. By Lemma 3.2.2 there is an invertible matrix $g \in \mathbb{M}_{2k}(A/I)[-\overline{\delta}][-\overline{\alpha}]$ such that

$$\overline{q} = g\overline{p}g^{-1}. \tag{3.21}$$

The following standard trick lets us lift an invertible matrix over A/I to an invertible matrix over A. Consider the invertible matrix

$$\begin{pmatrix} g & 1 \\ 0 & g^{-1} \end{pmatrix} \in \mathbb{M}_{4k}(A/I)[-\overline{\delta}, -\overline{\alpha}][-\overline{\alpha}, -\overline{\delta}]. \tag{3.22}$$

This matrix is a product of the following four matrices:

$$\begin{pmatrix} 1 & g \\ 0 & 1 \end{pmatrix} \in \mathbb{M}_{4k}(A/I)[-\overline{\delta}, -\overline{\alpha}][-\overline{\delta}, -\overline{\alpha}], \begin{pmatrix} 1 & 0 \\ -g^{-1} & 1 \end{pmatrix} \in \mathbb{M}_{4k}(A/I)[-\overline{\delta}, -\overline{\alpha}][-\overline{\delta}, -\overline{\alpha}],$$

$$\begin{pmatrix} 1 & g \\ 0 & 1 \end{pmatrix} \in \mathbb{M}_{4k}(A/I)[-\overline{\delta}, -\overline{\alpha}][-\overline{\delta}, -\overline{\alpha}], \begin{pmatrix} 0 & -1 \\ 1 & 0 \end{pmatrix} \in \mathbb{M}_{4k}(A/I)[-\overline{\delta}, -\overline{\alpha}][-\overline{\alpha}, -\overline{\delta}],$$

where each of them can be lifted from A/I to an invertible matrix over A with the same shift. Thus

$$\begin{pmatrix} g & 1 \\ 0 & g^{-1} \end{pmatrix} = \begin{pmatrix} 1 & g \\ 0 & 1 \end{pmatrix}\begin{pmatrix} 1 & 0 \\ -g^{-1} & 1 \end{pmatrix}\begin{pmatrix} 1 & g \\ 0 & 1 \end{pmatrix}\begin{pmatrix} 0 & -1 \\ 1 & 0 \end{pmatrix}.$$

Let

$$h \in \mathbb{M}_{4k}(A)[-\overline{\delta}, -\overline{\alpha}][-\overline{\alpha}, -\overline{\delta}]$$

be a matrix that lifts (3.22) with

$$h^{-1} \in \mathbb{M}_{4k}(A)[-\overline{\alpha}, -\overline{\delta}][-\overline{\delta}, -\overline{\alpha}].$$

Consider the enlargements of p and q as

$$\begin{pmatrix} p & 0 \\ 0 & 0 \end{pmatrix} \in \mathbb{M}_{4k}(A)(\overline{\alpha}, \overline{\delta})_0$$

and

$$\begin{pmatrix} q & 0 \\ 0 & 0 \end{pmatrix} \in \mathbb{M}_{4k}(A)(\overline{\delta}, \alpha)_0.$$

By Lemma 3.2.2, $h\begin{pmatrix} p & 0 \\ 0 & 0 \end{pmatrix}h^{-1}$ and $\begin{pmatrix} q & 0 \\ 0 & 0 \end{pmatrix}$ are graded equivalent to p and q, respectively, so replacing them does not change x. Replacing p and q with the new representatives, from (3.21), it follows that $\overline{p} = \overline{q}$. This means $p - q$ is a matrix over I, so (p, q) is an idempotent matrix over $D^\Gamma(A, I)$. Now

$$[(p, p)] - [(p, q)] \in K_0^{\mathrm{gr}}(D^\Gamma(A, I))$$

which maps to x under $\overline{\pi}_2$. This completes the proof. □

3.5 K_0^{gr} of nonunital rings

Let A be a ring which does not necessarily have identity. Let R be a ring with identity such that A is a two-sided ideal of R. Then $\mathcal{V}(A)$ is defined as

$$\mathcal{V}(A) := \ker\left(\mathcal{V}(R) \longrightarrow \mathcal{V}(R/A)\right).$$

It is easy to see that

$$\mathcal{V}(A) = \{\, [P] \mid P \text{ is a finitely generated projective } R\text{-module and } PA = P\,\},$$

where $[P]$ is the class of R-modules isomorphic to P and addition is defined via direct sum as before (compare this with (3.1)). One can show that this definition is independent of the choice of the ring R by interpreting P as an idempotent matrix of $\mathbb{M}_k(A)$ for a suitable $k \in \mathbb{N}$.

In order to define $K_0(A)$ for a nonunital ring A, consider the *unitisation ring* $\tilde{A} = \mathbb{Z} \times A$. The addition is component-wise and multiplication is defined as follows:

$$(n, a)(m, b) = (nm, ma + nb + ab), \tag{3.23}$$

where $m, n \in \mathbb{Z}$ and $a, b \in A$. This is a ring with $(1, 0)$ as the identity element and A a two-sided ideal of $\tilde{A}$ such that $\tilde{A}/A \cong \mathbb{Z}$. The canonical epimorphism $\tilde{A} \to \tilde{A}/A$ gives a natural homomorphism on the level of K_0 and then $K_0(A)$ is defined as

$$K_0(A) := \ker\left(K_0(\tilde{A}) \longrightarrow K_0(\tilde{A}/A)\right). \tag{3.24}$$

This construction extends the functor K_0 from the category of rings with identity to the category of rings (not necessarily with identity) to the category of abelian groups.

Remark 3.5.1 Notice that, for a ring A without unit, we didn't define $K_0(A)$ as the group completion of $\mathcal{V}(A)$ (as defined in the unital case). This is because the group completion is not in general a left exact functor. For a nonunital ring A, let R be a ring with identity containing A as a two-sided ideal. Then we have the following exact sequence of K-theory:

$$K_1(R) \longrightarrow K_1(R/A) \longrightarrow K_0(A) \longrightarrow K_0(R) \longrightarrow K_0(R/A),$$

(see §3.4 and [84, Theorem 2.5.4]). There is yet another construction, the *relative Grothendieck group of R and A*, $K_0(R, A)$, which one can show is isomorphic to $K_0(A)$ (see [8, §7] for comparisons between the groups $K_0(A)$ and $\mathcal{V}(A)^+$ in the nongraded setting).

A similar construction can be carried over to the graded setting as follows. Let A be a Γ-graded ring which does not necessarily have identity (Remark 1.1.14). Let R be a Γ-graded ring with identity such that A is a graded two-sided ideal of R. For example, consider $\tilde{A} = \mathbb{Z} \times A$ with multiplication given by (3.23). Moreover, $\tilde{A} = \mathbb{Z} \times A$ is Γ-graded with

$$\begin{aligned} \tilde{A}_0 &= \mathbb{Z} \times A_0, \\ \tilde{A}_\gamma &= 0 \times A_\gamma, \quad \text{for } \gamma \neq 0. \end{aligned} \tag{3.25}$$

Define

$$\mathcal{V}^{\mathrm{gr}}(A) = \{\, [P] \mid P \text{ is a graded finitely generated projective } R\text{-module} \\ \text{and } PA = P \,\},$$

where $[P]$ is the class of graded R-modules, graded isomorphic to P, and addition is defined via direct sum. Parallel to the nongraded setting we define

$$\mathcal{V}^{\mathrm{gr}}(A) := \ker\big(\mathcal{V}^{\mathrm{gr}}(R) \longrightarrow \mathcal{V}^{\mathrm{gr}}(R/A)\big). \tag{3.26}$$

As in the proof of Lemma 3.2.3(1), since $P \oplus Q \cong R^n(-\overline{\alpha})$, this gives an idempotent matrix p in $\mathbb{M}_n(R)(\overline{\alpha})_0$. However, since $PA = P$, we have

$$P \oplus QA = PA \oplus QA \cong A^n(-\overline{\alpha}).$$

This shows that $p \in \mathbb{M}_n(A)(\overline{\alpha})_0$. On the other hand, if

$$p \in \mathbb{M}_n(A)(\overline{\alpha})_0 \subseteq \mathbb{M}_n(R)(\overline{\alpha})_0$$

is an idempotent, then $pR^n(-\overline{\alpha})$ is a graded finitely generated projective R-module such that

$$pR^n(-\overline{\alpha})A = pA^n(-\overline{\alpha}) = pR^n(-\overline{\alpha}).$$

So $[pR^n(-\overline{\alpha})] \in \mathcal{V}^{\mathrm{gr}}(A)$. Now a repeat of Lemma 3.2.3 shows that $\mathcal{V}^{\mathrm{gr}}(A)$ is isomorphic to the monoid of equivalence classes of graded idempotent matrices over A as in §3.2. This shows that the construction of $\mathcal{V}^{\mathrm{gr}}(A)$ is independent of the choice of the graded ring R. It also shows that if A has identity, the two constructions (using the graded projective A-modules versus the graded projective R-modules) coincide.

To define the graded Grothendieck group for nonunital rings, we use a similar approach as in (3.24): Let A be a ring (possibly without identity) and let

$\tilde{A} = \mathbb{Z} \times A$ with the multiplication defined in (3.23), so that A is a graded two-sided ideal of $\tilde{A}$. The graded canonical epimorphism $\tilde{A} \to \tilde{A}/A$ gives a natural homomorphism on the level of K_0^{gr} and then $K_0^{\mathrm{gr}}(A)$ is defined as

$$\boxed{K_0^{\mathrm{gr}}(A) := \ker\left(K_0^{\mathrm{gr}}(\tilde{A}) \longrightarrow K_0^{\mathrm{gr}}(\tilde{A}/A)\right).} \tag{3.27}$$

Thus $K_0^{\mathrm{gr}}(A)$ is a $\mathbb{Z}[\Gamma]$-module. Since the graded homomorphism

$$\begin{aligned} \phi : \tilde{A} &\longrightarrow \tilde{A}/A \cong \mathbb{Z}, \\ (n, a) &\longmapsto n \end{aligned}$$

splits, we obtain the split exact sequence

$$0 \longrightarrow K_0^{\mathrm{gr}}(A) \longrightarrow K_0^{\mathrm{gr}}(\mathbb{Z} \times A) \longrightarrow K_0^{\mathrm{gr}}(\mathbb{Z}) \longrightarrow 0.$$

By Example 3.1.5, $K_0^{\mathrm{gr}}(\mathbb{Z}) \cong \mathbb{Z}[x, x^{-1}]$ and thus we get

$$K_0^{\mathrm{gr}}(\tilde{A}) \cong K_0^{\mathrm{gr}}(A) \oplus \mathbb{Z}[x, x^{-1}].$$

Interpreting this in the language of the reduced K_0^{gr} (§3.1.4), we have

$$\widetilde{K_0^{\mathrm{gr}}}(\tilde{A}) = K_0^{\mathrm{gr}}(A).$$

Let R and S be nonunital graded rings and $\phi : R \to S$ a nonunital graded homomorphism. Then

$$\begin{aligned} \tilde{\phi} : \tilde{R} &\longrightarrow \tilde{S}, \\ (r, n) &\longmapsto (\phi(r), n) \end{aligned}$$

is a unital graded homomorphism and the following commutative diagram shows that there is an order preserving $\mathbb{Z}[\Gamma]$-module homomorphism between their K_0^{gr}-groups.

$$\begin{array}{ccccc} K_0^{\mathrm{gr}}(R) & \hookrightarrow & K_0^{\mathrm{gr}}(\tilde{R}) & \longrightarrow & K_0^{\mathrm{gr}}(\tilde{R}/R) \\ \Big\downarrow{\scriptstyle \overline{\phi}} & & \Big\downarrow{\scriptstyle \overline{\tilde{\phi}}} & & \Big\| \\ K_0^{\mathrm{gr}}(S) & \hookrightarrow & K_0^{\mathrm{gr}}(\tilde{S}) & \longrightarrow & K_0^{\mathrm{gr}}(\tilde{S}/S) \end{array} \tag{3.28}$$

This construction extends the functor K_0^{gr} from the category of graded rings with identity to the category of graded rings (not necessarily with identity) to the category of $\mathbb{Z}[\Gamma]$-module.

In Example 3.7.9, we calculate the graded Grothendieck group of the nonunital ring of (countable) square matrices with finite number of nonzero entires using the idempotent representation.

Remark 3.5.2 UNITISATION OF A GRADED RING WITH IDENTITY

If a Γ-graded ring A has identity, then the ring $\tilde{A} = \mathbb{Z} \times A$ defined by the multiplication (3.23) is graded isomorphic to the cartesian product ring $\mathbb{Z} \times A$, where $\mathbb{Z}$ is a Γ-graded ring concentrated in degree zero (see Example 1.1.10). Indeed, the map

$$\begin{aligned} \tilde{A} &\longrightarrow \mathbb{Z} \times A \\ (n, a) &\longmapsto (n, a + n1_A) \end{aligned}$$

is a graded ring isomorphism of unital rings. This shows that if A has an identity, the two definitions of K_0^{gr} for A coincide. Note also that the unitisation ring $\tilde{A}$ is never a strongly graded ring.

Similar to the nongraded setting (see from example [84, Theorem 1.5.9]), one can prove that for a graded ideal A of the graded ring R,

$$K_0^{\mathrm{gr}}(R, A) \cong K_0^{\mathrm{gr}}(A)$$

as a $\mathbb{Z}[\Gamma]$-modules. This shows that $K_0^{\mathrm{gr}}(R, A)$ depends only on the structure of the nonunital ring A.

3.5.1 Graded inner automorphims

Let A be a graded ring and $f : A \to A$ an inner automorphism defined by $f(a) = rar^{-1}$, where r is a homogeneous and invertible element of A of degree δ. Clearly f is a graded automorphism. Then considering A as a graded left A-module via f, it is easy to observe, for any graded right A-module P, that there is a graded right A-module isomorphism

$$\begin{aligned} P(-\delta) &\longrightarrow P \otimes_f A, \\ p &\longmapsto p \otimes r, \end{aligned}$$

(with the inverse $p \otimes a \mapsto pr^{-1}a$). This induces an isomorphism between the functor

$$- \otimes_f A : \text{Pgr-}A \longrightarrow \text{Pgr-}A$$

and the δ-suspension functor

$$\mathcal{T}_\delta : \text{Pgr-}A \longrightarrow \text{Pgr-}A.$$

Recall that a Quillen's K_i-group, $i \geq 0$, is a functor from the category of exact categories with exact functors to the category of abelian groups (see Chapter 6). Moreover, isomorphic functors induce the same map on the K-groups [81, p.19]. Thus $K_i^{\mathrm{gr}}(f) = K_i^{\mathrm{gr}}(\mathcal{T}_\delta)$. Therefore if r is a homogeneous element of degree zero, *i.e.*, $\delta = 0$, then $K_i^{\mathrm{gr}}(f) : K_i^{\mathrm{gr}}(A) \to K_i^{\mathrm{gr}}(A)$ is the identity map.

3.6 K_0^{gr} is a pre-ordered module

3.6.1 Γ-pre-ordered modules

An abelian group G is called a *pre-ordered abelian group* if there is a relation, denoted by $\geq$, on G which is reflexive and transitive and respects the group structure. It follows that the set $G_+ := \{x \in G \mid x \geq 0\}$ forms a monoid. Conversely, any monoid C in G induces a pre-ordering on G, by defining $x \geq y$ if $x - y \in C$. It follows that with this pre-ordering $G_+ = C$, which is called the *positive cone* of the ordering. (Note that G_+ should not be confused with G^+ used for the group completion in §3.1.)

Example 3.6.1 Let V be a monoid and V^+ its completion (see §3.1). There exist a natural homomorphism $\phi : V \to V^+$ (see (3.4)), which makes V^+ a pre-ordered abelian group with the image of V under this homomorphism as a positive cone of V^+.

Let G have a pre-ordering. An element $u \in G$ is called an *order-unit* if $u \geq 0$ and for any $x \in G$, there is an $n \in \mathbb{N}$ such that $nu \geq x$.

For a ring R, by Example 3.6.1, the Grothendieck group $K_0(R)$ is a pre-ordered abelian group. Concretely, consider the set of isomorphism classes of finitely generated projective R-modules in $K_0(R)$. This set forms a monoid which is considered the positive cone of $K_0(R)$ and is denoted by $K_0(R)_+$. This monoid induces a pre-ordering on $K_0(R)$. We check that with this pre-ordering $[R]$ is an order-unit. Let $u \in K_0(R)$. Then $u = [P] - [Q]$, where P, Q are finitely generated projective R-modules. But there is a finitely generated projective module P' such that $P \oplus P' \cong R^n$ as right R-modules, for some $n \in \mathbb{N}$. Then $n[R] \geq u$. Indeed,

$$\begin{aligned} n[R] - u = n[R] - [P] + [Q] &= [R^n] - [P] + [Q] \\ &= [P \oplus P'] - [P] + [Q] = [P'] + [Q] \in K_0(R)_+. \end{aligned}$$

Example 3.6.2 Suppose R is a ring such that $K_0(R) \neq 0$, but the order unit $[R] = 0$. Since for any finitely generated projective module P, there are a finitely generated projective module Q and an $n \in \mathbb{N}$ such that $P \oplus Q \cong R^n$, it follows that

$$-[P] = -n[R] + [Q] = [Q] \in K_0(R)_+.$$

Thus $K_0(R) = K_0(R)_+$. Therefore it is possible that $x \in K_0(R)$ with $x > 0$ and $0 > x$ simultaneously. One instance of such a ring is constructed in Example 3.9.3.

The Grothendieck group of a ring as a pre-ordered group is studied extensively in [40, §15]. In particular it was established that for the so-called ultramatricial algebras R, the abelian group $K_0(R)$ along with its positive cone $K_0(R)_+$ and the order-unit $[R]$ is a complete invariant (see [40, Theorem 15.26] and the introduction to Chapter 5). This invariant is also called the *dimension group* in the literature as it coincides with an invariant called the dimension group by Elliot [36] to classify such algebras.

Since we will consider the graded Grothendieck groups, which have an extra $\mathbb{Z}[\Gamma]$-module structure, we need to adopt the above definitions on ordering to the graded setting. For this reason, here we define the category of Γ-pre-ordered modules.

Let Γ be a group and G be a (left) Γ-module. Let $\geq$ be a reflexive and transitive relation on G which respects the monoid and the module structures, *i.e.,* for $\gamma \in \Gamma$ and $x, y, z \in G$, if $x \geq y$, then $x + z \geq y + z$ and $\gamma x \geq \gamma y$. We call G a Γ-*pre-ordered module.* We call G a *pre-ordered module* when Γ is clear from the context. The *positive cone* of G is defined as $\{x \in G \mid x \geq 0\}$ and denoted by G_+. The set G_+ is a Γ-submonoid of G, *i.e.,* a submonoid which is closed under the action of Γ. In fact, G is a Γ-pre-ordered module if and only if there exists a Γ-submonoid of G. (Since G is a Γ-module, it can be considered as a $\mathbb{Z}[\Gamma]$-module.) An element $u \in G_+$ is called an *order-unit* if for any $x \in G$, there are $\alpha_1, \dots, \alpha_n \in \Gamma$, $n \in \mathbb{N}$, such that

$$\sum_{i=1}^{n} \alpha_i u \geq x. \tag{3.29}$$

As usual, in this setting, we only consider homomorphisms which preserve the pre-ordering, *i.e.,* a Γ-homomorphism $f : G \to H$, such that $f(G_+) \subseteq H_+$. We refer to these homomorphisms as the *order preserving homomorphisms.* We denote by $\mathcal{P}_\Gamma$ the category of pointed Γ-pre-ordered modules with pointed order preserving homomorphisms, *i.e.,* the objects are the pairs (G, u), where G is a Γ-pre-ordered module and u is an order-unit, and $f : (G, u) \to (H, v)$ is an order preserving Γ-homomorphism such that $f(u) = v$. We sometimes write $f : (G, G_+, u) \to (H, H_+, v)$ to emphasise the ordering of f. The focus of this book is to study the graded Grothendieck group as a functor from the category of graded rings to the category $\mathcal{P}_\Gamma$

Note that when Γ is a trivial group, we are in the classical setting of pre-ordered abelian groups. When $\Gamma = \mathbb{Z}$ and so $\mathbb{Z}[\Gamma] = \mathbb{Z}[x, x^{-1}]$, we simply write $\mathcal{P}$ for $\mathcal{P}_{\mathbb{Z}}$.

Example 3.6.3 $K_0^{\mathrm{gr}}(A)$ AS A Γ-PRE-ORDERED MODULE

Let A be a Γ-graded ring. Then $K_0^{\mathrm{gr}}(A)$ is a Γ-pre-ordered module with the set of isomorphic classes of graded finitely generated projective right A-modules as the positive cone of ordering. This monoid is denoted by $K_0^{\mathrm{gr}}(A)_+$ which is the image of $\mathcal{V}^{\mathrm{gr}}(A)$ under the natural homomorphism $\mathcal{V}^{\mathrm{gr}}(A)$. We consider $[A]$ as an order-unit of this group. Indeed, if $x \in K_0^{\mathrm{gr}}(A)$, then by Lemma 3.1.7(1) there are graded finitely generated projective modules P and P' such that $x = [P] - [P']$. But there is a graded module Q such that $P \oplus Q \cong A^n(\overline{\alpha})$, where $\overline{\alpha} = (\alpha_1, \dots, \alpha_n)$, $\alpha_i \in \Gamma$ (see (1.39)). Now

$$[A^n(\overline{\alpha})] - x = [P] + [Q] - [P] + [P'] = [Q] + [P'] = [Q \oplus P'] \in K_0^{\mathrm{gr}}(A)_+.$$

This shows that $\sum_{i=1}^n \alpha_i[A] = [A^n(\overline{\alpha})] \geq x$.

In Chapter 5 we will show that for the so-called graded ultramatricial algebras R, the $\mathbb{Z}[\Gamma]$-module $K_0^{\mathrm{gr}}(R)$ along with its positive cone $K_0^{\mathrm{gr}}(R)_+$ and the order-unit $[R]$ is a complete invariant (see Theorem 5.2.4). We denote this invariant $(K_0^{\mathrm{gr}}(R), K_0^{\mathrm{gr}}(R)_+, [R])$ and we call it the *graded dimension module* of R.

Theorem 3.6.4 *Let A and B be Γ-graded rings. If A is graded Morita equivalent to B, then there is an order preserving $\mathbb{Z}[\Gamma]$-module isomorphism*

$$K_0^{\mathrm{gr}}(A) \cong K_0^{\mathrm{gr}}(B).$$

Proof By Theorem 2.3.7, there is a graded $A-B$-bimodule Q such that the functor $- \otimes_A Q : \text{Gr-}A \to \text{Gr-}B$ is a graded equivalence. The restriction to $\text{Pgr-}A$ induces a graded equivalence $- \otimes_A Q : \text{Pgr-}A \to \text{Pgr-}B$. This in return induces an order preserving isomorphism $\phi : K_0^{\mathrm{gr}}(A) \to K_0^{\mathrm{gr}}(B)$ which is also a $\mathbb{Z}[\Gamma]$-module isomorphism by Equation (1.21). □

If Γ-graded rings A and B are graded Morita equivalent, *i.e.*, $\text{Gr-}A \approx \text{Gr-}B$, then by Theorem 2.3.7, $\text{Mod-}A \approx \text{Mod-}B$ and thus by Theorem 3.6.4, not only is $K_0^{\mathrm{gr}}(A) \cong K_0^{\mathrm{gr}}(B)$, but also $K_0(A) \cong K_0(B)$.

In a more specific case, we have the following.

Lemma 3.6.5 *Let A be a Γ-graded ring and P be a A-graded progenerator. Let $B = \mathrm{End}_A(P)$. Then $[P]$ is an order-unit in $K_0^{\mathrm{gr}}(A)$ and there is an order preserving $\mathbb{Z}[\Gamma]$-module isomorphism*

$$(K_0^{\mathrm{gr}}(B), [B]) \cong (K_0^{\mathrm{gr}}(A), [P]).$$

Proof Since P is a graded generator, by Theorem 2.2.2, there are $\alpha_i \in \Gamma$, $1 \leq i \leq n$, such that $[A] \leq \sum_i \alpha_i[P]$. Since $[A]$ is an order-unit, this immediately implies that $[P]$ is an order-unit. By (2.9), the functor $- \otimes_B P : \text{Gr-}B \to \text{Gr-}A$ is a graded equivalence. The restriction to Pgr-B induces a graded equivalence $- \otimes_B P : \text{Pgr-}B \to \text{Pgr-}A$. This in return induces an order preserving $\mathbb{Z}[\Gamma]$-module isomorphism $\phi : K_0^{\mathrm{gr}}(B) \to K_0^{\mathrm{gr}}(A)$, with

$$\phi([Q] - [L]) = [Q \otimes_B P] - [L \otimes_B P],$$

where Q and L are graded finitely generated projective B-modules. In particular $\phi([B]) = [P]$. This completes the proof. □

3.7 K_0^{gr} of graded division rings

In the case of graded division rings, using the description of graded free modules, one can compute the graded Grothendieck group completely.

Proposition 3.7.1 *Let A be a Γ-graded division ring with the support the subgroup Γ_A. Then the monoid of isomorphism classes of Γ-graded finitely generated projective A-modules is isomorphic to $\mathbb{N}[\Gamma/\Gamma_A]$. Consequently there is a canonical $\mathbb{Z}[\Gamma]$-module isomorphism*

$$\begin{aligned} K_0^{\mathrm{gr}}(A) &\longrightarrow \mathbb{Z}[\Gamma/\Gamma_A] \\ [A^n(\delta_1, \dots, \delta_n)] &\longmapsto \sum_{i=1}^{n} \Gamma_A + \delta_i. \end{aligned}$$

In particular if A is a (trivially graded) division ring, Γ a group and A considered as a Γ-graded division ring concentrated in degree zero, then $K_0^{\mathrm{gr}}(A) \cong \mathbb{Z}[\Gamma]$ and $[A^n(\delta_1, \dots, \delta_n)] \in K_0^{\mathrm{gr}}(A)$ corresponds to $\sum_{i=1}^{n} \delta_i$ in $\mathbb{Z}[\Gamma]$.

Proof By Proposition 1.3.16, $A(\delta_1) \cong_{\mathrm{gr}} A(\delta_2)$ as a graded A-module if and only if $\delta_1 - \delta_2 \in \Gamma_A$. Thus any graded free module of rank 1 is graded isomorphic to some $A(\delta_i)$, where $\{\delta_i\}_{i \in I}$ is a complete set of coset representative of the subgroup Γ_A in Γ, *i.e.*, $\{\Gamma_A + \delta_i, i \in I\}$ represents Γ/Γ_A. Since any graded finitely generated module M over A is graded free (Proposition 1.4.1), it follows that

$$M \cong_{\mathrm{gr}} A(\delta_{i_1})^{r_1} \oplus \cdots \oplus A(\delta_{i_k})^{r_k}, \tag{3.30}$$

where $\delta_{i_1}, \dots \delta_{i_k}$ are distinct elements of the coset representative. Now suppose

$$M \cong_{\mathrm{gr}} A(\delta_{i'_1})^{s_1} \oplus \cdots \oplus A(\delta_{i'_{k'}})^{s_{k'}}. \tag{3.31}$$

Considering the A_0-module $M_{-\delta_{i_1}}$, from (3.30) we have $M_{-\delta_{i_1}} \cong A_0^{r_1}$. Comparing this with $M_{-\delta_{i_1}}$ from (3.31), it follows that one of $\delta_{i'_j}$, $1 \leq j \leq k'$, say, $\delta_{i'_1}$, has to be δ_{i_1} and so $r_1 = s_1$ as A_0 is a division ring. Repeating the same argument for each δ_{i_j}, $1 \leq j \leq k$, we see $k = k'$, $\delta_{i'_j} = \delta_{i_j}$ and $r_j = s_j$, for all $1 \leq j \leq k$ (possibly after suitable permutation). Thus any graded finitely generated projective A-module can be written uniquely as $M \cong_{\mathrm{gr}} A(\delta_{i_1})^{r_1} \oplus \cdots \oplus A(\delta_{i_k})^{r_k}$, where $\delta_{i_1}, \ldots \delta_{i_k}$ are distinct elements of the coset representative. The (well-defined) map

$$\begin{aligned} \mathcal{V}^{\mathrm{gr}}(A) &\to \mathbb{N}[\Gamma/\Gamma_A] \qquad (3.32)\\ [A(\delta_{i_1})^{r_1} \oplus \cdots \oplus A(\delta_{i_k})^{r_k}] &\mapsto r_1(\Gamma_A + \delta_{i_1}) + \cdots + r_k(\Gamma_A + \delta_{i_k}), \end{aligned}$$

gives a $\mathbb{N}[\Gamma]$-monoid isomorphism between the monoid of isomorphism classes of Γ-graded finitely generated projective A-modules $\mathcal{V}^{\mathrm{gr}}(A)$ and $\mathbb{N}[\Gamma/\Gamma_A]$. The rest of the proof follows easily. □

Remark 3.7.2 One can use Equation (3.9) to calculate the graded K-theory of graded division algebras as well. One can also compute this group via the G_0^{gr}-theory of Artinian rings (see Corollary 3.10.2).

Example 3.7.3 Using Proposition 3.7.1, we calculate the graded K_0 of two types of graded field and we determine the action of $\mathbb{Z}[x, x^{-1}]$ on these groups. These are graded fields obtained from Leavitt path algebras of acyclic and C_n-comet graphs, respectively (see Theorem 1.6.19 and 1.6.21).

1 Let K be a field. Consider $A = K$ as a $\mathbb{Z}$-graded field with support $\Gamma_A = 0$, *i.e.*, A is concentrated in degree 0. By Proposition 3.7.1, $K_0^{\mathrm{gr}}(A) \cong \mathbb{Z}[x, x^{-1}]$ as a $\mathbb{Z}[x, x^{-1}]$-module. With this presentation $[A(i)]$ corresponds to x^i in $\mathbb{Z}[x, x^{-1}]$.

2 Let $A = K[x^n, x^{-n}]$ be a $\mathbb{Z}$-graded field with $\Gamma_A = n\mathbb{Z}$. By Proposition 3.7.1,

$$K_0^{\mathrm{gr}}(A) \cong \mathbb{Z}[\mathbb{Z}/n\mathbb{Z}] \cong \bigoplus_n \mathbb{Z}$$

is a $\mathbb{Z}[x, x^{-1}]$-module. The action of x on $(a_1, \ldots, a_n) \in \bigoplus_n \mathbb{Z}$ is

$$x(a_1, \ldots, a_n) = (a_n, a_1, \ldots, a_{n-1}).$$

With this presentation $[A]$ corresponds to $(1, 0, \ldots, 0)$ in $\bigoplus_n \mathbb{Z}$. Moreover, the map

$$U : K_0^{\mathrm{gr}}(A) \longrightarrow K_0(A)$$

induced by the forgetful functor (§1.2.7), gives a group homomorphism

$$U : \bigoplus_n \mathbb{Z} \longrightarrow \mathbb{Z},$$
$$(a_i) \longmapsto \sum_i a_i.$$

In fact, the sequence

$$\bigoplus_n \mathbb{Z} \xrightarrow{f} \bigoplus_n \mathbb{Z} \xrightarrow{U} \mathbb{Z} \longrightarrow 0,$$

where $f(a_1, \dots, a_n) = (a_n - a_1, a_1 - a_2, \dots, a_{n-1} - a_n)$, is exact. In §6.4 we systematically relate K_0^{gr} to K_0 for certain rings.

Example 3.7.4 Consider the Hamilton quaternion algebra

$$\mathbb{H} = \mathbb{R} \oplus \mathbb{R}i \oplus \mathbb{R}j \oplus \mathbb{R}k.$$

By Example 1.1.20, $\mathbb{H}$ is a $\mathbb{Z}_2 \times \mathbb{Z}_2$-graded division ring with $\Gamma_{\mathbb{H}} = \mathbb{Z}_2 \times \mathbb{Z}_2$. By Proposition 3.7.1, $K_0^{\text{gr}}(\mathbb{H}) \cong \mathbb{Z}$. In fact, since $\mathbb{H}$ is a strongly $\mathbb{Z}_2 \times \mathbb{Z}_2$-graded division ring, one can deduce the result using the Dade's theorem (see §3.1.3), *i.e.*, $K_0^{\text{gr}}(\mathbb{H}) \cong K_0(\mathbb{H}_0) = K_0(\mathbb{R}) \cong \mathbb{Z}$.

Example 3.7.5 Let (D, v) be a valued division algebra, where $v : D^* \to \Gamma$ is the valuation homomorphism. By Example 1.4.7, there is a Γ-graded division algebra $\text{gr}(D)$ associated with D, where $\Gamma_{\text{gr}(D)} = \Gamma_D$. By Proposition 3.7.1,

$$K_0^{\text{gr}}(\text{gr}(D)) \cong \mathbb{Z}[\Gamma/\Gamma_D].$$

The following example generalises Example 3.7.4 of the Hamilton quaternion algebra $\mathbb{H}$ as a $\mathbb{Z}_2 \times \mathbb{Z}_2$-graded ring.

Remark 3.7.6 We saw in Example 1.1.20 that $\mathbb{H}$ can be considered as a $\mathbb{Z}_2$-graded division ring. So $\mathbb{H}$ is also strongly $\mathbb{Z}_2$-graded, and

$$K_0^{\text{gr}}(\mathbb{H}) \cong K_0(\mathbb{H}_0) = K_0(\mathbb{C}) \cong \mathbb{Z}.$$

Then $Z(\mathbb{H}) = \mathbb{R}$, which we can consider as a trivially $\mathbb{Z}_2$-graded field, so by Proposition 3.7.1, $K_0^{\text{gr}}(\mathbb{R}) = \mathbb{Z} \oplus \mathbb{Z}$. We note that for both grade groups, $\mathbb{Z}_2$ and $\mathbb{Z}_2 \times \mathbb{Z}_2$, we have $K_0^{\text{gr}}(\mathbb{H}) \cong \mathbb{Z}$, but the $K_0^{\text{gr}}(\mathbb{R})$ are different. So the graded K-theory of a graded ring depends not only on the ring, but also on its grade group.

Example 3.7.7 Let K be a field and let $R = K[x^2, x^{-2}]$. Then R is a $\mathbb{Z}$-graded field, with support $2\mathbb{Z}$, where R can be written as $R = \bigoplus_{n \in \mathbb{Z}} R_n$, with $R_n = Kx^n$ if n is even and $R_n = 0$ if n is odd. Consider the shifted graded matrix ring

$A = \mathbb{M}_3(R)(0, 1, 1)$, which has support $\mathbb{Z}$. Then we will show that A is a graded central simple algebra over R.

It is clear that the centre of A is R, and A is finite dimensional over R. Recall that a graded ideal is generated by homogeneous elements. If J is a nonzero graded ideal of A, then using the elementary matrices, we can show that $J = A$ (see [50, §III, Exercise 2.9]), so A is graded simple.

By Theorem 1.6.21, A is the Leavitt path algebra of the following graph.

$$E : \bullet \longrightarrow \bullet \rightleftarrows \bullet$$

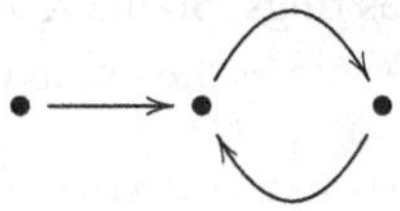

Thus by Theorem 1.6.15 A is a strongly $\mathbb{Z}$-graded ring.

Here, we also show that A is a strongly $\mathbb{Z}$-graded ring by checking the conditions of Proposition 1.1.15. Since $\mathbb{Z}$ is finitely generated, it is sufficient to show that $I_3 \in A_1A_{-1}$ and $I_3 \in A_{-1}A_1$. We have

$$I_3 = \begin{pmatrix} 0 & 1 & 0 \\ 0 & 0 & 0 \\ 0 & 0 & 0 \end{pmatrix}\begin{pmatrix} 0 & 0 & 0 \\ 1 & 0 & 0 \\ 0 & 0 & 0 \end{pmatrix} + \begin{pmatrix} 0 & 0 & 0 \\ x^2 & 0 & 0 \\ 0 & 0 & 0 \end{pmatrix}\begin{pmatrix} 0 & x^{-2} & 0 \\ 0 & 0 & 0 \\ 0 & 0 & 0 \end{pmatrix} + \begin{pmatrix} 0 & 0 & 0 \\ 0 & 0 & 0 \\ x^2 & 0 & 0 \end{pmatrix}\begin{pmatrix} 0 & 0 & x^{-2} \\ 0 & 0 & 0 \\ 0 & 0 & 0 \end{pmatrix}$$

and

$$I_3 = \begin{pmatrix} 0 & x^{-2} & 0 \\ 0 & 0 & 0 \\ 0 & 0 & 0 \end{pmatrix}\begin{pmatrix} 0 & 0 & 0 \\ x^2 & 0 & 0 \\ 0 & 0 & 0 \end{pmatrix} + \begin{pmatrix} 0 & 0 & 0 \\ 1 & 0 & 0 \\ 0 & 0 & 0 \end{pmatrix}\begin{pmatrix} 0 & 1 & 0 \\ 0 & 0 & 0 \\ 0 & 0 & 0 \end{pmatrix} + \begin{pmatrix} 0 & 0 & 0 \\ 0 & 0 & 0 \\ 1 & 0 & 0 \end{pmatrix}\begin{pmatrix} 0 & 0 & 1 \\ 0 & 0 & 0 \\ 0 & 0 & 0 \end{pmatrix}.$$

As in the previous examples, using Dade's theorem (see §3.1.3), we have $K_0^{\mathrm{gr}}(A) \cong K_0(A_0)$. Since $R_0 = K$, by (1.47) there is a ring isomorphism

$$A_0 = \begin{pmatrix} R_0 & R_1 & R_1 \\ R_{-1} & R_0 & R_0 \\ R_{-1} & R_0 & R_0 \end{pmatrix} = \begin{pmatrix} K & 0 & 0 \\ 0 & K & K \\ 0 & K & K \end{pmatrix} \cong K \times \mathbb{M}_2(K).$$

Then

$$K_0^{\mathrm{gr}}(A) \cong K_0(A_0) \cong K_0(K) \oplus K_0(\mathbb{M}_2(K)) \cong \mathbb{Z} \oplus \mathbb{Z},$$

since K_0 respects Cartesian products and Morita equivalence. Note that

$$K_0(A) = K_0(\mathbb{M}_3(R)(0,1,1)) \cong K_0(R) = K_0(K[x^2,x^{-2}]) \cong \mathbb{Z}.$$

The isomorphism $K_0(K[x^2,x^{-2}]) \cong \mathbb{Z}$ comes from the fundamental theorem of algebraic K-theory [84, Theorem 3.3.3] (see also [67, p. 484]),

$$K_0(K[x,x^{-1}]) \cong K_0(K) \cong \mathbb{Z},$$

and that $K[x^2,x^{-2}] \cong K[x,x^{-1}]$ as rings. So the K-theory of A is isomorphic to one copy of $\mathbb{Z}$, which is not the same as the graded K-theory of A.

Example 3.7.8 REDUCED K_0^{gr} OF GRADED CENTRAL SIMPLE ALGEBRAS

Recall from §3.1.4 that, for a Γ-graded ring A, $\widetilde{K_0^{\mathrm{gr}}}$ is the cokernel of the homomorphism

$$\begin{aligned} \phi : \mathbb{Z}[\Gamma] &\longrightarrow K_0^{\mathrm{gr}}(A), \\ \sum_\alpha n_\alpha \alpha &\longmapsto \sum_\alpha n_\alpha [A(\alpha)]. \end{aligned}$$

Let A be a Γ-graded division ring. We calculate $\widetilde{K_0^{\mathrm{gr}}}(\mathbb{M}_n(A))$, where $\mathbb{M}_n(A)$ is a Γ-graded ring (with no shift). The $\mathbb{Z}[\Gamma]$-module homomorphism ϕ above takes the form

$$\begin{aligned} \mathbb{Z}[\Gamma] &\longrightarrow K_0^{\mathrm{gr}}(\mathbb{M}_n(A)) \overset{\cong}{\longrightarrow} K_0^{\mathrm{gr}}(A) \overset{\cong}{\longrightarrow} \mathbb{Z}[\Gamma/\Gamma_A], \\ \alpha &\longmapsto [\mathbb{M}_n(A)(\alpha)] \longmapsto \bigoplus_n [A(\alpha)] \longmapsto n(\Gamma_A + \alpha), \end{aligned}$$

where the second map is induced by the Morita theory (see Proposition 2.1.2) and the third map is induced by Proposition 3.7.1. The cokernel of the composition maps gives $\widetilde{K_0^{\mathrm{gr}}}(\mathbb{M}_n(A))$, which one can immediately calculate:

$$\widetilde{K_0^{\mathrm{gr}}}(\mathbb{M}_n(A)) \cong \frac{\mathbb{Z}}{n\mathbb{Z}}[\frac{\Gamma}{\Gamma_A}].$$

In particular, for a graded division ring A, we have

$$\widetilde{K_0^{\mathrm{gr}}}(A) \cong 0.$$

Example 3.7.9 Let K be a field. Consider the following sequence of graded matrix rings:

$$K \subseteq \mathbb{M}_2(K)(0,1) \subseteq \mathbb{M}_3(K)(0,1,2) \subseteq \mathbb{M}_4(K)(0,1,2,3) \subseteq \cdots \tag{3.33}$$

where the inclusion comes from the nonunital (graded) ring homomorphism of (3.15). Let R be the graded ring

$$R = \bigcup_{i=1}^{\infty} \mathbb{M}_i(K)(\overline{\alpha}_i),$$

where $\overline{\alpha}_i = (0, 1, \dots, i-1)$. Note that R is a nonunital ring. (This is a nonunital graded ultramatricial algebra which will be studied in Chapter 5). We calculate $K_0^{\mathrm{gr}}(R)$. Since K_0^{gr} respects the direct limit (Theorem 3.2.4), we have

$$K_0^{\mathrm{gr}}(R) = K_0^{\mathrm{gr}}(\varinjlim R_i) = \varinjlim K_0^{\mathrm{gr}}(R_i),$$

where $R_i = \mathbb{M}_i(K)(\overline{\alpha}_i)$. The nonunital homomorphism

$$\phi_i : R_i \to R_{i+1},$$

(see (3.15)) induces the homomorphism $\overline{\phi}_i : K_0^{\mathrm{gr}}(R_i) \to K_0^{\mathrm{gr}}(R_{i+1})$ (see (3.28)). We will observe that each of these K-groups is $\mathbb{Z}[x, x^{-1}]$ and each map ϕ_i is an identity. First, note that the ring R_i, $i \in \mathbb{N}$, is unital, so using the Morita theory (see Proposition 2.1.2) and Example 3.7.3(1), $K_0^{\mathrm{gr}}(R_i) = K_0^{\mathrm{gr}}(K) \cong \mathbb{Z}[x, x^{-1}]$. Switching to the idempotent representation of the graded Grothendieck group, observe that the idempotent matrix p_i having 1 on the upper left corner and 0 everywhere else is in $R_{i0} = \mathbb{M}_i(K)(0, 1, \dots, i-1)_0$ and

$$p_i R_i \cong_{\mathrm{gr}} K \oplus K(1) \oplus \cdots \oplus K(i-1)$$

(see Example 2.1.3). Again Proposition 2.1.2 (see also Proposition 2.1.1) shows that $[p_i R_i] = [K \oplus K(1) \oplus \cdots \oplus K(i-1)]$ corresponds to $1 \in \mathbb{Z}[x, x^{-1}]$. Now $p_{i+1} := \phi_i(p_i)$ gives again a matrix with 1 on the upper left corner and 0 everywhere else. So $p_{i+1}R_{i+1} \cong_{\mathrm{gr}} K \oplus K(1) \oplus \cdots \oplus K(i)$ and consequently $[p_{i+1}R_{i+1}] = 1$. So $\overline{\phi}_i(1) = 1$. Since $\overline{\phi}_i$ respects the shift, this shows that $\overline{\phi}_i$ is the identity map. Thus $K_0^{\mathrm{gr}}(R) \cong \mathbb{Z}[x, x^{-1}]$.

A similar argument shows that for the graded ring

$$S = \bigcup_{i=1}^{\infty} \mathbb{M}_i(K)(\overline{\alpha}_i),$$

where $\overline{\alpha}_i = (0, 1, 1, \dots, 1)$, with one 0 and 1 repeated $i-1$ times, we have an ordered $\mathbb{Z}[x, x^{-1}]$-module isomorphism

$$K_0^{\mathrm{gr}}(R) \cong K_0^{\mathrm{gr}}(S).$$

Using a similar argument as in the proof of Theorem 1.6.19, one can show that R is isomorphic to the Leavitt path algebra associated with the infinite graph

$$E: \quad \cdots\cdots\to \bullet \longrightarrow \bullet \longrightarrow \bullet$$

whereas S is the Leavitt path algebra of a graph F consisting of infinite vertices, the one in the middle and the rest are connected to this vertex with an edge.

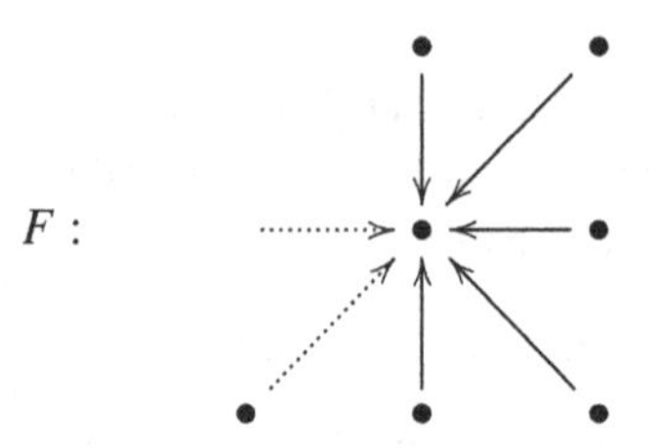

Clearly the rings R and S are not graded isomorphic, as the support of R is $\mathbb{Z}$ whereas the support of S is $\{-1, 0, 1\}$. See Remark 5.2.7 in regard to classification of graded ultramatricial algebras.

Example 3.7.10 K_0^{gr} RING OF GRADED FIELDS

If A is a Γ-graded field, then by §3.1.5, $K_0^{\mathrm{gr}}(A)$ is a ring. One can easily check that by Proposition 3.7.1 and (1.23), $K_0^{\mathrm{gr}}(A) \cong \Gamma/\Gamma_A$ as algebras.

3.8 K_0^{gr} of graded local rings

Recall from §1.1.8 that a Γ-graded ring A is called a graded local ring if the two-sided ideal M generated by noninvertible homogeneous elements of A is a proper ideal.

In Proposition 3.8.3, we explicitly calculate the graded Grothendieck group of graded local rings. There are two ways to do this. One can prove that a graded finitely generated projective module over a graded local ring is graded free with a unique rank, as in the case of graded division rings (see §1.4) and adopt the same approach to calculate the graded Grothendieck group (see §3.7). However, there is a more direct way to do so, which we develop here.

Recall the definition of a graded Jacobson radical of a graded ring A, $J^{\mathrm{gr}}(A)$, from §1.1.5. We also need a graded version of the Nakayama lemma which holds in this setting as well. That is, if $J \subseteq J^{\mathrm{gr}}(A)$ is a graded right ideal of A and P is a graded finitely generated right A-module, and Q is a graded submodule of P such that $P = Q + PJ$, then $P = Q$ (see [75, Corollary 2.9.2]).

We need the following two lemmas in order to calculate the K_0^{gr} of a graded local ring.

Lemma 3.8.1 *Let A be a Γ-graded ring, $J \subseteq J^{\mathrm{gr}}(A)$ a homogeneous two-sided ideal and P and Q be graded finitely generated projective A-modules. If*

$\overline{P} = P \otimes_A A/J = P/PJ$ and $\overline{Q} = Q \otimes_A A/J = Q/QJ$ are isomorphic as graded A/J-modules, then P and Q are isomorphic as graded A-modules.

Proof Let $\phi : \overline{P} \to \overline{Q}$ be a graded A/J-module isomorphism. Clearly ϕ is also an A-module isomorphism. Consider Diagram (3.34). Since π_2 is an epimorphism and P is a graded finitely generated projective, there is a graded homomorphism $\psi : P \to Q$ which makes the diagram commutative.

$$\begin{array}{ccccc} P & \xrightarrow{\pi_1} & \overline{P} & \longrightarrow & 0 \\ {\scriptstyle\psi}\downarrow & & \downarrow{\scriptstyle\phi} & & \\ Q & \xrightarrow{\pi_2} & \overline{Q} & \longrightarrow & 0 \end{array} \tag{3.34}$$

We will show that ψ is a graded A-module isomorphism. Since $\pi_2\psi$ is an epimorphism, $Q = \psi(P) + \ker \pi_2 = \psi(P) + QJ$. The Nakayama lemma then implies that $Q = \psi(P)$, *i.e.*, ψ is an epimorphism. Now since Q is graded projective, there is a graded homomorphism $i : Q \to P$ such that $\psi i = 1_Q$ and $P = i(Q) \oplus \ker \psi$. But

$$\ker \psi \subseteq \ker \pi_2\psi = \ker \phi\pi_1 = \ker \pi_1 = PJ.$$

Thus $P = i(Q) + PJ$. Again the Nakayama lemma implies $P = i(Q)$. Thus $\ker \psi = 0$, and so $\psi : P \to Q$ is an isomorphism. □

Lemma 3.8.2 *Let A and B be Γ-graded rings and $\phi : A \to B$ be a graded epimorphism such that $\ker \phi \subseteq J^{\mathrm{gr}}(A)$. Then the induced homomorphism*

$$\overline{\phi} : K_0^{\mathrm{gr}}(A) \to K_0^{\mathrm{gr}}(B)$$

is a $\mathbb{Z}[\Gamma]$-module monomorphism.

Proof Let $J = \ker(\phi)$. Without the loss of generality we can assume $B = A/J$. Let $x \in \ker \overline{\phi}$. Write $x = [P] - [Q]$ for two graded finitely generated projective A-modules P and Q. So $\phi(x) = [\overline{P}] - [\overline{Q}] = 0$, where $\overline{P} = P \otimes_A A/J = P/PJ$ and $\overline{Q} = Q \otimes_A A/J = Q/QJ$ are graded A/J-modules. Since $\phi(x) = 0$, by Lemma 3.1.7(3), $\overline{P} \bigoplus B^n(\overline{\alpha}) \cong_{\mathrm{gr}} \overline{Q} \bigoplus B^n(\overline{\alpha})$, where $\overline{\alpha} = (\alpha_1, \dots, \alpha_n)$. So

$$\overline{P \bigoplus A^n(\overline{\alpha})} \cong_{\mathrm{gr}} \overline{Q \bigoplus A^n(\overline{\alpha})}.$$

By Lemma 3.8.1, $P \bigoplus A^n(\overline{\alpha}) \cong_{\mathrm{gr}} Q \bigoplus A^n(\overline{\alpha})$. Therefore $x = [P] - [Q] = 0$ in $K_0^{\mathrm{gr}}(A)$. This completes the proof. □

For a Γ-graded ring A, recall that Γ_A is the support of A and

$$\Gamma_A^* = \{\alpha \in \Gamma \mid A_\alpha^* \neq \varnothing\}$$

is a subgroup of Γ_A.

Proposition 3.8.3 *Let A be a Γ-graded local ring. Then there is a $\mathbb{Z}[\Gamma]$-module isomorphism*

$$K_0^{\mathrm{gr}}(A) \cong \mathbb{Z}[\Gamma/\Gamma_A^*].$$

Proof Let $M = J^{\mathrm{gr}}(A)$ be the unique graded maximal ideal of A. Then by Lemma 3.8.2 the homomorphism

$$\begin{aligned} \phi : K_0^{\mathrm{gr}}(A) &\longrightarrow K_0^{\mathrm{gr}}(A/M) \\ [P] &\longmapsto [P \otimes_A A/M] \end{aligned} \tag{3.35}$$

is a monomorphism. Since A/M is a graded division ring, by Proposition 3.7.1, $K_0^{\mathrm{gr}}(A/M) \cong \mathbb{Z}[\Gamma/\Gamma_{A/M}]$. Observe that $\Gamma_{A/M} = \Gamma_A^*$. On the other hand, for the graded A-module $A(\alpha)$, where $\alpha \in \Gamma$, we have

$$A(\alpha) \otimes_A A/M \cong_{\mathrm{gr}} (A/M)(\alpha) \tag{3.36}$$

as the graded A/M-module. Since

$$K_0^{\mathrm{gr}}(A/M) \cong \mathbb{Z}[\Gamma/\Gamma_A^*]$$

is generated by $[(A/M)(\alpha_i)]$, where $\{\alpha_i\}_{i\in I}$ is a complete set of coset representative of Γ/Γ_A^* (see Proposition 3.7.1), from (3.35) and (3.36) we get $\phi([A(\alpha_i)] = [(A/M)(\alpha_i)]$. So ϕ is an epimorphism as well. This finishes the proof. □

Example 3.8.4 K_0^{gr} RING OF COMMUTATIVE GRADED LOCAL RINGS

If A is a commutative Γ-graded local ring, it follows from Proposition 3.8.3 and Example 3.7.10 that $K_0^{\mathrm{gr}}(A) \cong \Gamma/\Gamma_A^*$ as algebras.

3.9 K_0^{gr} of Leavitt path algebras

For a graph E, its associated path algebra $\mathcal{P}(E)$ is a positively graded ring. In §6.1, we will use Quillen's theorem on graded K-theory of such rings to calculate the graded Grothendieck group of paths algebras (see Theorem 6.1.2).

In this section we calculate the K_0 and K_0^{gr} of Leavitt path algebras. For one thing, they provide very nice examples. We first calculate the nongraded K_0 of the Leavitt path algebras in §3.9.1, and in §3.9.2 we determine their K_0^{gr}-groups. Throughout this section, a Leavitt path algebra $\mathcal{L}_K(E)$ is defined over the coefficient field K. We sometimes write $\mathcal{L}(E)$ instead of $\mathcal{L}_K(E)$.

3.9.1 K_0 of Leavitt path algebras

For a Leavitt path algebra $\mathcal{L}_K(E)$, the monoid $\mathcal{V}(\mathcal{L}_K(E))$ is studied in [5]. In particular using [5, Theorem 3.5], one can calculate the Grothendieck group of a Leavitt path algebra from the adjacency matrix of a graph (see [3, p.1998]). We present the calculation of the Grothendieck group of a Leavitt path algebra here.

Let F be a free abelian monoid generated by a countable set X. The nonzero elements of F can be written as $\sum_{t=1}^{n} x_t$, where $x_t \in X$. Let $I \subseteq \mathbb{N}$ and r_i, s_i be elements of F, where $i \in I$. We define an equivalence relation on F denoted by $\langle r_i = s_i \mid i \in I\rangle$ as follows: Define a binary relation $\to$ on $F\backslash\{0\}$ by

$$r_i + \sum_{t=1}^{n} x_t \to s_i + \sum_{t=1}^{n} x_t, i \in I$$

and generate the equivalence relation on F using this binary relation. Namely, $a \sim a$ for any $a \in F$ and for $a, b \in F\backslash\{0\}$, $a \sim b$ if there is a sequence

$$a = a_0, a_1, \ldots, a_n = b$$

such that for each $t = 0, \ldots, n-1$ either $a_t \to a_{t+1}$ or $a_{t+1} \to a_t$. We denote the quotient monoid by $F/\langle r_i = s_i \mid i \in I\rangle$. Completing the monoid (see §3.1.1), one can see that there is a canonical group isomorphism

$$\Big(\frac{F}{\langle r_i = s_i \mid i \in I\rangle}\Big)^+ \cong \frac{F^+}{\langle r_i - s_i \mid i \in I\rangle}. \tag{3.37}$$

Let E be a graph (as usual we consider only graphs with no sinks) and A_E be the *adjacency matrix* $(n_{ij}) \in \mathbb{Z}^{E^0 \oplus E^0}$, where n_{ij} is the number of edges from v_i to v_j. Clearly the adjacency matrix depends on the ordering we put on E^0. We usually fix an ordering on E^0.

Multiplying the matrix $A_E^t - I$ from the left defines a homomorphism

$$\mathbb{Z}^{E^0} \longrightarrow \mathbb{Z}^{E^0},$$

where $\mathbb{Z}^{E^0}$ is the direct sum of copies of $\mathbb{Z}$ indexed by E^0. The next theorem shows that the cokernel of this map gives the Grothendieck group of Leavitt path algebras.

Theorem 3.9.1 *Let E be finite graph with no sinks and $\mathcal{L}(E)$ be the Leavitt path algebra associated with E. Then*

$$K_0(\mathcal{L}(E)) \cong \operatorname{coker}(A_E^t - I : \mathbb{Z}^{E^0} \longrightarrow \mathbb{Z}^{E^0}). \tag{3.38}$$

Proof Let M_E be the abelian monoid generated by $\{v \mid v \in E^0\}$ subject to the relations

$$v = \sum_{\{\alpha \in E^1 | s(\alpha) = v\}} r(\alpha) \tag{3.39}$$

for every $v \in E^0$. The relations (3.39) can be then written as $A_E^t \overline{v}_i = I\overline{v}_i$, where $v_i \in E^0$ and $\overline{v}_i$ is the $(0, \dots, 1, 0, \dots)$ with 1 in the ith component. Therefore,

$$M_E \cong \frac{F}{\langle A_E^t \overline{v}_i = I\,\overline{v}_i, v_i \in E^0 \rangle},$$

where F is the free abelian monoid generated by the vertices of E. By [5, Theorem 3.5] there is a natural monoid isomorphism

$$\mathcal{V}(\mathcal{L}(E)) \cong M_E.$$

So, using (3.37) we have

$$K_0(\mathcal{L}(E)) \cong \mathcal{V}(\mathcal{L}(E))^+ \cong M_E^+ \cong \frac{F^+}{\langle (A_E^t - I)\overline{v}_i, v_i \in E^0 \rangle}. \tag{3.40}$$

Now F^+ is $\mathbb{Z}^{E^0}$ and it is easy to see that the denominator in (3.40) is the image of $A_E^t - I : \mathbb{Z}^{E^0} \longrightarrow \mathbb{Z}^{E^0}$. □

Example 3.9.2 Let E be the following graph:

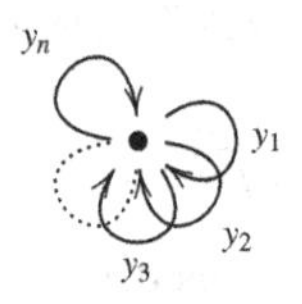

Then the Leavitt path algebra associated with E, $\mathcal{L}(E)$, is the algebra constructed by Leavitt in (1.60). By Theorem 3.9.1,

$$K_0(\mathcal{L}(E)) \cong \mathbb{Z}/(n-1)\mathbb{Z}.$$

Example 3.9.3 Here is an example of a ring R such that $K_0(R) \neq 0$ but $[R] = 0$. Let A be the Leavitt path algebra associated with the graph

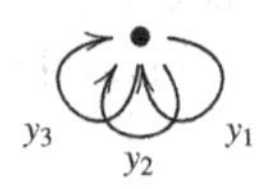

Thus as a right A-module, $A^3 \cong A$ (see Example 1.3.19). By Example 3.9.2, $K_0(A) = \mathbb{Z}/2\mathbb{Z}$ and $2[A] = 0$. The ring $R = \mathbb{M}_2(A)$ is Morita equivalent to A (using the assignment $P_{\mathbb{M}_2(A)} \mapsto P \otimes_{\mathbb{M}_2(A)} A^2$, see Proposition 2.1.1). Thus $K_0(R) \cong K_0(A) \cong \mathbb{Z}/2\mathbb{Z}$. Under this assignment, $[R] = [\mathbb{M}_2(A)]$ is sent to $2[A]$ which is zero, thus $[R] = 0$.

3.9.2 Action of $\mathbb{Z}$ on K_0^{gr} of Leavitt path algebras

Recall that Leavitt path algebras have a natural $\mathbb{Z}$-graded structure (see §1.6.4). The graded Grothendieck group as a possible invariant for these algebras was first considered in [48]. In the case of finite graphs with no sinks, there is a good description of the action of $\mathbb{Z}$ on the graded Grothendieck group which we give here.

Let E be a finite graph with no sinks. Set $\mathcal{A} = \mathcal{L}(E)$, which is a strongly $\mathbb{Z}$-graded ring by Theorem 1.6.13. For any $u \in E^0$ and $i \in \mathbb{Z}$, $u\mathcal{A}(i)$ is a graded finitely generated projective right $\mathcal{A}$-module and any graded finitely generated projective $\mathcal{A}$-module is generated by these modules up to isomorphism, *i.e.*,

$$\boxed{\mathcal{V}^{\text{gr}}(\mathcal{A}) = \big\langle [u\mathcal{A}(i)] \mid u \in E^0, i \in \mathbb{Z} \big\rangle.} \tag{3.41}$$

By §3.1, $K_0^{\text{gr}}(\mathcal{A})$ is the group completion of $\mathcal{V}^{\text{gr}}(\mathcal{A})$. The action of $\mathbb{N}[x, x^{-1}]$ on $\mathcal{V}^{\text{gr}}(\mathcal{A})$ and thus the action of $\mathbb{Z}[x, x^{-1}]$ on $K_0^{\text{gr}}(\mathcal{A})$ are defined on generators

$$x^j[u\mathcal{A}(i)] = [u\mathcal{A}(i + j)],$$

where $i, j \in \mathbb{Z}$. We first observe that for $i \geq 0$,

$$\boxed{x[u\mathcal{A}(i)] = [u\mathcal{A}(i + 1)] = \sum_{\{\alpha \in E^1 \mid s(\alpha) = u\}} [r(\alpha)\mathcal{A}(i)].} \tag{3.42}$$

First notice that for $i \geq 0$, $\mathcal{A}_{i+1} = \sum_{\alpha \in E^1} \alpha \mathcal{A}_i$. It follows that

$$u\mathcal{A}_{i+1} = \bigoplus_{\{\alpha \in E^1 \mid s(\alpha) = u\}} \alpha \mathcal{A}_i$$

as $\mathcal{A}_0$-modules. Using the fact that $\mathcal{A}_n \otimes_{\mathcal{A}_0} \mathcal{A} \cong \mathcal{A}(n)$, $n \in \mathbb{Z}$, and the fact that $\alpha\mathcal{A}_i \cong r(\alpha)\mathcal{A}_i$ as an $\mathcal{A}_0$-module, we get

$$u\mathcal{A}(i + 1) \cong \bigoplus_{\{\alpha \in E^1 \mid s(\alpha) = u\}} r(\alpha)\mathcal{A}(i)$$

as graded $\mathcal{A}$-modules. This gives (3.42).

Recall that for a Γ-graded ring A, $K_0^{\text{gr}}(A)$ is a pre-ordered abelian group with the set of isomorphic classes of graded finitely generated projective right A-modules as the positive cone of ordering, denoted by $K_0^{\text{gr}}(A)_+$ (*i.e.*, the image of $\mathcal{V}^{\text{gr}}(A)$ under the natural homomorphism $\mathcal{V}^{\text{gr}}(A) \to K_0^{\text{gr}}(A)$). Moreover, $[A]$ is an order-unit. We call the triple, $(K_0^{\text{gr}}(A), K_0^{\text{gr}}(A)_+, [A])$ the *graded dimension group* (see [40, §15] for some background on dimension groups).

In [48] it was conjectured that the graded dimension group is a complete invariant for Leavitt path algebras. Namely, for graphs E and F, $\mathcal{L}(E) \cong_{\text{gr}} \mathcal{L}(F)$

if and only if there is an order preserving $\mathbb{Z}[x, x^{-1}]$-module isomorphism

$$\phi : K_0^{\mathrm{gr}}(\mathcal{L}(E)) \to K_0^{\mathrm{gr}}(\mathcal{L}(F)) \tag{3.43}$$

such that $\phi([\mathcal{L}(E)] = \mathcal{L}(F)$.

3.9.3 K_0^{gr} of a Leavitt path algebra via its 0-component ring

In this section we calculate the graded Grothendieck group of Leavitt path algebras. For a finite graph with no sinks, $\mathcal{L}_K(E)$ is strongly graded (Theorem 1.6.13) and thus $K_0^{\mathrm{gr}}(\mathcal{L}(E)) \cong K_0(\mathcal{L}(E)_0)$ (§3.1.3). It is known that $\mathcal{L}(E)_0$ is an ultramatricial algebra, *i.e.*, the union of an increasing countable chain of a finite product of matrix algebras over the field K. We recall the description of $\mathcal{L}(E)_0$ in the setting of finite graphs with no sinks (see the proof of Theorem 5.3 in [5] for the general case). We will then use it to calculate $K_0^{\mathrm{gr}}(\mathcal{L}(E))$.

Let A_E be the adjacency matrix of E. Let $L_{0,n}$ be the K-linear span of all elements of the form pq^* with $r(p) = r(q)$ and $|p| = |q| \leq n$. Then

$$\mathcal{L}(E)_0 = \bigcup_{n=0}^{\infty} L_{0,n}, \tag{3.44}$$

with the transition inclusion

$$\begin{aligned} L_{0,n} &\longrightarrow L_{0,n+1}, \\ pq^* &\longmapsto \sum_{\{\alpha | s(\alpha)=v\}} p\alpha(q\alpha)^*, \end{aligned} \tag{3.45}$$

where $r(p) = r(q) = v$ and extended linearly. Note that since E does not have sinks, for any $v \in E_0$, the set $\{\alpha | s(\alpha) = v\}$ is not empty.

For a fixed $v \in E^0$, let $L_{0,n}^v$ be the K-linear span of all elements of the form pq^* with $|p| = |q| = n$ and $r(p) = r(q) = v$. Arrange the paths of length n with the range v in a fixed order $p_1^v, p_2^v, \ldots, p_{k_n^v}^v$, and observe that the correspondence of $p_i^v p_j^{v*}$ to the matrix unit $\mathbf{e}_{ij}$ gives rise to a ring isomorphism $L_{0,n}^v \cong \mathbb{M}_{k_n^v}(K)$. Moreover, $L_{0,n}^v$, $v \in E^0$ form a direct sum. This implies that

$$L_{0,n} \cong \bigoplus_{v \in E^0} \mathbb{M}_{k_n^v}(K),$$

where k_n^v, $v \in E^0$, is the number of paths of length n with the range v. The inclusion map $L_{0,n} \to L_{0,n+1}$ of (3.45) can be represented by

$$A_E^t : \bigoplus_{v \in E^0} \mathbb{M}_{k_n^v}(K) \longrightarrow \bigoplus_{v \in E^0} \mathbb{M}_{k_{n+1}^v}(K). \tag{3.46}$$

This means $(A_1, \dots, A_l) \in \bigoplus_{v \in E^0} \mathbb{M}_{k_n^v}(K)$ is sent to

$$\Big(\sum_{j=1}^{l} n_{j1}A_j, \dots, \sum_{j=1}^{l} n_{jl}A_j\Big) \in \bigoplus_{v \in E^0} \mathbb{M}_{k_{n+1}^v}(K),$$

where n_{ji} is the number of edges connecting v_j to v_i and

$$\sum_{j=1}^{l} k_j A_j = \begin{pmatrix} A_1 & & & & & \\ & \ddots & & & & \\ & & A_1 & & & \\ & & & \ddots & & \\ & & & & A_l & \\ & & & & & \ddots \\ & & & & & & A_l \end{pmatrix}$$

in which each matrix is repeated k_j times down the leading diagonal and if $k_j = 0$, then A_j is omitted. This shows that $\mathcal{L}(E)_0$ is an *ultramatricial algebra, i.e.,* it is isomorphic to the union of an increasing chain of a finite product of matrix algebras over a field K. These algebras will be studied in Chapter 5.

Writing $\mathcal{L}(E)_0 = \varinjlim_n L_{0,n}$, since the Grothendieck group K_0 respects the direct limit, we have

$$K_0(\mathcal{L}(E)_0) \cong \varinjlim_n K_0(L_{0,n}).$$

Since the K_0 of (Artinian) simple algebras are $\mathbb{Z}$, the ring homomorphism

$$L_{0,n} \longrightarrow L_{0,n+1}$$

induces the group homomorphism

$$\mathbb{Z}^{E^0} \xrightarrow{A_E^t} \mathbb{Z}^{E^0},$$

where $A_E^t : \mathbb{Z}^{E^0} \to \mathbb{Z}^{E^0}$ is multiplication from the left which is induced by the homomorphism (3.46).

For a finite graph E with no sinks, with n vertices and adjacency matrix A, by Theorem 1.6.13, $K_0^{\mathrm{gr}}(\mathcal{L}(E)) \cong K_0(\mathcal{L}(E)_0)$. Thus $K_0^{\mathrm{gr}}(\mathcal{L}(E))$ is the direct limit of the ordered direct system

$$\mathbb{Z}^n \xrightarrow{A^t} \mathbb{Z}^n \xrightarrow{A^t} \mathbb{Z}^n \xrightarrow{A^t} \cdots, \tag{3.47}$$

where the ordering in $\mathbb{Z}^n$ is defined point-wise.

In general, the direct limit of the system, $\varinjlim_A \mathbb{Z}^n$, where $A \in \mathbb{M}_n(\mathbb{Z})$, is an ordered group and can be described as follows. Consider the pair (a, k), where $a \in \mathbb{Z}^n$ and $k \in \mathbb{N}$, and define the equivalence relation $(a, k) \sim (b, k')$ if $A^{k''-k}a = A^{k''-k'}b$ for some $k'' \in \mathbb{N}$. Let $[a, k]$ denote the equivalence class of (a, k). Clearly $[A^n a, n+k] = [a, k]$. Then it is not difficult to show that the direct limit $\varinjlim_A \mathbb{Z}^n$ is the abelian group consists of equivalent classes $[a, k]$, $a \in \mathbb{Z}^n$, $k \in \mathbb{N}$, with addition defined by

$$[a, k] + [b, k'] = [A^{k'}a + A^k b, k + k']. \tag{3.48}$$

The positive cone of this ordered group is the set of elements $[a, k]$, where $a \in \mathbb{Z}^{+n}$, $k \in \mathbb{N}$. Moreover, there is an automorphism $\delta_A : \varinjlim_A \mathbb{Z}^n \to \varinjlim_A \mathbb{Z}^n$ defined by $\delta_A([a, k]) = [Aa, k]$.

There is another presentation for $\varinjlim_A \mathbb{Z}^n$ which is sometimes easier to work with. Consider the set

$$\Delta_A = \{\, v \in A^n\mathbb{Q}^n \mid A^k v \in \mathbb{Z}^n, \text{ for some } k \in \mathbb{N}\,\}. \tag{3.49}$$

The set Δ_A forms an ordered abelian group with the usual addition of vectors and the positive cone

$$\Delta_A^+ = \{\, v \in A^n\mathbb{Q}^n \mid A^k v \in \mathbb{Z}^{+n}, \text{ for some } k \in \mathbb{N}\,\}. \tag{3.50}$$

Moreover, there is an automorphism $\delta_A : \Delta_A \to \Delta_A$ defined by $\delta_A(v) = Av$. The map

$$\begin{aligned} \phi : \Delta_A &\to \varinjlim_A \mathbb{Z}^n \\ v &\mapsto [A^k v, k], \end{aligned} \tag{3.51}$$

where $k \in \mathbb{N}$ such that $A^k v \in \mathbb{Z}^n$, is an isomorphism which respects the action of A and the ordering, *i.e.*, $\phi(\Delta_A^+) = (\varinjlim_A \mathbb{Z}^n)^+$ and $\phi(\delta_A(v)) = \delta_A\phi(v)$.

Example 3.9.4 Let E be the following graph:

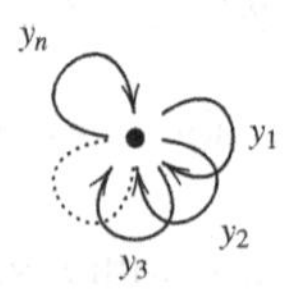

The nongraded K_0 of $\mathcal{L}(E)$ was computed in Example 3.9.2. The graph E has no sinks, and so by (3.47),

$$K_0^{\mathrm{gr}}(\mathcal{L}(E)) \cong \varinjlim \mathbb{Z},$$

of the inductive system $\mathbb{Z} \xrightarrow{n} \mathbb{Z} \xrightarrow{n} \mathbb{Z} \xrightarrow{n} \cdots$. This gives that

$$K_0^{\mathrm{gr}}(\mathcal{L}(E)) \cong \mathbb{Z}[1/n].$$

Example 3.9.5 For the graph

$$E: \quad \bullet \rightleftarrows \bullet$$

with the adjacency $A_E = \begin{pmatrix} 1 & 2 \\ 1 & 0 \end{pmatrix}$, the ring of homogeneous element of degree zero, $\mathcal{L}(E)_0$, is the direct limit of the system

$$K \oplus K \xrightarrow{A_E^t} \mathbb{M}_2(K) \oplus \mathbb{M}_2(K) \xrightarrow{A_E^t} \mathbb{M}_4(K) \oplus \mathbb{M}_4(K) \xrightarrow{A_E^t} \cdots$$

$$(a,b) \mapsto \begin{pmatrix} a & 0 \\ 0 & b \end{pmatrix} \oplus \begin{pmatrix} a & 0 \\ 0 & a \end{pmatrix}.$$

So $K_0^{\mathrm{gr}}(\mathcal{L}(E))$ is the direct limit of the direct system

$$\mathbb{Z}^2 \xrightarrow{A_E^t} \mathbb{Z}^2 \xrightarrow{A_E^t} \mathbb{Z}^2 \xrightarrow{A_E^t} \cdots.$$

Since $\det(A_E^t) = -2$, one can easily calculate that

$$K_0^{\mathrm{gr}}(\mathcal{L}(E)) \cong \mathbb{Z}[1/2] \bigoplus \mathbb{Z}[1/2].$$

Moreover, $[\mathcal{L}(E)] \in K_0^{\mathrm{gr}}(\mathcal{L}(E))$ is represented by $(1,1) \in \mathbb{Z}[1/2] \bigoplus \mathbb{Z}[1/2]$. Adopting (3.49) for the description of $K_0^{\mathrm{gr}}(\mathcal{L}(E))$, since the action of x on $K_0^{\mathrm{gr}}(\mathcal{L}(E))$ is represented by the action of A_E^t from the left, we have

$$x(a,b) = (a+b, 2a).$$

Moreover, considering (3.50) for the positive cone, $(A_E^t)^k(a,b)$ is eventually positive, if $v(a,b) > 0$, where $v = (2,1)$ is the Perron eigenvector of A_E (see [51, Lemma 7.3.8]). It follows that

$$K_0^{\mathrm{gr}}(\mathcal{L}(E))^+ = \Delta_{A_E^t}^+ = \{(a,b) \in \mathbb{Z}[1/2] \oplus \mathbb{Z}[1/2] \mid 2a + b > 0\} \cup \{(0,0)\}.$$

Example 3.9.6 REDUCED K_0^{gr} OF STRONGLY GRADED RINGS

When A is a strongly graded ring, the graded K-groups coincide with K-groups of its 0-component ring (see (3.8)). However this example shows that this is not the case for the reduced graded Grothendieck groups.

Let A be the Leavitt algebra generated by $2n$ symbols (which is associated with a graph with one vertex and n-loops) (see 1.3.19). By Theorem 1.6.13, this is a strongly graded ring. The homomorphism (3.13), *i.e.*,

$$\phi : \mathbb{Z}[\Gamma] \longrightarrow K_0^{\mathrm{gr}}(A),$$

$$\sum_\alpha n_\alpha \alpha \longmapsto \sum_\alpha n_\alpha [A(\alpha)]$$

takes the form

$$\begin{aligned} \phi : \mathbb{Z}[x, x^{-1}] &\longrightarrow K_0^{\mathrm{gr}}(A) \cong \mathbb{Z}[1/n], \\ \sum_i n_i x^i &\longmapsto \sum_i n_i [A(i)]. \end{aligned}$$

This shows that ϕ is surjective (see (3.41)) and thus $\widetilde{K_0^{\mathrm{gr}}}(A)$ is trivial. On the other hand, by Example 3.9.4, $K_0(A_0) \cong K_0^{\mathrm{gr}}(A) \cong \mathbb{Z}[1/n]$. But

$$\begin{aligned} \phi_0 : \mathbb{Z} &\longrightarrow K_0(A_0) \cong \mathbb{Z}[1/n], \\ n &\longmapsto n[A_0]. \end{aligned}$$

This shows that $\widetilde{K_0}(A_0)$ is a nontrivial torsion group $\mathbb{Z}[1/n]/\mathbb{Z}$. Thus

$$\widetilde{K_0^{\mathrm{gr}}}(A) \ncong \widetilde{K_0}(A_0).$$

Remark 3.9.7 K_0^{gr} OF WEYL ALGEBRAS

Let $A = K\langle x, y\rangle/\langle xy - yx - 1\rangle$ be a Weyl algebra, where K is an algebraically closed field of characteristic 0. By Example 1.6.4, this is a $\mathbb{Z}$-graded ring. The graded Grothendieck group of this ring is calculated in [87]. It is shown that $K_0^{\mathrm{gr}}(A) \cong \bigoplus_{\mathbb{Z}} \mathbb{Z}$ (*i.e.*, a direct sum of a countably many $\mathbb{Z}$), $\widetilde{K_0^{\mathrm{gr}}}(A) = 0$ and $K_0(A) = 0$.

3.10 G_0^{gr} of graded rings

Recall that for a Γ-graded ring A with identity, the graded Grothendieck group K_0^{gr} was defined as the group completion of the monoid (see (3.2)),

$$\mathcal{V}^{\mathrm{gr}}(A) = \{\, [P] \mid P \text{ is graded finitely generated projective A-module} \,\}.$$

If, instead of isomorphism classes of graded finitely generated projective A-modules, we consider the isomorphism classes of all graded finitely generated A-modules, the group completion of this monoid is denoted by $G_0^{\mathrm{gr}}(A)$.

If A is graded Noetherian, then the category of graded finitely generated (right) modules over A, gr-A, is an abelian category (but not necessarily the category Pgr-A). Several of the K-theory techniques work only over such categories (such as Dévissage and localisations (§6.2.1)). For this reason, it is beneficial to develop the G_0^{gr}-theory.

For a Γ-graded ring A, $G_0^{\mathrm{gr}}(A)$ can equivalently be defined as the free abelian group generated by isomorphism classes $[M]$, where M is a graded finitely generated right A-module, subject to the relation $[M] = [K] + [N]$ if there is an exact sequence

$$0 \longrightarrow K \longrightarrow M \longrightarrow N \longrightarrow 0.$$

This definition is extended to exact categories in §3.12.

3.10.1 G_0^{gr} of graded Artinian rings

The theory of composition series for modules is used to calculate the G_0 group of Artinian rings. A similar theory in the graded setting is valid and we briefly recall the concepts.

Let M be a nonzero graded right A-module. A finite chain of graded submodules of M

$$M = M_0 \supset M_1 \supset \cdots \supset M_n = 0$$

is called *a graded composition series of length n* for M if M_i/M_{i+1} is a graded simple A-module, where $0 \leq i \leq n-1$. The graded simple modules M_i/M_{i+1} are called *graded composition factors* of the series. Two graded composition series

$$M = M_0 \supset M_1 \supset \cdots \supset M_n = 0,$$
$$M = N_0 \supset N_1 \supset \cdots \supset N_p = 0$$

are called *equivalent* if $n = p$ and for a suitable permutation $\sigma \in S_n$

$$M_i/M_{i+1} \cong N_{\sigma(i)}/N_{\sigma(i)+1},$$

i.e., there is a one to one correspondence between composition factors of these two chains such that the corresponding factors are isomorphic as graded A-modules.

We also need the *graded Jordan–Hölder theorem* which is valid with a similar proof as in the nongraded case. Namely, if a graded module M has a graded composition series, then all graded composition series of M are equivalent.

If a module is graded Artinian and Noetherian, then it has a graded composition series. In particular, if A is a graded right Artinian ring, then any graded finitely generated right A-module has a graded composition series. The proofs of these statements are similar to the nongraded case (see for example [4, §11]).

Let A be a Γ-graded right Artinian ring and let V_1 be a graded right simple A-module. Suppose V_2 is a graded right simple module which is not graded isomorphic to $V_1(\gamma)$ for any $\gamma \in \Gamma$. Continuing in this fashion, using the graded Jordan–Hölder theorem one can prove that there are a finite number of graded simple modules $\{V_1, \dots, V_s\}$ such that any graded simple module is isomorphic to some shift of one of V_i. We call $\{V_1, \dots, V_s\}$ a *basic set of graded simple right A-modules.*

Theorem 3.10.1 *Let A be a graded Artinian ring and $\{V_1, \dots, V_s\}$ a basic set of graded simple right A-modules. Then*

$$G_0^{\mathrm{gr}}(A) \cong \bigoplus_{i=1}^{s} \frac{\langle\, V_i(\gamma) \mid \gamma \in \Gamma \,\rangle}{\langle\, V_i(\alpha) - V_i(\beta) \mid V_i(\alpha) \cong_{\mathrm{gr}} V_i(\beta), \alpha, \beta \in \Gamma \,\rangle}, \tag{3.52}$$

as $\mathbb{Z}[\Gamma]$-modules. Here $\langle\, V_i(\gamma) \mid \gamma \in \Gamma \,\rangle$ is the free abelian group on generators $V_i(\gamma)$ and the action of Γ on the generators defined as $\alpha.V_i(\gamma) = V_i(\alpha + \gamma)$.

Proof Let M be a graded finitely generated A-module with a graded composition series

$$M = M_0 \supset M_1 \supset \cdots \supset M_n = 0, \tag{3.53}$$

and composition factors M_i/M_{i+1}. From the exact sequences

$$0 \longrightarrow M_{i+1} \longrightarrow M_i \longrightarrow M_i/M_{i+1} \longrightarrow 0$$

we get

$$[M] = [M] - 0 = \sum_{i=0}^{n-1} ([M_i] - [M_{i+1}]) = \sum_{i=0}^{n-1} [M_i/M_{i+1}].$$

Since M_i/M_{i+1}, $0 \leq i \leq n-1$, are graded simple, we can write

$$[M] = \sum_{t=1}^{s} \sum_{\gamma \in \Gamma} r_{t,\gamma}(M)[V_t(\gamma)] \in G_0^{\mathrm{gr}}(A),$$

where $r_{t,\gamma}(M)$ is the multiplicity of $V_t(\gamma)$ in the composite factors of M (all but a finite number of which are nonzero). Since any other graded composition series is equivalent to the above composition series, the graded Jordan–Hölder theorem guarantees that $r_{t,\gamma}(M)$ are independent of the choice of the graded composition series. This defines a homomorphism from the free abelian group generated by isomorphism classes of graded finitely generated right A-modules to the right hand side of (3.52).

Consider an exact sequence of graded finitely generated modules

$$0 \to K \xrightarrow{\phi} M \xrightarrow{\psi} N \to 0$$

and the graded composition series

$$K = K_0 \supset K_1 \supset \cdots \supset K_n = 0,$$
$$N = N_0 \supset N_1 \supset \cdots \supset N_p = 0.$$

Then one obtains a graded composition series

$$M = N'_0 \supset N'_1 \supset \cdots \supset N'_p = K_0 \supset K_1 \supset \cdots \supset K_n = 0,$$

where $N'_i = \psi^{-1}(N_i)$. This shows that the homomorphism above extends to the well-defined homomorphism

$$f : G_0^{\mathrm{gr}}(A) \longrightarrow \bigoplus_{i=1}^{s} \frac{\langle V_i(\gamma) \mid \gamma \in \Gamma \rangle}{\langle V_i(\alpha) - V_i(\beta) \mid V_i(\alpha) \cong_{\mathrm{gr}} V_i(\beta) \rangle}.$$

The fact that f is group isomorphism is not difficult to observe.

To show that f is a $\mathbb{Z}[\Gamma]$-module, note that if (3.53) is a composition series for M, then clearly

$$M(\alpha) = M_0(\alpha) \supset M_1(\alpha) \supset \cdots \supset M_n(\alpha) = 0$$

is a composition series for $M(\alpha)$, where $\alpha \in \Gamma$. Since

$$M_i(\alpha)/M_{i+1}(\alpha) = M_i/M_{i+1}(\alpha),$$

it follows that

$$f(\alpha.[M]) = f([M(\alpha)]) = \sum_{t=1}^{s} \sum_{\gamma \in \Gamma} r_{t,\gamma}(M)[V_t(\gamma + \alpha)]$$

$$= \alpha. \sum_{t=1}^{s} \sum_{\gamma \in \Gamma} r_{t,\gamma}(M)[V_t(\gamma)] = \alpha f([M]).$$

Thus f is a $\mathbb{Z}[\Gamma]$-module as well. □

We obtain the graded Grothendieck group of a graded division ring (Proposition 3.7.1) as a consequence of Theorem 3.10.1.

Corollary 3.10.2 *Let A be a Γ-graded division ring with the support Γ_A. Then*

$$K_0^{\mathrm{gr}}(A) \cong \mathbb{Z}[\Gamma/\Gamma_A].$$

Proof First note that A is a graded left and right Artinian ring with $\{A\}$ as a basic set of graded simple modules. By Proposition 1.4.1, $G_0^{\mathrm{gr}}(A) = K_0^{\mathrm{gr}}(A)$. Thus by Theorem 3.10.1

$$K_0^{\mathrm{gr}}(A) \cong \frac{\langle A(\gamma) \mid \gamma \in \Gamma \rangle}{\langle A(\alpha) - A(\beta) \mid A(\alpha) \cong_{\mathrm{gr}} A(\beta) \rangle}. \tag{3.54}$$

By Corollary 1.3.17 $A(\alpha) \cong_{\mathrm{gr}} A(\beta)$ if and only if $\alpha - \beta \in \Gamma_A$. It is now easy to show that (3.54) reduces to $K_0^{\mathrm{gr}}(A) \cong \mathbb{Z}[\Gamma/\Gamma_A]$. □

3.11 Symbolic dynamics and K_0^{gr}

One of the central objects in the theory of symbolic dynamics is a *shift of finite type* (*i.e.,* a topological Markov chain). Every finite directed graph E with no sinks and sources gives rise to a shift of finite type X_E by considering the set of bi-infinite paths and the natural shift of the paths to the left. This is called an *edge shift*. Conversely, any shift of finite type is conjugate to an edge shift (for a comprehensive introduction to symbolic dynamics see [51]). Several invariants have been proposed in order to classify shifts of finite type, among them Krieger's dimension group. In this section we see that Krieger's invariant can be expressible as the graded Grothendieck group of a Leavitt path algebra.

We briefly recall the objects of our interest. Let $\mathcal{A}$ be a finite alphabet (*i.e.,* a finite set). A *full shift space* is defined as

$$\mathcal{A}^{\mathbb{Z}} := \{ (a_i)_{i\in\mathbb{Z}} \mid a_i \in \mathcal{A} \},$$

and a *shift map* $\sigma : \mathcal{A}^{\mathbb{Z}} \to \mathcal{A}^{\mathbb{Z}}$ is defined as

$$\sigma((a_i)_{i\in\mathbb{Z}}) = (a_{i+1})_{i\in\mathbb{Z}}.$$

Moreover, a *subshift* $X \subseteq \mathcal{A}^{\mathbb{Z}}$ is a closed σ-invariant subspace of $\mathcal{A}^{\mathbb{Z}}$.

Given a finite graph E (see §1.6.3 for terminologies related to graphs), a *subshift of finite type associated with E* is defined as

$$X_E := \{ (e_i)_{i\in\mathbb{Z}} \in (E^1)^{\mathbb{Z}} \mid r(e_i) = s(e_{i+1}) \}.$$

We say X_E is essential if the graph E has no sinks and sources. Moreover, X_E is called *irreducible* if the adjacency matrix A_E is irreducible. For a square nonnegative integer matrix A, we denote by X_A the subshift of finite type associated with the graph with the adjacency matrix A. Finally, two shifts of finite type X_A and X_B are called *conjugate* (or *topologically conjugate of subshifts*) and denoted by $X_A \cong X_B$, if there exists a homeomorphism $h : X_A \to X_B$ such that $\sigma_B h = h \sigma_A$.

The notion of the shift equivalence for matrices was introduced by Williams [97] (see also [51, §7]) in an attempt to provide a computable machinery for determining the conjugacy between two shifts of finite type. Two square nonnegative integer matrices A and B are called *elementary shift equivalent*, and denoted by $A \sim_{ES} B$, if there are nonnegative matrices R and S such that $A = RS$ and $B = SR$. The equivalence relation $\sim_S$ on square nonnegative integer matrices generated by elementary shift equivalence is called *strong shift equivalence*.

Example 3.11.1 Let $A = \begin{pmatrix} 1 & 2 \\ 1 & 0 \end{pmatrix}$ and $A^T = \begin{pmatrix} 1 & 1 \\ 2 & 0 \end{pmatrix}$. We show that A is strongly shift equivalent to A^T. We have

$$A = \begin{pmatrix} 1 & 1 & 0 \\ 0 & 0 & 1 \end{pmatrix} \begin{pmatrix} 1 & 1 \\ 0 & 1 \\ 1 & 0 \end{pmatrix}$$

$$\begin{pmatrix} 1 & 1 \\ 0 & 1 \\ 1 & 0 \end{pmatrix} \begin{pmatrix} 1 & 1 & 0 \\ 0 & 0 & 1 \end{pmatrix} = \begin{pmatrix} 1 & 1 & 1 \\ 0 & 0 & 1 \\ 1 & 1 & 0 \end{pmatrix} = E_1.$$

Moreover, we have

$$E_1 = \begin{pmatrix} 1 & 1 & 1 \\ 0 & 0 & 1 \\ 1 & 1 & 0 \end{pmatrix} = \begin{pmatrix} 0 & 1 & 1 \\ 1 & 0 & 0 \\ 0 & 0 & 1 \end{pmatrix} \begin{pmatrix} 0 & 0 & 1 \\ 0 & 0 & 1 \\ 1 & 1 & 0 \end{pmatrix}$$

$$\begin{pmatrix} 0 & 0 & 1 \\ 0 & 0 & 1 \\ 1 & 1 & 0 \end{pmatrix} \begin{pmatrix} 0 & 1 & 1 \\ 1 & 0 & 0 \\ 0 & 0 & 1 \end{pmatrix} = \begin{pmatrix} 0 & 0 & 1 \\ 0 & 0 & 1 \\ 1 & 1 & 1 \end{pmatrix} = E_2.$$

Finally,

$$E_2 = \begin{pmatrix} 0 & 0 & 1 \\ 0 & 0 & 1 \\ 1 & 1 & 1 \end{pmatrix} = \begin{pmatrix} 1 & 0 \\ 1 & 0 \\ 1 & 1 \end{pmatrix} \begin{pmatrix} 0 & 0 & 1 \\ 1 & 1 & 0 \end{pmatrix}$$

$$\begin{pmatrix} 0 & 0 & 1 \\ 1 & 1 & 0 \end{pmatrix} \begin{pmatrix} 1 & 0 \\ 1 & 0 \\ 1 & 1 \end{pmatrix} = \begin{pmatrix} 1 & 1 \\ 2 & 0 \end{pmatrix} = A^T.$$

This shows that

$$A \sim_{ES} E_1 \sim_{ES} E_2 \sim_{ES} A^T.$$

Thus $A \sim_S A^T$.

Besides elementary and strongly shift equivalence, there is a weaker notion, called shift equivalence, defined as follows. The nonnegative integer square matrices A and B are called *shift equivalent* if there are nonnegative matrices R and S such that $A^l = RS$ and $B^l = SR$, for some $l \in \mathbb{N}$, and $AR = RB$ and $SA = BS$. Clearly, strongly shift equivalence implies shift equivalence, but the converse, an open question for almost 20 years, does not hold [51].

Before stating Williams' main theorem, we need to recall the concept of out-splitting and in-splitting of a graph.

Definition 3.11.2 Let $E = (E^0, E^1, r, s)$ be a finite graph. For each $v \in E^0$ which is not a sink, partition $s^{-1}(v)$ into disjoint nonempty subsets $\mathcal{E}_v^1, \dots, \mathcal{E}_v^{m(v)}$, where $m(v) \geq 1$. (If v is a sink then put $m(v) = 0$.) Let $\mathcal{P}$ denote the resulting partition of E^1. We form the *out-split graph $E_s(\mathcal{P})$ from E using* $\mathcal{P}$ as follows. Let

$$E_s(\mathcal{P})^0 = \{\, v^i \mid v \in E^0, 1 \le i \le m(v) \,\} \cup \{\, v \mid m(v) = 0 \,\},$$
$$E_s(\mathcal{P})^1 = \{\, e^j \mid e \in E^1, 1 \le j \le m(r(e)) \,\} \cup \{\, e \mid m(r(e)) = 0 \,\},$$

and define $r_{E_s(\mathcal{P})}, s_{E_s(\mathcal{P})} : E_s(\mathcal{P})^1 \to E_s(\mathcal{P})^0$ for $e \in \mathcal{E}^i_{s(e)}$ and $1 \le j \le m(r(e))$ by

$$s_{E_s(\mathcal{P})}(e^j) = s(e)^i \text{ and } s_{E_s(\mathcal{P})}(e) = s(e)^i,$$
$$r_{E_s(\mathcal{P})}(e^j) = r(e)^j \text{ and } r_{E_s(\mathcal{P})}(e) = r(e).$$

Definition 3.11.3 Let $E = (E_0, E_1, r, s)$ be a finite graph. For each $v \in E^0$ which is not a source, partition the set $r^{-1}(v)$ into disjoint nonempty subsets $\mathcal{E}^v_1, \ldots, \mathcal{E}^v_{m(v)}$, $m(v) \ge 1$. (If v is a source then put $m(v) = 0$.) Let $\mathcal{P}$ denote the resulting partition of E^1. We form the *in-split graph $E_r(\mathcal{P})$ from E using* $\mathcal{P}$ as follows. Let

$$E_r(\mathcal{P})^0 = \{\, v_i \mid v \in E^0, 1 \le i \le m(v) \,\} \cup \{\, v \mid m(v) = 0 \,\},$$
$$E_r(\mathcal{P})^1 = \{\, e_j \mid e \in E^1, 1 \le j \le m(s(e)) \,\} \cup \{\, e \mid m(s(e)) = 0 \,\},$$

and define $r_{E_r(\mathcal{P})}, s_{E_r(\mathcal{P})} : E_r(\mathcal{P})^1 \to E_r(\mathcal{P})^0$ for $e \in \mathcal{E}^{r(e)}_i$ and $1 \le j \le m(s(e))$ by

$$s_{E_r(\mathcal{P})}(e_j) = s(e)_j \text{ and } s_{E_r(\mathcal{P})}(e) = s(e),$$
$$r_{E_r(\mathcal{P})}(e_j) = r(e)_i \text{ and } r_{E_r(\mathcal{P})}(e) = r(e)_i.$$

Example 3.11.4 OUT-SPLITTING OF A GRAPH

Consider the graph

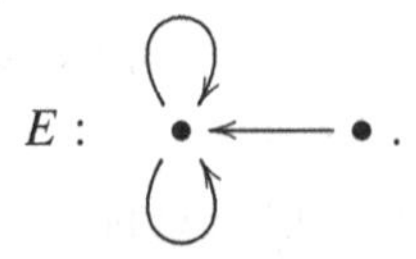

Let $\mathcal{P}$ be the partition of the edges of E containing only one edge in each partition. Then the out-split graph of E using $\mathcal{P}$ is

$E_s(\mathcal{P})$:

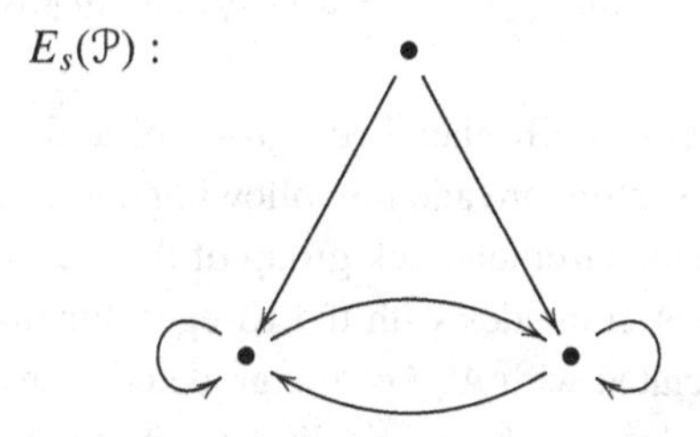

Theorem 3.11.5 (Williams [97, 51]) *Let A and B be two square nonnegative integer matrices and let E and F be two essential graphs.*

(1) *X_A is conjugate to X_B if and only if A is strongly shift equivalent to B.*
(2) *X_E is conjugate to X_F if and only if E can be obtained from F by a sequence of in/out-splittings and their converses.*

Krieger in [56] defined an invariant for classifying the irreducible shifts of finite type up to shift equivalence. Later Wagoner systematically used this invariant to relate it to higher K-groups. Surprisingly, Krieger's dimension group and Wagoner's dimension module in symbolic dynamics turn out to be expressible as the graded Grothendieck groups of Leavitt path algebras. Here, we briefly describe this relation.

In general, a nonnegative integral $n \times n$ matrix A gives rise to a stationary system. This in turn gives a direct system of order free abelian groups with A acting as an order preserving group homomorphism as follows:

$$\mathbb{Z}^n \xrightarrow{A} \mathbb{Z}^n \xrightarrow{A} \mathbb{Z}^n \xrightarrow{A} \cdots,$$

where the ordering in $\mathbb{Z}^n$ is defined point-wise (*i.e.*, the positive cone is $\mathbb{N}^n$). The direct limit of this system, $\Delta_A := \varinjlim_A \mathbb{Z}^n$ (*i.e.*, the K_0 of the stationary system) along with its positive cone, Δ^+, and the automorphism which induced by A on the direct limit, $\delta_A : \Delta_A \to \Delta_A$, is the invariant considered by Krieger, now known as *Krieger's dimension group*. Following [51], we denote this triple by $(\Delta_A, \Delta_A^+, \delta_A)$.

The following theorem was proved by Krieger ([56, Theorem 4.2], and [51, Theorem 7.5.8], see also [51, §7.5] for a detailed algebraic treatment).

Theorem 3.11.6 *Let A and B be two square nonnegative integer matrices. Then A and B are shift equivalent if and only if*

$$(\Delta_A, \Delta_A^+, \delta_A) \cong (\Delta_B, \Delta_B^+, \delta_B).$$

Wagoner noted that the induced structure on Δ_A by the automorphism δ_A makes Δ_A a $\mathbb{Z}[x, x^{-1}]$-module which was systematically used in [92, 93] (see also [25, §3]).

Recall that the graded Grothendieck group of a $\mathbb{Z}$-graded ring has a natural $\mathbb{Z}[x, x^{-1}]$-module structure and the following observation (Theorem 3.11.7) shows that the graded Grothendieck group of the Leavitt path algebra associated with a matrix A coincides with the Krieger dimension group of the shift of finite type associated with A^t, *i.e.*, the graded dimension group of a Leavitt path algebra coincides with Krieger's dimension group,

$$(K_0^{\mathrm{gr}}(\mathcal{L}(E)), (K_0^{\mathrm{gr}}(\mathcal{L}(E))^+) \cong (\Delta_{A^t}, \Delta_{A^t}^+).$$

This will provide a link between the theory of Leavitt path algebras and symbolic dynamics.

Theorem 3.11.7 *Let E be a finite graph with no sinks with the adjacency matrix A. Then there is an isomorphism $\phi : K_0^{\mathrm{gr}}(\mathcal{L}(E)) \longrightarrow \Delta_{A^t}$ such that $\phi(x\alpha) = \delta_{A^t}\phi(\alpha)$, $\alpha \in \mathcal{L}(E)$, $x \in \mathbb{Z}[x, x^{-1}]$ and $\phi(K_0^{\mathrm{gr}}(\mathcal{L}(E))^+) = \Delta_{A^t}^+$.*

Proof Since $\mathcal{L}(E)$ is strongly graded by Theorem 1.6.13, there is an ordered isomorphism

$$K_0^{\mathrm{gr}}(\mathcal{L}(E)) \to K_0(\mathcal{L}(E)_0).$$

Thus the ordered group $K_0^{\mathrm{gr}}(\mathcal{L}(E))$ coincides with the ordered group Δ_{A^t} (see (3.47)). We only need to check that their module structures are compatible. It is enough to show that the action of x on K_0^{gr} coincides with the action of A^t on $K_0(\mathcal{L}(E)_0)$, *i.e.*, $\phi(x\alpha) = \delta_{A^t}\phi(\alpha)$.

Set $\mathcal{A} = \mathcal{L}(E)$. Since graded finitely generated projective modules are generated by $u\mathcal{A}(i)$, where $u \in E^0$ and $i \in \mathbb{Z}$, it suffices to show that $\phi(x[u\mathcal{A}]) = \delta_{A^t}\phi([u\mathcal{A}])$. Since the image of $u\mathcal{A}$ in $K_0(\mathcal{A}_0)$ is $[u\mathcal{A}_0]$, and $\mathcal{A}_0 = \bigcup_{n=0}^{\infty} L_{0,n}$, (see (3.44)) using the presentation of K_0 given in (3.48), we have

$$\phi([u\mathcal{A}]) = [u\mathcal{A}_0] = [uL_{0,0}, 1] = [u, 1].$$

Thus

$$\delta_{A^t}\phi([u\mathcal{A}]) = \delta_{A^t}([u, 1]) = [A^t u, 1] = \sum_{\{\alpha \in E^1 | s(\alpha) = u\}} [r(\alpha), 1].$$

On the other hand,

$$\phi(x[u\mathcal{A}]) = \phi([u\mathcal{A}(1)]) = \phi\Big(\sum_{\{\alpha\in E^1 \mid s(\alpha)=u\}} [r(\alpha)\mathcal{A}]\Big)$$

$$= \sum_{\{\alpha\in E^1 \mid s(\alpha)=u\}} [r(\alpha)\mathcal{A}_0] = \sum_{\{\alpha\in E^1 \mid s(\alpha)=u\}} [r(\alpha)L_{0,0}, 1] = \sum_{\{\alpha\in E^1 \mid s(\alpha)=u\}} [r(\alpha), 1]. \tag{3.55}$$

Thus $\phi(x[u\mathcal{A}]) = \delta_{A^t}\phi([u\mathcal{A}])$. This finishes the proof. □

It is easy to see that two matrices A and B are shift equivalent if and only if A^t and B^t are shift equivalent. Combining this with Theorem 3.11.7 and the fact that Krieger's dimension group is a complete invariant for shift equivalent, we have the following corollary.

Corollary 3.11.8 *Let E and F be finite graphs with no sinks and A_E and A_F be their adjacency matrices, respectively. Then A_E is shift equivalent to A_F if and only if there is an order preserving $\mathbb{Z}[x, x^{-1}]$-module isomorphism $K_0^{gr}(\mathcal{L}(E)) \cong K_0^{gr}(\mathcal{L}(F))$.*

3.12 K_1^{gr}-theory

For a Γ-graded ring A, the category of finitely generated Γ-graded projective right A-modules, $\text{Pgr}^\Gamma\text{-}A$, is an exact category. Thus, using Quillen's Q construction one defines

$$K_i^{gr}(A) := K_i(\text{Pgr}^\Gamma\text{-}A), \quad i \geq 0.$$

Moreover, the shift functors (1.16) induce auto-equivalences (in fact, automorphisms) $\mathcal{T}_\alpha : \text{Pgr-}A \to \text{Pgr-}A$. These in return give a group homomorphism $\Gamma \to \text{Aut}(K_i^{gr}(A))$ or, equivalently, a $\mathbb{Z}[\Gamma]$-module on $K_i^{gr}(A)$. This general construction will be used in Chapter 6.

In this section we concretely construct the graded K_1-group, using a graded version of Bass' construction of K_1-group [13]. For a concise introduction of the (nongraded) groups K_0 and K_1 see [58].

Definition 3.12.1 (KK_0 AND K_1 OF AN EXACT CATEGORY) Let $\mathcal{P}$ be an *exact category*, *i.e.*, a full additive subcategory of an abelian category $\mathcal{A}$ such that, if

$$0 \longrightarrow P_1 \longrightarrow P \longrightarrow P_2 \longrightarrow 0$$

is an exact sequence in $\mathcal{A}$ and $P_1, P_2 \in \mathcal{P}$, then $P \in \mathcal{P}$ (*i.e.*, $\mathcal{P}$ is closed under extension). Moreover, we assume $\mathcal{P}$ has a small skeleton, *i.e.*, $\mathcal{P}$ has a full subcategory $\mathcal{P}_0$ which is small and $\mathcal{P}_0 \hookrightarrow \mathcal{P}$ is an equivalence.

The groups $K_0(\mathcal{P})$ and $K_1(\mathcal{P})$ are defined as follows.

1 $K_0(\mathcal{P})$ is the free abelian group generated by objects of $\mathcal{P}_0$, subject to the relation $[P] = [P_1] + [P_2]$ if there is an exact sequence

$$0 \longrightarrow P_1 \longrightarrow P \longrightarrow P_2 \longrightarrow 0$$

in $\mathcal{P}$.

2 $K_1(\mathcal{P})$ is the free abelian group generated by pairs (P, f), where P is an object of $\mathcal{P}_0$ and $f \in \operatorname{Aut}(P)$, subject to the relations

$$[P, f] + [P, g] = [P, fg]$$

and

$$[P, f] = [P_1, g] + [P_2, h],$$

if there is a commutative diagram in $\mathcal{P}_0$

$$\begin{array}{ccccccccc} 0 & \longrightarrow & P_1 & \xrightarrow{i} & P & \xrightarrow{\pi} & P_2 & \longrightarrow & 0 \\ & & \downarrow g & & \downarrow f & & \downarrow h & & \\ 0 & \longrightarrow & P_1 & \xrightarrow{i} & P & \xrightarrow{\pi} & P_2 & \longrightarrow & 0. \end{array}$$

Note that from the relations of K_1 it follows that

$$[P, fg] = [P, gf],$$
$$[P, fg] = [P \oplus P, f \oplus g].$$

If Γ is a group and $\mathcal{T}_\alpha : \mathcal{P} \to \mathcal{P}$, $\alpha \in \Gamma$, are auto-equivalences such that $\mathcal{T}_\beta \mathcal{T}_\alpha \cong \mathcal{T}_{\alpha+\beta}$, then $K_0(\mathcal{P})$ and $K_1(\mathcal{P})$ have a Γ-module structure.

One can easily see that $K_0^{\mathrm{gr}}(A) = K_0(\operatorname{Pgr}^\Gamma\text{-}A)$. Following Bass, one defines

$$K_1^{\mathrm{gr}}(A) = K_1(\operatorname{Pgr}^\Gamma\text{-}A).$$

Since the category Gr-A is an abelian category and Pgr-A is an exact category, the main theorems of K-theory are valid for the graded Grothendieck group, such as Dévissage, the Resolution theorem and the localisation exact sequences of K-theory (see §6.2.1, [84, Chapter 3] and [81, 95]).

Example 3.12.2 Let F be a field and $F[x_1, \dots, x_r]$ be the polynomial ring with r variables, which is considered as a $\mathbb{Z}$-grade ring with support $\mathbb{N}$, namely, $\deg(x_i) = 1$, $1 \le i \le r$. Then we prove in Chapter 6 (see Theorem 6.1.1) that

$$K_1^{\mathrm{gr}}(F[x_1, \dots, x_r]) = F^* \otimes_{\mathbb{Z}} \mathbb{Z}[x, x^{-1}].$$

Example 3.12.3 K_1^{gr} OF GRADED DIVISION ALGEBRAS

Let A be a Γ-graded division ring. Then A_0 is a division ring and $\Omega := \Gamma_A$ is a group (§1.1.4). By (3.9), for $i = 1$, and the description of K_1-group of division ring due to Dieudonné ([35, § 20]), we have

$$K_1^{\Gamma}(A) \cong \bigoplus_{\Gamma/\Omega} K_1^{\Omega}(A) \cong \bigoplus_{\Gamma/\Omega} K_1(A_0) = \bigoplus_{\Gamma/\Omega} A_0^*/[A_0^*, A_0^*], \tag{3.56}$$

where A_0^* is the group of invertible elements of division ring A_0 and $[A_0^*, A_0^*]$ is the multiplicative commutator subgroup. Representing $\bigoplus_{\Gamma/\Omega} K_1(A_0)$ as the additive group of the group ring $K_1(A_0)[\Gamma/\Omega]$ (see Example 3.1.11), the action of Γ can be described as follows: for $\beta \in \Gamma$,

$$\beta\Big(\bigoplus_{\Omega+\alpha\in\Gamma/\Omega} K_1(A_0)(\Omega + \alpha)\Big) = \bigoplus_{\Omega+\alpha\in\Gamma/\Omega} K_1(A_0)(\Omega + \alpha + \beta).$$

As a concrete example, let $A = K[x^n, x^{-n}]$ be a $\mathbb{Z}$-graded field with $\Gamma_A = n\mathbb{Z}$. Then by (3.56),

$$K_1^{gr}(A) \cong \bigoplus_n K^*$$

is a $\mathbb{Z}[x, x^{-1}]$-module. The action of x on $(a_1, \dots, a_n) \in \bigoplus_n K^*$ is

$$x(a_1, \dots, a_n) = (a_n, a_1, \dots, a_{n-1}).$$

Compare this with the computation of $K_0^{gr}(A)$ in Example 3.7.3(2).

Remark 3.12.4 THE MATRIX DESCRIPTION OF THE K_1^{gr}-GROUP

Let A be a strongly Γ-graded ring. Then the map

$$\begin{aligned} K_1^{gr}(A) &\longrightarrow K_1(A_0), \\ [P, f] &\longmapsto [P_0, f_0] \end{aligned} \tag{3.57}$$

is an isomorphism of groups (see §1.5). Here, for a graded isomorphism $f : P \to P$, we denote by f_α the restriction of f to P_α, where $\alpha \in \Gamma$. Note that $K_1^{gr}(A)$ is a $\mathbb{Z}[\Gamma]$-graded module, where the action of Γ on the generators is defined by $\alpha[P, f] = [P(\alpha), f]$.

Since $K_1(A_0)$ has a matrix description ([84, Theorem 3.1.7]), from (3.57) we get a matrix representation

$$K_1^{gr}(A) \cong K_1(A_0) \cong \mathrm{GL}(A_0)/E(A_0).$$

We don't know whether, for an arbitrary graded ring, one can give a matrix description for $K_1^{gr}(A)$. For some work in this direction see [100].

4
Graded Picard groups

Let A be a commutative ring. If M is a finitely generated projective A-module of constant rank 1, then there is an A-module N such that $M \otimes_A N \cong A$. In fact this is an equivalent condition. The module M above is called an *invertible module*. The isomorphism classes of invertible modules with the tensor product form an abelian group, denoted by $\mathrm{Pic}(A)$ and called the *Picard group* of A. On the other hand, since A is commutative, $K_0(A)$ is a ring with multiplication defined by the tensor product and one can prove that there is an exact sequence

$$1 \longrightarrow \mathrm{Pic}(A) \stackrel{\phi}{\longrightarrow} K_0(A)^*, \tag{4.1}$$

where $K_0(A)^*$ is the group of invertible elements of $K_0(A)$ and $\phi([A]) = [A]$.

When A is a graded commutative ring, a parallel construction, using graded modules, gives the graded Picard group $\mathrm{Pic}^{\mathrm{gr}}(A)$. As one expects when A is a strongly graded commutative ring, using Dade's Theorem 1.5.1 one immediately gets

$$\mathrm{Pic}^{\mathrm{gr}}(A) \cong \mathrm{Pic}(A_0). \tag{4.2}$$

However, when A is a noncommutative ring, the situation is substantially more involved. One needs to define the invertible bimodules in order to define the Picard group. Moreover, since for bimodules M and N, $M \otimes_A N$ is not necessarily isomorphic to $N \otimes_A M$ as bimodules, the Picard group is not abelian. Even when A is strongly graded, the identities such as (4.2) do not hold in the noncommutative setting (see [43]).

By Morita theory, an auto-equivalence of Mod-A gives rise to an invertible A-bimodule. This shows that the isomorphism classes of auto-equivalence of Mod-A under composition form a group which is isomorphic to $\mathrm{Pic}(A)$ ([61, Theorem 18.29 and Corollary 18.29]). In this chapter we study the graded version of the Picard group.

4.1 $\mathrm{Pic}^{\mathrm{gr}}$ of a graded commutative ring

Let A be a commutative Γ-graded ring. A graded A-module P is called a *graded invertible* module if there is a graded module Q such that $P \otimes_A Q \cong_{\mathrm{gr}} A$ as graded A-modules. It is clear that if P is a graded invertible module, so is $P(\alpha)$ for any $\alpha \in \Gamma$. The *graded Picard group*, $\mathrm{Pic}^{\mathrm{gr}}(A)$, is defined as the set of graded isomorphism classes of graded invertible A-modules with tensor product as multiplication. This is a well-defined binary operation and makes $\mathrm{Pic}^{\mathrm{gr}}(A)$ an abelian group with the isomorphism class of A as an identity element. Since a graded invertible module is an invertible module, we have a group homomorphism

$$\begin{aligned}\mathrm{Pic}^{\mathrm{gr}}(A) &\longrightarrow \mathrm{Pic}(A),\\ [P] &\longmapsto [P],\end{aligned}$$

where $[P]$ represents the isomorphism class of P in either group.

Since for any $\alpha, \beta \in \Gamma$, $A(\alpha) \otimes_A A(\beta) \cong_{\mathrm{gr}} A(\alpha + \beta)$ (see §1.2.6), the map

$$\begin{aligned}\phi : \Gamma &\longrightarrow \mathrm{Pic}^{\mathrm{gr}}(A),\\ \alpha &\longmapsto [A(\alpha)]\end{aligned} \tag{4.3}$$

is a group homomorphism. We use this map in the next lemma to calculate the graded Picard group of graded fields.

For a Γ-graded ring A, recall from §1.1.3 that Γ_A is the support of A and $\Gamma_A^* = \{\alpha \in \Gamma \mid A_\alpha^* \neq \varnothing\}$. Moreover, for a graded field A, Γ_A is a subgroup of Γ.

Proposition 4.1.1 *Let A be a Γ-graded field with support Γ_A. Then*

$$\mathrm{Pic}^{\mathrm{gr}}(A) \cong \Gamma/\Gamma_A.$$

Proof Consider the map $\phi : \Gamma \to \mathrm{Pic}^{\mathrm{gr}}(A)$ from (4.3). Since any graded invertible module is graded projective and graded projective modules over graded fields are graded free (Proposition 1.4.1), it follows that the graded invertible modules must be of the form $A(\alpha)$ for some $\alpha \in \Gamma$. This shows that ϕ is an epimorphism. Now if $\phi(\alpha) = [A(\alpha)] = [A]$, then $A(\alpha) \cong_{\mathrm{gr}} A$, which implies by Corollary 1.3.17 that $\alpha \in \Gamma_A$. Conversely, if $\alpha \in \Gamma_A$, then there is an element of degree α, which has to be invertible, as A is a graded field. Thus $A(\alpha) \cong_{\mathrm{gr}} A$, again by Corollary 1.3.17. This shows that the kernel of ϕ is Γ_A. This completes the proof. □

The graded Grothendieck group of a graded local ring was determined in Proposition 3.8.3. Here we determine its graded Picard group.

Proposition 4.1.2 *Let A be a commutative Γ-graded local ring with support Γ_A. Then*

$$\operatorname{Pic}^{\mathrm{gr}}(A) \cong \Gamma/\Gamma_A^*.$$

Proof Let M be the unique graded maximal ideal of A. By Lemma 3.8.1, if for graded projective A-modules P and Q, $\overline{P} = P/PM$ is isomorphic to $\overline{Q} = Q/QM$ as graded A/M-modules, then P is isomorphic to Q as graded A-modules. This immediately implies that the natural map

$$\begin{aligned} \phi : \operatorname{Pic}^{\mathrm{gr}}(A) &\longrightarrow \operatorname{Pic}^{\mathrm{gr}}(A/M), \\ [P] &\longmapsto [\overline{P}] \end{aligned}$$

is a monomorphism. But by the proof of Proposition 4.1.1, any graded invertible A/M-module is of the form $(A/M)(\alpha)$ for some $\alpha \in \Gamma$. Since

$$\phi([A(\alpha)]) = [(A/M)(\alpha)],$$

ϕ is an isomorphism. Since $\Gamma_{A/M} = \Gamma_A^*$, by Proposition 4.1.1, $\operatorname{Pic}^{\mathrm{gr}}(A/M) = \Gamma/\Gamma_A^*$. This completes the proof. □

In the next theorem (Theorem 4.1.5) we will be using the calculus of exterior algebras in the graded setting. Recall that if M is an A-module, the nth *exterior power* of M is the quotient of the tensor product of n copies of M over A, denoted by $\bigotimes^n M$ (or $T_n(M)$ as in Example 1.1.3), by the submodule generated by $m_1 \otimes \cdots \otimes m_n$, where $m_i = m_j$ for some $1 \leq i \neq j \leq n$. The nth exterior power of M is denoted by $\bigwedge^n M$. We set $\bigwedge^0 M = A$ and clearly $\bigwedge^1 M = M$.

If A is a commutative Γ-graded ring and M is a graded A-module, then $\bigotimes^n M$ is a graded A-module (§1.2.6) and the submodule generated by $m_1 \otimes \cdots \otimes m_n$, where $m_i = m_j$ for some $1 \leq i \neq j \leq n$, coincides with the submodule generated by $m_1 \otimes \cdots \otimes m_n$, where all m_i are homogeneous and $m_i = m_j$ for some $1 \leq i \neq j \leq n$, and $m_1 \otimes \cdots \otimes m_n + m_1' \otimes \cdots \otimes m_n'$, where all m_i and m_i' are homogeneous, $m_i = m_i'$ for all $1 \leq i \leq n$ except for two indices i, j, where $i \neq j$ and $m_i = m_j'$ and $m_j = m_i'$. Thus this submodule is a graded submodule of $\bigotimes^n M$ and therefore $\bigwedge^n M$ is a graded A-module as well. We will use the isomorphism

$$\bigwedge^n (M \oplus N) \cong_{\mathrm{gr}} \bigoplus_{r=0}^{n} \Big(\bigwedge^r M \otimes_A \bigwedge^{n-r} N\Big), \tag{4.4}$$

which is valid in the nongraded setting, and one also checks that it respects the grading.

Lemma 4.1.3 *Let A be a commutative Γ-graded ring. Then*

$$\bigwedge^n A^m(\alpha_1, \dots, \alpha_m) \cong_{\mathrm{gr}} \bigoplus_{1 \leq i_1 < i_2 < \cdots < i_n \leq m} A(\alpha_{i_1} + \alpha_{i_2} + \cdots + \alpha_{i_n}). \tag{4.5}$$

Proof We prove the lemma by induction on m. For $m = 1$ and $n = 1$ we clearly have $\bigwedge^1 A(\alpha_1) \cong_{\mathrm{gr}} A(\alpha_1)$. For $n \geq 2$, since $\bigwedge^2 A = 0$ it follows that $\bigwedge^n A(\alpha) = 0$. This shows that (4.5) is valid for $m = 1$. Now by induction, by (4.4), since $\bigwedge^n A(\alpha) = 0$ for any $\alpha \in \Gamma$ and $n \geq 2$, we have

$$\begin{aligned}
\bigwedge^n A^m(\alpha_1, \dots, \alpha_m) &\cong_{\mathrm{gr}} \bigwedge^n \big(A(\alpha_1) \oplus A^{m-1}(\alpha_2, \dots, \alpha_m)\big) \\
&\cong_{\mathrm{gr}} A \otimes_A \bigwedge^n A^{m-1}(\alpha_2, \dots, \alpha_m) \bigoplus A(\alpha_1) \otimes_A \bigwedge^{n-1} A^{m-1}(\alpha_2, \dots, \alpha_m) \\
&\cong_{\mathrm{gr}} \bigoplus_{2 \leq i_1 < i_2 < \cdots < i_n \leq m} A(\alpha_{i_1} + \alpha_{i_2} + \cdots + \alpha_{i_n}) \\
&\qquad \bigoplus \Big(A(\alpha_1) \otimes_A \bigoplus_{2 \leq i_1 < i_2 < \cdots < i_{n-1} \leq m} A(\alpha_{i_1} + \alpha_{i_2} + \cdots + \alpha_{i_n})\Big) \\
&\cong_{\mathrm{gr}} \bigoplus_{2 \leq i_1 < i_2 < \cdots < i_n \leq m} A(\alpha_{i_1} + \alpha_{i_2} + \cdots + \alpha_{i_n}) \\
&\qquad \bigoplus_{2 \leq i_1 < i_2 < \cdots < i_{n-1} \leq m} A(\alpha_1 + \alpha_{i_1} + \alpha_{i_2} + \cdots + \alpha_{i_n}) \\
&\cong_{\mathrm{gr}} \bigoplus_{1 \leq i_1 < i_2 < \cdots < i_n \leq m} A(\alpha_{i_1} + \alpha_{i_2} + \cdots + \alpha_{i_n}).
\end{aligned}$$

□

The following corollary is immediate and will be used in Theorem 4.1.5.

Corollary 4.1.4 *Let A be a commutative Γ-graded ring. Then*

$$\bigwedge^n A^n(\alpha_1, \dots, \alpha_n) \cong_{\mathrm{gr}} A(\alpha_1 + \alpha_2 + \cdots + \alpha_n).$$

Recall from §3.1.5 that if A is a Γ-graded commutative ring, then $K_0^{\mathrm{gr}}(A)$ is a commutative ring. Denote by $K_0^{\mathrm{gr}}(A)^*$ the group of invertible elements of this ring. The following theorem establishes the graded version of the exact sequence (4.1).

Theorem 4.1.5 *Let A be a commutative Γ-graded ring. Then there is an exact sequence*

$$1 \longrightarrow \operatorname{Pic}^{\mathrm{gr}}(A) \xrightarrow{\phi} K_0^{\mathrm{gr}}(A)^*,$$

where $\phi([P]) = [P]$.

Proof It is clear that $\phi : \operatorname{Pic}^{\mathrm{gr}}(A) \to K_0^{\mathrm{gr}}(A)^*$ is a well-defined group homomorphism. We only need to show that ϕ is injective. Suppose $\phi([P]) = \phi([Q])$. Thus $[P] = [Q]$ in $K_0^{\mathrm{gr}}(A)$. By Lemma 3.1.7, $P \oplus A^n(\overline{\alpha}) \cong_{\mathrm{gr}} Q \oplus A^n(\overline{\alpha})$, where $\overline{\alpha} = (\alpha_1, \dots, \alpha_n)$. Since P and Q are graded invertible, they are in particular invertible, and so are of constant rank 1. Thus $\bigwedge^n(P) = 0$, for $n \geq 2$. Now by (4.4)

$$\begin{aligned}
\bigwedge^{n+1}(P \oplus A^n(\overline{\alpha})) &\cong_{\mathrm{gr}} \bigoplus_{i+j=n+1} \bigwedge^{i} A^n(\overline{\alpha}) \otimes \bigwedge^{j} P \\
&\cong_{\mathrm{gr}} \bigwedge^{n} A^n(\overline{\alpha}) \otimes \bigwedge^{1} P \\
&\cong_{\mathrm{gr}} A(\alpha_1 + \cdots + \alpha_n) \otimes P \cong_{\mathrm{gr}} P(\alpha_1 + \cdots + \alpha_n),
\end{aligned}$$

thanks to Corollary 4.1.4. Similarly

$$\bigwedge^{n+1}(Q \oplus A^n(\overline{\alpha})) \cong_{\mathrm{gr}} Q(\alpha_1 + \cdots + \alpha_n).$$

Thus

$$P(\alpha_1 + \cdots + \alpha_n) \cong_{\mathrm{gr}} Q(\alpha_1 + \cdots + \alpha_n),$$

which implies $P \cong_{\mathrm{gr}} Q$. So ϕ is an injective map. □

4.2 $\operatorname{Pic}^{\mathrm{gr}}$ of a graded noncommutative ring

When A is a noncommutative ring, the definition of the (graded) Picard group is more involved (see [13, Chapter 2], [16], [31, §55], and [38] for nongraded Picard groups of noncommutative rings). Note that if P and Q are $A-A$-bimodules then $P \otimes_A Q$ is not necessarily isomorphic to $Q \otimes_A P$ as A-bimodules. This is an indication that the Picard group, in this setting, is not necessarily an abelian group.

Let A and B be Γ-graded rings and P be a graded $A-B$-bimodule. Then P is called a *graded invertible* A–B-bimodule, if there is a graded B–A-bimodule Q

such that $P \otimes_B Q \cong_{\mathrm{gr}} A$ as A–A-bimodules and $Q \otimes_A P \cong_{\mathrm{gr}} B$ as B–B-bimodules and the following diagrams are commutative.

$$\begin{array}{ccc} P \otimes_B Q \otimes_A P & \longrightarrow & A \otimes_A P \\ \downarrow & & \downarrow \\ P \otimes_B B & \longrightarrow & P \end{array} \qquad \begin{array}{ccc} Q \otimes_A P \otimes_B Q & \longrightarrow & B \otimes_B Q \\ \downarrow & & \downarrow \\ Q \otimes_A A & \longrightarrow & Q. \end{array} \tag{4.6}$$

As in the commutative case (§4.1), for a noncommutative graded ring A, the *graded Picard group*, $\mathrm{Pic}^{\mathrm{gr}}(A)$, is defined as the set of graded isomorphism classes of graded invertible $A-A$-bimodules with tensor product as multiplications. The graded isomorphism class of the graded bimodule P is denoted by $[P]$. Since P is invertible, it has an inverse $[Q]$ in $\mathrm{Pic}^{\mathrm{gr}}(A)$ and Diagram (4.6) guarantees that $([P][Q])[P] = [P]([Q][P]) = [P]$.

Theorem 4.2.1 *Let A and B be Γ-graded rings. If A is graded Morita equivalent to B, then*

$$\mathrm{Pic}^{\mathrm{gr}}(A) \cong \mathrm{Pic}^{\mathrm{gr}}(B).$$

Proof Let $\text{Gr-}A \approx_{\mathrm{gr}} \text{Gr-}B$. Then there is a graded equivalence $\phi : \text{Gr-}A \to \text{Gr-}B$ with an inverse $\psi : \text{Gr-}B \to \text{Gr-}A$. By Theorem 2.3.7 (and its proof), $\psi(B) = P$ is a graded $B-A$-bimodule, $\phi \cong - \otimes_A P^*$ and $\psi \cong - \otimes_B P$. Now one can easily check that the map

$$\begin{aligned} \mathrm{Pic}^{\mathrm{gr}}(A) &\longrightarrow \mathrm{Pic}^{\mathrm{gr}}(B), \\ [M] &\longmapsto [P \otimes_A M \otimes_A P^*] \end{aligned}$$

is an isomorphism of groups. □

In fact, in view of the fact that the Picard group coincides with the group of isomorphism classes of auto-equivalences of a given module category, Theorem 4.2.1 can be established directly.

Let A be a Γ-graded ring. Consider the group $\mathrm{Aut}^{\mathrm{gr}}(A)$ of all the graded automorphisms and its subgroup $\mathrm{Inn}^{\mathrm{gr}}_{A_0}(A)$ which consists of graded inner automorphisms induced by the homogeneous elements of degree zero as follows:

$$\mathrm{Inn}^{\mathrm{gr}}_{A_0}(A) = \{ f : A \longrightarrow A \mid f(x) = uxu^{-1}, u \in A_0^* \}.$$

The graded A-bimodule structure on A induced by $f, g \in \mathrm{Aut}^{\mathrm{gr}}(A)$ will be denoted by ${}_fA_g$. Namely, A acts on ${}_fA_g$ as follows: $a.x.b = f(a)xg(b)$. For

$f, g, h \in \mathrm{Aut}^{\mathrm{gr}}(A)$, one can prove the following graded A-bimodule isomorphisms:

$$\begin{aligned} {}_fA_g &\cong_{\mathrm{gr}} {}_{hf}A_{hg}, \\ {}_fA_1 \otimes_A {}_gA_1 &\cong_{\mathrm{gr}} {}_{fg}A_1, \\ {}_fA_1 \otimes_A {}_1A_f &\cong_{\mathrm{gr}} {}_1A_f \otimes_A {}_fA_1 \cong_{\mathrm{gr}} A. \end{aligned} \tag{4.7}$$

The following theorem provides two exact sequences between the groups Γ, $\mathrm{Inn}^{\mathrm{gr}}_{A_0}(A)$ and $\mathrm{Aut}^{\mathrm{gr}}(A)$ and the graded Picard group. The first sequence belongs to the graded setting, whereas the second sequence is a graded version of a similar result in the nongraded setting (see [31, Theorem 55.11]).

Recall that $C(A)$ stands for the centre of the ring A which is a graded subring of A (when the grade group Γ is abelian, see Example 1.1.25). Moreover, it is easy to see that the map

$$\begin{aligned} \phi : \Gamma &\longrightarrow \mathrm{Pic}^{\mathrm{gr}}(A), \\ \alpha &\longmapsto [A(\alpha)] \end{aligned} \tag{4.8}$$

(which was considered in the case of commutative graded rings in (4.3)) is well-defined and is a homomorphism.

Theorem 4.2.2 *Let A be a Γ-graded ring. Then the following sequences are exact:*

(1) *the sequence* $1 \longrightarrow \Gamma^*_{C(A)} \longrightarrow \Gamma \stackrel{\phi}{\longrightarrow} \mathrm{Pic}^{\mathrm{gr}}(A)$, *where* $\phi(\alpha) = [A(\alpha)]$.

(2) *the sequence* $1 \longrightarrow \mathrm{Inn}^{\mathrm{gr}}_{A_0}(A) \longrightarrow \mathrm{Aut}^{\mathrm{gr}}(A) \stackrel{\phi}{\longrightarrow} \mathrm{Pic}^{\mathrm{gr}}(A)$, *where* $\phi(f) = [{}_fA_1]$.

Proof (1) Consider the group homomorphism ϕ defined in (4.8). If $u \in C(A)^*$ is a homogeneous element of degree α, then the map $\psi : A \to A(\alpha)$, $a \mapsto ua$, is a graded A-bimodule isomorphism. This shows that $\Gamma^*_{C(A)} \subseteq \ker(\phi)$. On the other hand, if $\phi(\alpha) = [A(\alpha)] = 1_{\mathrm{Pic}^{\mathrm{gr}}(A)} = [A]$, then there is a graded A-bimodule isomorphism $\psi : A \to A(\alpha)$. From this it follows that there is an invertible homogeneous element $u \in C(A)$ of degree α such that $\psi(x) = ux$ (see also Corollary 1.3.17 and Proposition 1.3.16). This completes the proof.

(2) The fact that the map ϕ is well-defined and is a homomorphism follows from (4.7). We only need to show that $\ker\phi$ coincides with $\mathrm{Inn}^{\mathrm{gr}}_{A_0}(A)$. Let $f \in \mathrm{Inn}^{\mathrm{gr}}_{A_0}(A)$, so that $f(x) = uxu^{-1}$, for any $x \in A$, where $u \in A_0^*$. Then the A-graded bimodules ${}_fA_1$ and A are isomorphic. Indeed, for the map

$$\begin{aligned} \theta : {}_fA_1 &\longrightarrow A, \\ x &\longmapsto u^{-1}x \end{aligned}$$

we have

$$\theta(a.x.b) = \theta(uau^{-1}xb) = au^{-1}xb = a.\theta(x).b,$$

which gives a graded A-bimodule isomorphisms. This shows that

$$\mathrm{Inn}^{\mathrm{gr}}_{A_0}(A) \subseteq \ker\phi.$$

Conversely, suppose $\phi(f) = [{}_fA_1] = [A]$. Thus there is a graded A-bimodule isomorphism $\theta : {}_fA_1 \to A$ such that

$$\theta(a.x.b) = \theta(f(a)xb) = a\theta(x)b. \tag{4.9}$$

Since θ is bijective, there is a $u \in A$ such that $\theta(u) = 1$. Also, since θ is graded and $1 \in A_0$ it follows that $u \in A_0$. But $A = \theta(A) = \theta(A1) = Au$ and similarly $A = uA$. This implies $u \in A_0^*$. Plugging $x = a = 1$ in (4.9), we have $\theta(b) = ub$, for any $b \in A$. Using this identity, by plugging $x = b = 1$ in (4.9), we obtain $uf(a) = au$. Since u is invertible, we get $f(a) = u^{-1}au$, so $f \in \mathrm{Inn}^{\mathrm{gr}}_{A_0}(A)$ and we are done. □

For a graded noncommutative ring A the problem of determining the Picard group $\mathrm{Pic}(A)$ is difficult. There are several cases in the literature where $\mathrm{Picent}(A) := \mathrm{Pic}_R(A)$ is determined. Here R is a centre of A and $\mathrm{Pic}_R(A)$ is a subgroup of $\mathrm{Pic}(A)$ consisting of all isomorphism classes of invertible bimodules P of A which are centralised by R, (*i.e.*, $rp = pr$, for all $r \in R$ and $p \in P$) *i.e.*, P is a $A \otimes_R A^{\mathrm{op}}$-left module (see [38] and [31, §55]). For a graded division algebra A (*i.e.*, a graded division ring which is finite dimensional over its centre) we will be able to determine $\mathrm{Pic}^{\mathrm{gr}}(A)$. In fact, since graded division algebras are graded Azumaya algebras, we first determine $\mathrm{Pic}^{\mathrm{gr}}_R(A)$, where A is a graded Azumaya algebra over its centre R. The approach follows the idea in Bass [16, §3, Corollary 4.5], where it is shown that for an Azumaya algebra A over R, the Picard group $\mathrm{Pic}_R(A)$ coincides with $\mathrm{Pic}(R)$.

Recall that a graded algebra A over a commutative graded ring R is called a *graded Azumaya algebra* if A is a graded faithfully projective R-module and the natural map

$$\phi : A \otimes_R A^{\mathrm{op}} \longrightarrow \mathrm{End}_R(A),$$
$$a \otimes b \longmapsto \phi(a \otimes b)(x) = axb,$$

where $a, x \in A$ and $b \in A^{\mathrm{op}}$, is a graded R-isomorphism. This implies that A is an Azumaya algebra over R. Conversely, if A is a graded algebra over graded ring R and A is an Azumaya algebra over R, then A is faithfully projective as an R-module and the natural map $\phi : A \otimes_R A^{\mathrm{op}} \to \mathrm{End}_R(A)$ is naturally graded. Thus A is also a graded Azumaya algebra.

A graded division algebra is a graded central simple algebra. We first define these types of ring and show that they are graded Azumaya algebras.

A graded algebra A over a graded commutative ring R is said to be a *graded central simple algebra* over R if A is graded simple ring, *i.e.*, A does not have any proper graded two-sided ideals, $C(A) \cong_{\mathrm{gr}} R$ and A is finite dimensional as an R-module. Note that since A is graded simple, the centre of A (identified with R), is a graded field. Thus A is graded free as a graded module over its centre by Proposition 1.4.1, so the dimension of A over R is uniquely defined.

Proposition 4.2.3 *Let A be a Γ-graded central simple algebra over a graded field R. Then A is a graded Azumaya algebra over R.*

Proof Since A is graded free of finite dimension over R, it follows that A is faithfully projective over R. Consider the natural graded R-algebra homomorphism $\psi : A \otimes_R A^{\mathrm{op}} \to \operatorname{End}_R(A)$ defined by $\psi(a \otimes b)(x) = axb$ where $a, x \in A$, $b \in A^{\mathrm{op}}$. Since graded ideals of A^{op} coincide with graded ideals of A, A^{op} is also graded simple. Thus $A \otimes A^{\mathrm{op}}$ is also graded simple (see [90, Chapter 2], so ψ is injective. Hence the map is surjective by dimension count, using Theorem 1.4.3. This shows that A is an Azumaya algebra over R, as required. □

Let A be a graded algebra over a graded commutative ring R. For a graded A-bimodule P centralised by R, define

$$P^A = \{\, p \in P \mid ap = pa, \text{ for all } a \in A \,\}.$$

Denote by $\text{Gr-}A_R\text{-Gr}$ the category of graded A-bimodules centralised by R. When A is a graded Azumaya algebra, the functors

$$\begin{aligned} \text{Gr-}A_R\text{-Gr} &\longrightarrow \text{Gr-}R, & \qquad \text{Gr-}R &\longrightarrow \text{Gr-}A_R\text{-Gr}, \\ P &\longmapsto P^A & N &\longmapsto N \otimes_R A \end{aligned}$$

are inverse equivalences of categories for which graded projective modules correspond to graded projective modules and, moreover, invertible modules correspond to invertible modules (see [26, Proposition III.4.1] and [55, Theorem 5.1.1] for the nongraded version). This immediately implies that

$$\operatorname{Pic}_R^{\mathrm{gr}}(A) \cong \operatorname{Pic}^{\mathrm{gr}}(R). \tag{4.10}$$

Lemma 4.2.4 *Let A be a Γ-graded division algebra with centre R. Then*

$$\operatorname{Pic}_R^{\mathrm{gr}}(A) \cong \Gamma/\Gamma_R.$$

Proof By Proposition 4.2.3, A is a graded Azumaya algebra over R. By (4.10), $\operatorname{Pic}_R^{\mathrm{gr}}(A) \cong \operatorname{Pic}^{\mathrm{gr}}(R)$. Since R is a graded field, by Proposition 4.1.1, $\operatorname{Pic}^{\mathrm{gr}}(R) = \Gamma/\Gamma_R$. □

Example 4.2.5 Recall from Example 1.1.20 that the quaternion algebra $\mathbb{H} = \mathbb{R} \oplus \mathbb{R}i \oplus \mathbb{R}j \oplus \mathbb{R}k$ is a graded division algebra having two different gradings, *i.e.*, $\mathbb{Z}_2$ and $\mathbb{Z}_2 \times \mathbb{Z}_2$, respectively. Since the centre $\mathbb{R}$ is concentrated in degree 0 in either grading, by Lemma 4.2.4, $\mathrm{Pic}^{\mathrm{gr}}_{\mathbb{R}}(\mathbb{H}) = \mathbb{Z}_2$ or $\mathrm{Pic}^{\mathrm{gr}}_{\mathbb{R}}(\mathbb{H}) = \mathbb{Z}_2 \oplus \mathbb{Z}_2$, depending on the grading. This also shows that, in contrast to the graded Grothendieck group, $\mathrm{Pic}^{\mathrm{gr}}_{\mathbb{R}}(\mathbb{H}) \not\cong \mathrm{Pic}(\mathbb{H}_0)$, although $\mathbb{H}$ is a strongly graded ring in either grading.

Example 4.2.6 Let (D, v) be a tame valued division algebra over a henselian field F, where $v : D^* \to \Gamma$ is the valuation homomorphism. By Example 1.4.7, there is a Γ-graded division algebra $\mathrm{gr}(D)$ associated with D with the centre $\mathrm{gr}(F)$, where $\Gamma_{\mathrm{gr}(D)} = \Gamma_D$ and $\Gamma_{\mathrm{gr}(F)} = \Gamma_F$. Since by Lemma 4.2.3 graded division algebras are graded Azumaya, by Lemma 4.2.4,

$$\mathrm{Pic}^{\mathrm{gr}}_{\mathrm{gr}(F)}(\mathrm{gr}(D)) = \Gamma/\Gamma_F.$$

Let A be a strongly graded Γ-graded ring. Then for any $\alpha \in \Gamma$, A_α is a finitely generated projective invertible A_0-bimodule (see Theorem 1.5.12) and the map

$$\begin{aligned} \psi : \Gamma &\longrightarrow \mathrm{Pic}(A_0), \\ \alpha &\longmapsto [A_\alpha] \end{aligned} \tag{4.11}$$

is a group homomorphism. We then have a natural commutative diagram (see Theorem 4.2.2)

$$\begin{array}{ccccccc} 1 & \longrightarrow & C(A) \cap A^*_\alpha & \longrightarrow & \Gamma & \xrightarrow{\phi} & \mathrm{Pic}^{\mathrm{gr}}(A) \\ & & \downarrow & & \| & & \downarrow (-)_0 \\ 1 & \longrightarrow & C_A(A_0) \cap A^*_\alpha & \longrightarrow & \Gamma & \xrightarrow{\psi} & \mathrm{Pic}(A_0) \end{array} \tag{4.12}$$

Remark 4.2.7 RELATING $\mathrm{Pic}^{\mathrm{gr}}(A)$ TO $\mathrm{Pic}(A_0)$ FOR A STRONGLY GRADED RING A

Example 4.2.5 shows that for a strongly graded ring A, $\mathrm{Pic}^{\mathrm{gr}}(A)$ is not necessarily isomorphic to $\mathrm{Pic}(A_0)$. However, one can relate these two groups with an exact sequence as follows:

$$1 \longrightarrow H^1(\Gamma, Z(A_0)^*) \longrightarrow \mathrm{Pic}^{\mathrm{gr}}(A) \xrightarrow{\Psi} \mathrm{Pic}(A_0)^\Gamma \longrightarrow H^2(\Gamma, Z(A_0)^*).$$

Here Γ acts on $\mathrm{Pic}(A_0)$ by

$$\gamma[P] = [A_\gamma \otimes_{A_0} P \otimes_{A_0} A_{-\gamma}],$$

so the notation $\mathrm{Pic}(A_0)^\Gamma$ refers to the group of Γ-invariant elements of $\mathrm{Pic}(A_0)$. Moreover, $Z(A_0)^*$ denotes the units of the centre of A_0 and H^1, H^2 denote the first and second cohomology groups. The map Ψ is defined by $\Psi([P]) = [P_0]$ (see [68] for details. Also see [17]).

We include a result from [77] which shows that if A_0 has IBN, then A has gr-IBN, based on a condition on $\mathrm{Pic}(A_0)$. In general, one can produce an example of strongly graded ring A such that A_0 has IBN whereas A is a non-IBN ring.

For a ring R, define

$$\mathrm{Pic}_n(R) = \{\, [X] \in \mathrm{Pic}(R) \mid X^n \cong R^n \text{ as right } R\text{-module}\,\}.$$

This is a subgroup of $\mathrm{Pic}(R)$ and $\mathrm{Pic}_n(R)$ and $\mathrm{Pic}_m(R)$ are subgroups of $\mathrm{Pic}_{nm}(R)$. Thus

$$\mathrm{Pic}_\infty(R) = \bigcup_{n\geq 1} \mathrm{Pic}_n(R)$$

is a subgroup of $\mathrm{Pic}(R)$.

Proposition 4.2.8 *Let A be a strongly Γ-graded ring such that $\{[A_\alpha] \mid \alpha \in \Gamma\} \subseteq \mathrm{Pic}_\infty(A_0)$. If A_0 has IBN then A has gr-IBN. Moreover, if Γ is finite, then A has IBN if and only if A_0 has IBN.*

Proof Suppose that $A^n(\overline{\alpha}) \cong_{\mathrm{gr}} A^m(\overline{\beta})$ as graded right A-modules, where $\overline{\alpha} = (\alpha_1, \dots, \alpha_n)$ and $\overline{\beta} = (\beta_1, \dots, \beta_m)$. Using Dade's Theorem 1.5.1, we have

$$A_{\alpha_1} \oplus \cdots \oplus A_{\alpha_n} \cong A_{\beta_1} \oplus \cdots \oplus A_{\beta_m}, \tag{4.13}$$

as right A_0-modules. Since each $[A_{\alpha_i}]$ and $[A_{\beta_j}]$ are in some $\mathrm{Pic}_{n_i}(A_0)$ and $\mathrm{Pic}_{n_j}(A_0)$, respectively, there is a large enough t such that $[A_{\alpha_i}], [A_{\beta_j}] \in \mathrm{Pic}_t(A_0)$, for all $1 \leq i \leq n$ and $1 \leq j \leq m$. Thus $A^t_{\alpha_i} \cong A^t_0$ and $A^t_{\beta_j} \cong A^t_0$, for all i and j. Replacing this into (4.13) we get $A_0^{nt} \cong A_0^{mt}$. Since A_0 has IBN, it follows that $n = m$.

Now suppose Γ is finite. If A_0 has IBN, then we will prove that A has IBN. From what we proved above, it follows that A has gr-IBN. If $A^n \cong A^m$, then $F(A^n) \cong_{\mathrm{gr}} F(A^m)$, where F is the adjoint to the forgetful functor (see §1.2.7). Since $F(A^n) = F(A)^n$ and $F(A) = \bigoplus_{\gamma\in\Gamma} A(\gamma)$, using the fact that A has gr-IBN, it follows immediately that $n = m$. Conversely, suppose A has IBN. If $A_0^n \cong A_0^m$, then tensoring with A over A_0 we have

$$A^n \cong_{\mathrm{gr}} A_0^n \otimes_{A_0} A \cong A_0^m \otimes_{A_0} A \cong_{\mathrm{gr}} A^m.$$

So $n = m$ and we are done. □

Here is another application of Picard groups, relating the ideal theory of

A_0 to A. This proposition is taken from [78, Theorem 3.4]. Recall the group homomorphism

$$\psi : \Gamma \longrightarrow \mathrm{Pic}(A_0)$$
$$\alpha \longmapsto [A_\alpha],$$

from (4.11). For nonzero element $a \in A$, define the *length* of a, $l(a) = n$, where n is the number of nonzero homogeneous elements appearing in the decomposition of a into the homogeneous sum.

Proposition 4.2.9 *Let A be a strongly Γ-graded ring such that the homomorphism $\psi : \Gamma \to \mathrm{Pic}(A_0)$ is injective. If A_0 is a simple ring then so is A.*

Proof Let I be a nonzero ideal of A and x a nonzero element of I which has the minimum length among all elements of I. Suppose x_α is the homogeneous element of degree α appearing in the decomposition of x into the homogeneous sum. If $x_\alpha \neq 0$, then multiplying x with a homogeneous element of degree $-\alpha$, we can assume that $x_0 \neq 0$. Since A_0 is a simple ring, we have $\sum_i y_i x_0 z_i = 1$, where $y_i, z_i \in A_0$. Consider $\sum_i y_i x z_i \in I$, and note that its length is equal to the length of x. Replacing x with this element, we can assume $x_0 = 1$. If $x = x_0 = 1$ then $I = A$ and we are done. Otherwise, suppose there is $x_\alpha \neq 0$ in the decomposition of x. Note that for each $b \in A_0$, $bx - xb \in I$ with $l(bx - xb) < l(x)$. The minimality of x implies that $bx = xb$ and thus $bx_\alpha = x_\alpha b$ for any $b \in A_0$, *i.e.*, $x_\alpha \in C_A(A_0)$. We show that $x_\alpha \in A_\alpha^*$. Note that $A_{-\alpha}x_\alpha$ and $x_\alpha A_{-\alpha}$ are nonzero (nongraded) two-sided ideals of A_0. Again since A_0 is simple, this implies that x_α is invertible. Therefore $x_\alpha \in C_A(A_0) \cap A_\alpha^*$. This is a contradiction with the assumption that ψ is injective (see 4.12). This finishes the proof. □

5

Graded ultramatricial algebras classification via K_0^{gr}

Let F be a field. An F-algebra is called an *ultramatricial* algebra if it is isomorphic to the union of an increasing chain of a finite product of matrix algebras over F. When F is the field of complex numbers, these algebras are also called *locally semisimple* algebras (or LS-algebras for short), as they are isomorphic to a union of a chain of semisimple $\mathbb{C}$-algebras. An important example of such rings is the group ring $\mathbb{C}[S_\infty]$, where S_∞ is the infinite symmetric group. These rings appeared in the setting of C^*-algebras and then von Neumann regular algebras in the work of Grimm, Bratteli, Elliott, Goodearl, Handelman and many others after them.

Despite their simple constructions, the study of ultramatricial algebras is far from over. As is noted in [91]: "The current state of the theory of LS-algebras and its applications should be considered as the initial one; one has discovered the first fundamental facts and noted a general circle of questions. To estimate it in perspective, one must consider the enormous number of diverse and profound examples of such algebras. In addition one can observe the connections with a large number of areas of mathematics."

One of the sparkling examples of the Grothendieck group as a complete invariant is in the setting of ultramatricial algebras (and AF C^*-algebras). It is by now a classical result that the K_0-group along with its positive cone K_{0+} (the dimension group) and the position of identity is a complete invariant for such algebras ([40, Bratteli–Elliott Theorem 15.26]). To be precise, let R and S be (unital) ultramatricial algebras. Then $R \cong S$ if and only if there is an order isomorphism $(K_0(R), [R]) \cong (K_0(S), [S])$, that is, an isomorphism from $K_0(R)$ to $K_0(S)$ which sends the positive cone onto the positive cone and $[R]$ to $[S]$. To emphasise the ordering, this isomorphism is also denoted by $(K_0(R), K_0(R)_+, [R]) \cong (K_0(S), K_0(S)_+, [S])$ (see §3.6.1).

The theory has also been worked out for the nonunital ultramatricial algebras (see Remark 5.2.7 and [41, Chapter XII]). Two valuable surveys on ultramatricial algebras and their relations with other branches of mathematics are [91, 98]. The lecture notes by Effros [37] also give an excellent detailed account of this theory.

In this chapter we initiate the graded version of this theory. We define the graded ultramatricial algebras. We then show that the graded Grothendieck group, equipped with its module structure and its ordering, is a complete invariant for such algebras (Theorem 5.2.4). When the grade group is considered to be trivial, we retrieve the Bratteli–Elliott Theorem (Corollary 5.2.6).

When the grading is present, K_0^{gr} seems to capture more details than K_0. The following example demonstrates this: the Grothendieck group classifies matricial algebras, however, it is not a complete invariant if we enlarge the class of algebras to include matrices over Laurent rings. Consider the following graphs and their associated Leavitt path algebras (see Theorems 1.6.19 and 1.6.21).

$$F \qquad \bullet \longrightarrow \bullet \longrightarrow \bullet, \qquad \mathcal{L}_K(F) \cong \mathbb{M}_3(K),$$

$$E \qquad \bullet \longrightarrow \bullet \rightleftarrows \bullet, \qquad \mathcal{L}_K(E) \cong \mathbb{M}_3(K[x,x^{-1}]).$$

Since K_0 is Morita invariant, one can calculate that

$$\big(K_0(\mathcal{L}(F)), K_0(\mathcal{L}(F))_+, [\mathcal{L}(F)]\big) \cong (\mathbb{Z}, \mathbb{N}, 3)$$

and

$$\big(K_0(\mathcal{L}(E)), K_0(\mathcal{L}(E))_+, [\mathcal{L}(E)]\big) \cong (\mathbb{Z}, \mathbb{N}, 3).$$

However,

$$\mathbb{M}_3(K) \not\cong \mathbb{M}_3(K[x,x^{-1}]).$$

But, as we will prove in this Chapter, K_0^{gr} is a complete invariant for this class of algebras (Theorem 5.1.3).

5.1 Graded matricial algebras

We begin with the graded version of matricial algebras.

Definition 5.1.1 Let A be a Γ-graded field. A Γ-*graded matricial* A-algebra is a graded A-algebra of the form

$$\mathbb{M}_{n_1}(A)(\overline{\delta}_1) \times \cdots \times \mathbb{M}_{n_l}(A)(\overline{\delta}_l),$$

where $\overline{\delta}_i = (\delta_1^{(i)}, \dots, \delta_{n_i}^{(i)})$, $\delta_j^{(i)} \in \Gamma$, $1 \leq j \leq n_i$ and $1 \leq i \leq l$.

If Γ is a trivial group, then Definition 5.1.1 reduces to the definition of matricial algebras ([40, §15]). Note that if R is a graded matricial A-algebra, then R_0 is a matricial A_0-algebra (see §1.4.1).

In general, when two graded finitely generated projective modules represent the same element in the graded Grothendieck group, then they are not necessarily graded isomorphic but rather graded stably isomorphic (Lemma 3.1.7(3)). However, for graded matricial algebras one can prove that the graded stably isomorphic modules are in fact graded isomorphic. Later, in Lemma 5.1.5 we show this is also the case in the larger category of graded ultramatricial algebras.

Lemma 5.1.2 *Let A be a Γ-graded field and R be a Γ-graded matricial A-algebra. Let P and Q be graded finitely generated projective R-modules. Then $[P] = [Q]$ in $K_0^{\mathrm{gr}}(R)$, if and only if $P \cong_{\mathrm{gr}} Q$.*

Proof Since the functor K_0^{gr} respects the direct sum, it suffices to prove the statement for a graded matricial algebra of the form $R = \mathbb{M}_n(A)(\overline{\delta})$. Let P and Q be graded finitely generated projective R-modules such that $[P] = [Q]$ in $K_0^{\mathrm{gr}}(R)$. By Proposition 2.1.2, $R = \mathbb{M}_n(A)(\overline{\delta})$ is graded Morita equivalent to A. So there are equivalent functors ψ and ϕ such that $\psi\phi \cong 1$ and $\phi\psi \cong 1$, which also induce an isomorphism $K_0^{\mathrm{gr}}(\psi) : K_0^{\mathrm{gr}}(R) \to K_0^{\mathrm{gr}}(A)$ such that $[P] \mapsto [\psi(P)]$. Now since $[P] = [Q]$, it follows that $[\psi(P)] = [\psi(Q)]$ in $K_0^{\mathrm{gr}}(A)$. But since A is a graded field, by the proof of Proposition 3.7.1, any graded finitely generated projective A-module can be written uniquely as a direct sum of an appropriately shifted A. Writing $\psi(P)$ and $\psi(Q)$ in this form, the homomorphism (3.32) shows that $\psi(P) \cong_{\mathrm{gr}} \psi(Q)$. Now applying the functor ϕ to this we obtain $P \cong_{\mathrm{gr}} Q$. □

Let A be a Γ-graded field (with the support Γ_A) and let $\mathcal{C}$ be a category consisting of Γ-graded matricial A-algebras as objects and A-graded algebra homomorphisms as morphisms. We consider the quotient category $\mathcal{C}^{\mathrm{out}}$ obtained from $\mathcal{C}$ by identifying homomorphisms which differ up to a degree zero graded inner automorphism. That is, the graded homomorphisms $\phi, \psi \in \mathrm{Hom}_{\mathcal{C}}(R, S)$ represent the same morphism in $\mathcal{C}^{\mathrm{out}}$ if there is an inner automorphism $\theta : S \to S$, defined by $\theta(s) = xsx^{-1}$, where $\deg(x) = 0$, such that $\phi = \theta\psi$. The following theorem shows that K_0^{gr} as a graded dimension group (see Example 3.6.3) "classifies" the category of $\mathcal{C}^{\mathrm{out}}$. This is a graded analogue of a similar result for matricial algebras (see [40, Lemma 15.23]). Recall from §3.6.1 that $\mathcal{P}$ is a category with objects consisting of the pairs (G, u), where G is a Γ-pre-ordered module and u is an order-unit, and $f : (G, u) \to (H, v)$ is an order preserving Γ-homomorphism such that $f(u) = v$.

Theorem 5.1.3 *Let A be a Γ-graded field and $\mathcal{C}^{\mathrm{out}}$ be the category consisting of Γ-graded matricial A-algebras as objects and A-graded algebra homomorphisms modulo graded inner automorphisms as morphisms. Then*

$$K_0^{\mathrm{gr}} : \mathcal{C}^{\mathrm{out}} \to \mathcal{P}$$

is a fully faithful functor. Namely,

(1) *(well-defined and faithful) For any graded matricial A-algebras R and S and $\phi, \psi \in \mathrm{Hom}_{\mathcal{C}}(R, S)$, we have $\phi(r) = x\psi(r)x^{-1}$, $r \in R$, for some invertible homogeneous element x of degree 0 in S, if and only if $K_0^{\mathrm{gr}}(\phi) = K_0^{\mathrm{gr}}(\psi)$.*

(2) *(full) For any graded matricial A-algebras R and S and the morphism*

$$f : (K_0^{\mathrm{gr}}(R), [R]) \to (K_0^{\mathrm{gr}}(S), [S])$$

in $\mathcal{P}$, there is a $\phi \in \mathrm{Hom}_{\mathcal{C}}(R, S)$ such that $K_0^{\mathrm{gr}}(\phi) = f$.

Proof (1) (well-defined) Let $\phi, \psi \in \mathrm{Hom}_{\mathcal{C}}(R, S)$ such that $\phi = \theta\psi$, where $\theta(s) = xsx^{-1}$ for some invertible homogeneous element x of S of degree 0. Then

$$K_0^{\mathrm{gr}}(\phi) = K_0^{\mathrm{gr}}(\theta\psi) = K_0^{\mathrm{gr}}(\theta)K_0^{\mathrm{gr}}(\psi) = K_0^{\mathrm{gr}}(\psi),$$

since $K_0^{\mathrm{gr}}(\theta)$ is the identity map (see §3.5.1).

(faithful) The rest of the proof is similar to the nongraded version with an extra attention given to the grading (see [40, p.218]). Let $K_0^{\mathrm{gr}}(\phi) = K_0^{\mathrm{gr}}(\psi)$. Let $R = \mathbb{M}_{n_1}(A)(\overline{\delta}_1) \times \cdots \times \mathbb{M}_{n_l}(A)(\overline{\delta}_l)$ and set $g_{jk}^{(i)} = \phi(\mathbf{e}_{jk}^{(i)})$ and $h_{jk}^{(i)} = \psi(\mathbf{e}_{jk}^{(i)})$ for $1 \leq i \leq l$ and $1 \leq j, k \leq n_i$, where $\mathbf{e}_{jk}^{(i)}$ are the graded matrix units basis for $\mathbb{M}_{n_i}(A)$. Since ϕ and ψ are graded homomorphisms,

$$\deg(\mathbf{e}_{jk}^{(i)}) = \deg(g_{jk}^{(i)}) = \deg(h_{jk}^{(i)}) = \delta_j^{(i)} - \delta_k^{(i)}.$$

Since $\mathbf{e}_{jj}^{(i)}$ are pairwise graded orthogonal idempotents (of degree 0) in R and $\sum_{i=1}^{l}\sum_{j=1}^{n_i}\mathbf{e}_{jj}^{(i)} = 1$ (*i.e.*, graded full matrix units), the same is also the case for $g_{jj}^{(i)}$ and $h_{jj}^{(i)}$. Then

$$[g_{11}^{(i)}S] = K_0^{\mathrm{gr}}(\phi)([\mathbf{e}_{11}^{(i)}R]) = K_0^{\mathrm{gr}}(\psi)([\mathbf{e}_{11}^{(i)}R]) = [h_{11}^{(i)}S].$$

By Lemma 5.1.2, $g_{11}^{(i)}S \cong_{\mathrm{gr}} h_{11}^{(i)}S$. Thus there are homogeneous elements x_i and y_i of degree 0 such that $x_iy_i = g_{11}^{(i)}$ and $y_ix_i = h_{11}^{(i)}$ (see §1.3.2). Let

$$x = \sum_{i=1}^{l}\sum_{j=1}^{n_i} g_{j1}^{(i)}x_ih_{1j}^{(i)} \quad \text{and} \quad y = \sum_{i=1}^{l}\sum_{j=1}^{n_i} h_{j1}^{(i)}y_ig_{1j}^{(i)}.$$

Note that x and y are homogeneous elements of degree zero. One checks easily that $xy = yx = 1$. Now for $1 \leq i \leq l$ and $1 \leq j, k \leq n_i$, we have

$$\begin{aligned} xh_{jk}^{(i)} &= \sum_{s=1}^{l} \sum_{t=1}^{n_s} g_{t1}^{(s)} x_s h_{1t}^{(s)} h_{jk}^{(i)} \\ &= g_{j1}^{(i)} x_i h_{1j}^{(i)} h_{jk}^{(i)} = g_{jk}^{(i)} g_{k1}^{(i)} x_i h_{1k}^{(i)} \\ &= \sum_{s=1}^{l} \sum_{t=1}^{n_s} g_{jk}^{(i)} g_{t1}^{(s)} x_s h_{1t}^{(i)} = g_{jk}^{(i)} x. \end{aligned}$$

Let $\theta : S \to S$ be the graded inner automorphism $\theta(s) = xsy$. Then

$$\theta\psi(e_{jk}^{(i)})) = xh_{jk}^{(i)} y = g_{jk}^{(i)} = \phi(e_{jk}^{(i)}).$$

Since $\mathbf{e}_{jk}^{(i)}$, $1 \leq i \leq l$ and $1 \leq j, k \leq n_i$, form a homogeneous A-basis for R, $\theta\psi = \phi$.

(2) Let $R = \mathbb{M}_{n_1}(A)(\overline{\delta}_1) \times \cdots \times \mathbb{M}_{n_l}(A)(\overline{\delta}_l)$. Consider

$$R_i = \mathbb{M}_{n_i}(A)(\overline{\delta}_i), 1 \leq i \leq l.$$

Each R_i is a graded finitely generated projective R-module, so $f([R_i])$ is in the positive cone of $K_0^{\mathrm{gr}}(S)$, *i.e.*, there is a graded finitely generated projective S-module P_i such that $f([R_i]) = [P_i]$. Then

$$[S] = f([R]) = f([R_1] + \cdots + [R_l]) = [P_1] + \cdots + [P_l] = [P_1 \oplus \cdots \oplus P_l].$$

Since S is a graded matricial algebra, by Lemma 5.1.2, $P_1 \oplus \cdots \oplus P_l \cong_{\mathrm{gr}} S$ as right S-modules. So there are homogeneous orthogonal idempotents $g_1, \ldots, g_l$ in S such that $g_1 + \cdots + g_l = 1$ and $g_i S \cong_{\mathrm{gr}} P_i$ (see §1.3.2). Note that each of $R_i = \mathbb{M}_{n_i}(A)(\overline{\delta}_i)$ is a graded simple algebra. Set $\overline{\delta}_i = (\delta_1^{(i)}, \ldots, \delta_n^{(i)})$ (here $n = n_i$). Let $\mathbf{e}_{jk}^{(i)}$, $1 \leq j, k \leq n$, be the graded matrix units of R_i and consider the graded finitely generated projective (right) R_i-module

$$V = \mathbf{e}_{11}^{(i)} R_i = A(\delta_1^{(i)} - \delta_1^{(i)}) \oplus A(\delta_2^{(i)} - \delta_1^{(i)}) \oplus \cdots \oplus A(\delta_n^{(i)} - \delta_1^{(i)}).$$

Then (1.44) shows that

$$R_i \cong_{\mathrm{gr}} V(\delta_1^{(i)} - \delta_1^{(i)}) \oplus V(\delta_1^{(i)} - \delta_2^{(i)}) \oplus \cdots \oplus V(\delta_1^{(i)} - \delta_n^{(i)}),$$

as a graded R-module. Thus

$$\begin{aligned} [P_i] = [g_i S] = f([R_i]) = f([V(\delta_1^{(i)} - \delta_1^{(i)})]) + f([V(\delta_1^{(i)} - \delta_2^{(i)})]) + \cdots \\ + f([V(\delta_1^{(i)} - \delta_n^{(i)})]). \quad (5.1) \end{aligned}$$

There is a graded finitely generated projective S-module Q such that

$$f([V]) = f([V(\delta_1^{(i)} - \delta_1^{(i)})]) = [Q].$$

Since f is a $\mathbb{Z}[\Gamma]$-module homomorphism, for $1 \leq k \leq n$,

$$f([V(\delta_1^{(i)} - \delta_k^{(i)})]) = f((\delta_1^{(i)} - \delta_k^{(i)})[V]) = (\delta_1^{(i)} - \delta_k^{(i)})f([V])$$
$$= (\delta_1^{(i)} - \delta_k^{(i)})[Q] = [Q(\delta_1^{(i)} - \delta_k^{(i)})].$$

From (5.1) and Lemma 5.1.2 now follows

$$g_i S \cong_{\mathrm{gr}} Q(\delta_1^{(i)} - \delta_1^{(i)}) \oplus Q(\delta_1^{(i)} - \delta_2^{(i)}) \oplus \cdots \oplus Q(\delta_1^{(i)} - \delta_n^{(i)}). \quad (5.2)$$

Let

$$g_{jk}^{(i)} \in \mathrm{End}(g_i S) \cong_{\mathrm{gr}} g_i S g_i$$

map the jth summand of the right hand side of (5.2) to its k-th summand and everything else to zero. Observe that $\deg(g_{jk}^{(i)}) = \delta_j^{(i)} - \delta_k^{(i)}$ and $g_{jk}^{(i)}$, $1 \leq j, k \leq n$, form the matrix units. Moreover, $g_{11}^{(i)} + \cdots + g_{nn}^{(i)} = g_i$ and

$$g_{11}^{(i)} S = Q(\delta_1^{(i)} - \delta_1^{(i)}) = Q.$$

Thus $[g_{11}^{(i)} S] = [Q] = f([V]) = f([e_{11}^{(i)} R_i])$.

Now for any $1 \leq i \leq l$, define the A-algebra homomorphism

$$R_i \longrightarrow g_i S g_i,$$
$$e_{jk}^{(i)} \longmapsto g_{jk}^{(i)}.$$

This is a graded homomorphism, and induces a graded homomorphism $\phi : R \to S$ such that $\phi(\mathbf{e}_{jk}^{(i)}) = g_{jk}^{(i)}$. Clearly

$$K_0^{\mathrm{gr}}(\phi)([\mathbf{e}_{11}^{(i)} R_i]) = [\phi(\mathbf{e}_{11}^{(i)}) S] = [g_{11}^{(i)} S] = f([\mathbf{e}_{11}^{(i)} R_i]),$$

for $1 \leq i \leq l$. Now $K_0^{\mathrm{gr}}(R)$ is generated by $[\mathbf{e}_{11}^{(i)} R_i]$, $1 \leq i \leq l$, as $\mathbb{Z}[\Gamma]$-module. This implies that $K_0^{\mathrm{gr}}(\phi) = f$. □

Remark 5.1.4 In Theorem 5.1.3, both parts (1) and (2) are valid when the ring S is a graded ultramatricial algebra as well. In fact, in the proofs of (1) and (2), the only property of S which is used is that if $[P] = [Q]$ in $K_0^{\mathrm{gr}}(S)$, then $P \cong_{\mathrm{gr}} Q$, where P and Q are graded finitely generated projective S-modules. Lemma 5.1.5 below shows that this is the case for ultramatricial algebras.

Lemma 5.1.5 *Let A be a Γ-graded field and R be a Γ-graded ultramatricial A-algebra. Let P and Q be graded finitely generated projective R-modules. Then $[P] = [Q]$ in $K_0^{\mathrm{gr}}(R)$ if and only if $P \cong_{\mathrm{gr}} Q$.*

Proof We will use the description of K_0^{gr} based on homogeneous idempotents to prove this lemma (see §3.2). Let p and q be homogeneous idempotents matrices over R corresponding to the graded finitely generated R-modules P and

Q, respectively (see Lemma 3.2.3(2)). Suppose $[P] = [Q]$ in $K_0^{\mathrm{gr}}(R)$. We will show that p and q are graded equivalent in R, which by Lemma 3.2.3(3), implies that $P \cong_{\mathrm{gr}} Q$. Since by Definition 5.2.1, $R = \bigcup_{i\in I} R_i$, there is a $j \in I$ such that p and q are homogeneous idempotent matrices over R_j. But since $[p] = [q]$ in $K_0^{\mathrm{gr}}(R)$, there is an $n \in \mathbb{N}$, such that $p \oplus 1_n$ is graded equivalent to $q \oplus 1_n$ in R (see §3.2.1). So there is a $k \in I$, $k \geq j$, such that $p \oplus 1_n$ and $q \oplus 1_n$ are graded equivalent in R_k. Thus $[p\oplus 1_n] = [q\oplus 1_n]$ in $K_0^{\mathrm{gr}}(R_k)$. This implies that $[p] = [q]$ in $K_0^{\mathrm{gr}}(R_k)$. Since R_k is a graded matricial algebra, by Lemma 5.1.5, p is graded equivalent to q in R_k. So p is graded equivalent to q in R and consequently $P \cong_{\mathrm{gr}} Q$ as an R-module. The converse is immediate. □

5.2 Graded ultramatricial algebras, classification via K_0^{gr}

The direct limit of a direct system of graded rings is a ring with a graded structure (see Example 1.1.9). In this section, we study a particular case of such graded rings, namely the direct limit of graded matricial algebras. Recall from Definition 5.1.1 that a Γ-graded matricial algebra over the graded field A is of the form

$$\mathbb{M}_{n_1}(A)(\overline{\delta}_1) \times \cdots \times \mathbb{M}_{n_l}(A)(\overline{\delta}_l),$$

for some shifts $\overline{\delta}_i$, $1 \leq i \leq l$.

Definition 5.2.1 Let A be a Γ-graded field. A ring R is called a Γ-*graded ultramatricial* A-algebra if $R = \bigcup_{i=1}^{\infty} R_i$, where $R_1 \subseteq R_2 \subseteq \cdots$ is a sequence of graded matricial A-subalgebras. Here the inclusion respects the grading, *i.e.*, $R_{i\alpha} \subseteq R_{i+1\alpha}$. Moreover, under the inclusion $R_i \subseteq R_{i+1}$, we have $1_{R_i} = 1_{R_{i+1}}$.

Clearly R is also a Γ-graded A-algebra with $R_\alpha = \bigcup_{i=1}^{\infty} R_{i\alpha}$. If Γ is a trivial group, then Definition 5.2.1 reduces to the definition of ultramatricial algebras ([40, §15]). Note that if R is a graded ultramatricial A-algebra, then R_0 is an ultramatricial A_0-algebra (see §1.4.1).

Example 5.2.2 Let A be a ring. We identify $\mathbb{M}_n(A)$ as a subring of $\mathbb{M}_{2n}(A)$ under the monomorphism

$$X \in \mathbb{M}_n(A) \longmapsto \begin{pmatrix} X & 0 \\ 0 & X \end{pmatrix} \in \mathbb{M}_{2n}(A). \tag{5.3}$$

With this identification, we have a sequence

$$\mathbb{M}_2(A) \subseteq \mathbb{M}_4(A) \subseteq \cdots.$$

Now let A be a Γ-graded field, $\alpha_1, \alpha_2 \in \Gamma$ and consider the sequence of graded subalgebras

$$\mathbb{M}_2(A)(\alpha_1, \alpha_2) \subseteq \mathbb{M}_4(A)(\alpha_1, \alpha_2, \alpha_1, \alpha_2) \subseteq \cdots,$$

with the same embedding as (5.3). Then

$$R = \bigcup_{i=1}^{\infty} \mathbb{M}_{2^i}(A)((\alpha_1, \alpha_2)^{2^{i-1}}),$$

where $(\alpha_1, \alpha_2)^k$ stands for k copies of (α_1, α_2), is a graded ultramatricial algebra.

Example 5.2.3 Let $A = K[x^3, x^{-3}]$ be a $\mathbb{Z}$-graded field with support $3\mathbb{Z}$, where K is a field. Consider the following sequence of graded matricial algebras with the embedding as in (5.3):

$$\mathbb{M}_2(A)(0,1) \subseteq \mathbb{M}_4(A)(0,1,1,2) \subseteq \mathbb{M}_8(A)(0,1,1,2,0,1,1,2) \subseteq \cdots. \tag{5.4}$$

Let R be a graded ultramatricial algebra constructed as in Example 5.2.2, from the union of matricial algebras of sequence (5.4). We calculate $K_0^{\mathrm{gr}}(R)$. Since K_0^{gr} respects the direct limit (Theorem 3.2.4), we have

$$K_0^{\mathrm{gr}}(R) = K_0^{\mathrm{gr}}(\varinjlim R_i) = \varinjlim K_0^{\mathrm{gr}}(R_i),$$

where R_i corresponds to the ith algebra in the sequence (5.4). Since all the matricial algebras R_i are strongly graded, we get

$$\varinjlim K_0^{\mathrm{gr}}(R_i) = \varinjlim K_0(R_{i_0}).$$

Recall that (see Proposition 1.4.1) if

$$T \cong_{\mathrm{gr}} \mathbb{M}_m(K[x^n, x^{-n}])(p_1, \dots, p_m),$$

then letting d_l, $0 \leq l \leq n-1$, be the number of i such that p_i represents $\bar{l}$ in $\mathbb{Z}/n\mathbb{Z}$, we have

$$T \cong_{\mathrm{gr}} \mathbb{M}_m(K[x^n, x^{-n}])(0, \dots, 0, 1, \dots, 1, \dots, n-1, \dots, n-1),$$

where $0 \leq l \leq n-1$ occurs d_l times. It is now easy to see that

$$T_0 \cong \mathbb{M}_{d_0}(K) \times \cdots \times \mathbb{M}_{d_{n-1}}(K).$$

Using this, the 0-component rings of the sequence (5.4) take the form

$$K \oplus K \subseteq K \oplus \mathbb{M}_2(K) \oplus K \subseteq \mathbb{M}_2(K) \oplus \mathbb{M}_4(K) \oplus \mathbb{M}_2(K)$$

$$(x, y) \mapsto (x, \begin{pmatrix} y & 0 \\ 0 & x \end{pmatrix}, y)$$

$$(x, y, z) \mapsto (\begin{pmatrix} x & 0 \\ 0 & x \end{pmatrix}, \begin{pmatrix} y & 0 \\ 0 & y \end{pmatrix}, \begin{pmatrix} z & 0 \\ 0 & z \end{pmatrix}).$$

Thus

$$K_0^{\mathrm{gr}}(R) \cong \varinjlim K_0(R_{i_0}) \cong \mathbb{Z}[1/2] \oplus \mathbb{Z}[1/2] \oplus \mathbb{Z}[1/2].$$

We are now in a position to classify the graded ultramatricial algebras via the K_0^{gr}-group. The following theorem shows that the *dimension module* is a complete invariant for the category of graded ultramatricial algebras.

Theorem 5.2.4 *Let R and S be Γ-graded ultramatricial algebras over a graded field A. Then $R \cong_{\mathrm{gr}} S$ as graded A-algebras if and only if there is an order preserving $\mathbb{Z}[\Gamma]$-module isomorphism*

$$(K_0^{\mathrm{gr}}(R), [R]) \cong (K_0^{\mathrm{gr}}(S), [S]).$$

Proof One direction is clear. Let $R = \bigcup_{i=1}^{\infty} R_i$ and $S = \bigcup_{i=1}^{\infty} S_i$, where $R_1 \subseteq R_2 \subseteq \cdots$ and $S_1 \subseteq S_2 \subseteq \cdots$ are sequences of graded matricial A-algebras. Let $\phi_i : R_i \to R$ and $\psi_i : S_i \to S$, $i \in \mathbb{N}$, be inclusion maps. In order to show that the isomorphism

$$f : (K_0^{\mathrm{gr}}(R), [R]) \longrightarrow (K_0^{\mathrm{gr}}(S), [S])$$

between the graded Grothendieck groups give rise to an isomorphism between the rings R and S, we will find a sequence $n_1 < n_2 < \cdots$ of positive numbers and graded A-module injections $\rho_k : R_{n_k} \to S$ such that ρ_{k+1} is an extension of ρ_k and $\bigcup_{k=1}^{\infty} R_{n_k} = R$. To achieve this we repeatedly use Theorem 5.1.3 (and Remark 5.1.4) (*i.e.*, a "local" version of this theorem) and the fact that since R_n, $n \in \mathbb{N}$, is a finite dimensional A-algebra, for any A-graded homomorphism $\rho : R_n \to S$, we have $\rho(R_n) \subseteq S_i$, for some positive number i. Throughout the proof, for simplicity, we write $\overline{\theta}$ for the $\mathbb{Z}[\Gamma]$-homomorphism $K_0^{\mathrm{gr}}(\theta)$ induced by a graded A-algebra homomorphism $\theta : R \to S$.

We first prove two auxiliary facts.

I. If $\sigma : S_k \to R_n$ is a graded A-algebra homomorphism such that

$$\overline{\phi}_n \overline{\sigma} = f^{-1} \overline{\psi}_k, \tag{5.5}$$

(see Diagram (5.6)) then there exist an integer $j > k$ and a graded A-algebra homomorphism $\rho : R_n \to S_j$ such that $\psi_j \rho \sigma = \psi_k$ and $\overline{\psi}_j \overline{\rho} = \overline{f}\,\overline{\phi}_n$.

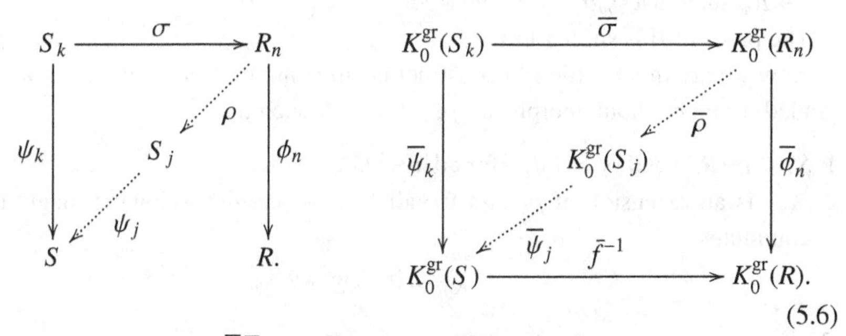

(5.6)

Proof of I. Consider $\overline{f}\,\overline{\phi}_n : K_0^{\mathrm{gr}}(R_n) \to K_0^{\mathrm{gr}}(S)$. By Theorem 5.1.3 (and Remark 5.1.4), there is an A-graded homomorphism $\rho' : R_n \to S$ such that $\overline{\rho'} = \overline{f}\,\overline{\phi}_n$. Since R_n (a matricial A-algebra) is a finite dimensional A-algebra, $\rho'(R_n) \subseteq S_i$ for some i. Thus ρ' gives a graded homomorphism $\rho'' : R_n \to S_i$ such that $\psi_i \rho'' = \rho'$ (recall that ψ_i is just an inclusion). Moreover,

$$\overline{\psi}_i \overline{\rho''} = \overline{\rho'} = \overline{f}\,\overline{\phi}_n. \tag{5.7}$$

Then, using (5.5),

$$\overline{\psi}_i \overline{\rho''}\,\overline{\sigma} = \overline{f}\,\overline{\phi}_n \overline{\sigma} = \overline{\psi}_k.$$

Theorem 5.1.3 implies that there is a graded inner automorphism θ of S such that

$$\theta \psi_i \rho'' \sigma = \psi_k. \tag{5.8}$$

The restriction of θ on S_i gives $\theta|_{S_i} : S_i \to S$. Since S_i is finite dimensional A-algebra, it follows that there is a j such that $\theta(S_i) \subseteq S_j$. So θ gives a graded homomorphism $\theta' : S_i \to S_j$ such that $\psi_j \theta' = \theta \psi_i$. Set

$$\rho := \theta' \rho'' : R_n \to S_j. \tag{5.9}$$

We get, using (5.8),

$$\psi_j \rho \sigma = \psi_j \theta' \rho'' \sigma = \theta \psi_i \rho'' \sigma = \psi_k.$$

This gives the first part of I. Using Theorem 5.1.3, (5.7) and (5.9) we have

$$\overline{\psi}_j \overline{\rho} = \overline{\psi}_j \overline{\theta}' \overline{\rho}'' = \overline{\theta}\,\overline{\psi}_i \overline{\rho}'' = \overline{\psi}_i \overline{\rho}'' = \overline{f}\,\overline{\phi}_n.$$

This completes the proof of I.

The second auxiliary fact we need is to replace the R_is and S_is in I as follows.

II. If $\rho : R_n \to S_k$ is a graded A-algebra such that

$$\overline{\psi}_k \overline{\rho} = \overline{f}\,\overline{\phi}_n,$$

then there exist an integer $m > n$ and a graded A-algebra homomorphism $\sigma : S_k \to R_m$ such that $\phi_m \sigma \rho = \phi_n$ and $\overline{\phi}_m \overline{\sigma} = \bar{f}^{-1}\overline{\psi}_k$.

The proof of II is similar to I.

Now we are in a position to construct positive numbers $n_1 < n_2 < \cdots$ and graded A-algebra homomorphisms $\rho_k : R_{n_k} \to S$ such that

1 $S_k \subseteq \rho_k(R_{n_k})$ and $\overline{\rho}_k = \bar{f}\,\overline{\phi}_{n_k}$, for all $k \in \mathbb{N}$.
2 ρ_{k+1} is an extension of ρ_k and for all $k \in \mathbb{N}$, i.e, the following diagram commutes:

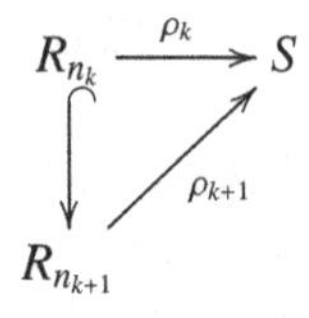

Moreover, ρ_k is injective for all $k \in \mathbb{N}$.

Consider the morphism $\bar{f}^{-1}\overline{\psi}_1 : K_0^{\mathrm{gr}}(S_1) \to K_0^{\mathrm{gr}}(R)$. By Theorem 5.1.3, there is a graded A-algebra homomorphism $\sigma' : S_1 \to R$ such that $\overline{\sigma}' = \bar{f}^{-1}\overline{\psi}_1$. Since S_1 is a finite dimensional A-algebra, $\sigma'(S_1) \subseteq R_{n_1}$ for some positive number n_1. So σ' gives a graded A-algebra homomorphism $\sigma : S_1 \to R_{n_1}$ such that $\phi_{n_1}\sigma = \sigma'$ and $\overline{\phi}_{n_1}\overline{\sigma} = \bar{f}^{-1}\overline{\psi}_1$. Thus σ satisfies the conditions of part I. Therefore there is a $j > 1$ and a graded A-algebra homomorphism $\rho : R_{n_1} \to S_j$ (see Diagram (5.10)) such that $\psi_j \rho \sigma = \psi_1$ and $\overline{\psi}_j \overline{\rho} = \bar{f}\,\overline{\phi}_{n_1}$.

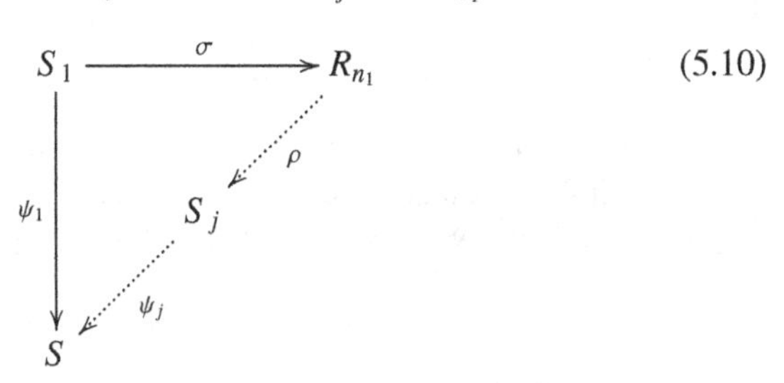

(5.10)

So $\rho_1 = \psi_j \rho : R_{n_1} \to S$ is a graded A-homomorphism such that $\rho_1 \sigma = \psi_1$ and $\overline{\rho}_1 = \bar{f}\,\overline{\phi}_{n_1}$. But

$$S_1 = \psi_1(S_1) = \rho_1\sigma(S_1) \subseteq \rho_1(R_{n_1}).$$

This proves (1). We now proceed by induction. Suppose there are $\{n_1, \ldots, n_k\}$ and $\{\rho_1, \ldots, \rho_k\}$, for some positive integer k such that (1) and (2) above are satisfied. Since R_{n_k} is a finite dimensional A-algebra, there is an $i \geq k + 1$ such that $\rho_k(R_{n_k}) \subseteq S_i$. So ρ_k gives a graded A-homomorphism $\rho' : R_{n_k} \to S_i$ such that $\psi_i \rho' = \rho_k$ and $\overline{\psi}_i \overline{\rho'} = \bar{f}\,\overline{\phi}_{n_k}$. By II, there are an $n_{k+1} > n_k$ and a $\sigma : S_i \to R_{n_{k+1}}$ such that $\phi_{n_{k+1}}\sigma\rho' = \phi_{n_k}$ and $\overline{\phi}_{n_{k+1}}\overline{\sigma} = \bar{f}^{-1}\overline{\phi}_i$. Since $\phi_{n_{k+1}}\sigma\rho' = \phi_{n_k}$, and

ϕ_{n_k} (being an inclusion) is injective, ρ' is injective and so $\rho_k = \psi_i\rho'$ is injective. Since $\sigma : S_i \to R_{n_{k+1}}$ satisfies the conditions of Part I, there is a $j > i$ such that $\rho : R_{n_{k+1}} \to S_j$ such that $\psi_j\rho\sigma = \psi_i$ and $\overline{\psi}_j\overline{\rho} = \overline{f}\,\overline{\phi}_{n_{k+1}}$. Thus for $\rho_{k+1} = \psi_j\rho : R_{n_{k+1}} \to S$ we have $\overline{\rho}_{k+1} = \overline{\psi}_j\overline{\rho} = \overline{f}\,\overline{\phi}_{n_{k+1}}$. Since $\rho_{k+1}\sigma = \psi_j\rho\sigma = \psi_i$ and $i \geq k+1$, we have

$$S_{k+1} = \psi_i(S_{k+1}) = \rho_{k+1}\sigma(S_{k+1}) \subseteq \rho_{k+1}\sigma(S_i) \subseteq \rho_{k+1}(R_{n_{k+1}}).$$

Finally, since $\phi_{n_{k+1}}\sigma\rho' = \phi_{n_k}$ and $\rho_{k+1}\sigma\rho' = \psi_i\rho' = \rho_k$, it follows that ρ_{k+1} is an extension of ρ_k. Then by induction (1) and (2) follow.

We have $n_1 < n_2 < \cdots$ and $k \leq n_k$. So $\bigcup R_{n_k} = R$. So ρ_k induces an injection $\rho : R \to S$ such that $\rho\phi_{n_k} = \rho_k$, for any k. But

$$S_k \subseteq \rho_k(R_{n_k}) = \rho(R_{n_k}) \subseteq \rho(R).$$

It follows that ρ is an epimorphism as well. This finishes the proof. □

Theorem 5.2.5 *Let R and S be Γ-graded ultramatricial algebras over a graded field A. Then R and S are graded Morita equivalent if and only if there is an order preserving $\mathbb{Z}[\Gamma]$-module isomorphism $K_0^{\mathrm{gr}}(R) \cong K_0^{\mathrm{gr}}(S)$.*

Proof Let R be graded Morita equivalent to S. Then by Theorem 2.3.8, there is an R-graded progenerator P, such that $S \cong_{\mathrm{gr}} \mathrm{End}_R(P)$. Now, using Lemma 3.6.5, we have

$$(K_0^{\mathrm{gr}}(S), [S]) \cong (K_0^{\mathrm{gr}}(R), [P]).$$

In particular $K_0^{\mathrm{gr}}(R) \cong K_0^{\mathrm{gr}}(S)$ as ordered $\mathbb{Z}[\Gamma]$-modules. (Note that the ultramatricial assumptions on rings are not used in this direction.)

For the converse, suppose $K_0^{\mathrm{gr}}(S) \cong K_0^{\mathrm{gr}}(R)$ as ordered $\mathbb{Z}[\Gamma]$-modules. Denote the image of $[S] \in K_0^{\mathrm{gr}}(S)$ under this isomorphism by $[P]$. Then

$$(K_0^{\mathrm{gr}}(S), [S]) \cong (K_0^{\mathrm{gr}}(R), [P]).$$

Since $[P]$ is an order-unit (see (3.29)), there is an $\overline{\alpha} = (\alpha_1, \ldots, \alpha_n)$, $\alpha_i \in \Gamma$, such that

$$\sum_{i=1}^{n} \alpha_i[P] = [P^n(\overline{\alpha})] \geq [R].$$

This means that there is a graded finitely generated projective A-module, Q such that $[P^n(\overline{\alpha})] = [R \oplus Q]$. Since R is a graded ultramatricial algebra, by Lemma 5.1.5, $P^n(\overline{\alpha}) \cong_{\mathrm{gr}} R \oplus Q$. Using Theorem 2.2.2, it follows that P is a graded progenerator. Let $T = \mathrm{End}_R(P)$. Using Lemma 3.6.5, we get an order preserving $\mathbb{Z}[\Gamma]$-module isomorphism

$$(K_0^{\mathrm{gr}}(T), [T]) \cong (K_0^{\mathrm{gr}}(R), [P]) \cong (K_0^{\mathrm{gr}}(S), [S]).$$

Observe that T is also a graded ultramatricial algebra. By the main Theorem 5.2.4, $\mathrm{End}_R(P) = T \cong_{\mathrm{gr}} S$. By Theorem 2.3.8, this implies that R and S are graded Morita equivalent. □

Considering a trivial group as the grade group, we retrieve the classical Bratteli–Elliott theorem (see the introduction to this chapter).

Corollary 5.2.6 (The Bratteli–Elliott theorem) *Let R and S be ultramatricial algebras over a field F. Then R and S are Morita equivalent if and only if there is an order preserving isomorphism $K_0(R) \cong K_0(S)$. Moreover, $R \cong S$ as F-algebras if and only if there is an order preserving isomorphism*

$$(K_0(R), [R]) \cong (K_0(S), [S]).$$

Proof The corollary follows by considering Γ to be a trivial group in Theorems 5.2.4 and 5.2.5. □

Remark 5.2.7 Theorem 5.2.4 can also be extended to the nonunital rings. If graded rings R and S are not unital, then instead of the order units in Theorem 5.2.4, one considers the generating interval

$$\{0 \leq x \leq [\tilde{R}] \mid x \in K_0^{\mathrm{gr}}(R)\},$$

where $[\tilde{R}]$ is the order unit in the group $K_0^{\mathrm{gr}}(\tilde{R})$. Then the order preserving map $f : K_0^{\mathrm{gr}}(R) \to K_0^{\mathrm{gr}}(S)$ should be *contractive*, *i.e.*, preserving the generating intervals (see [41, Chapter XII] for the nongraded version of this theorem).

As a demonstration, in Example 3.7.9, there is an order preserving $\mathbb{Z}[x, x^{-1}]$-module isomorphism $K_0^{\mathrm{gr}}(R) \cong K_0^{\mathrm{gr}}(S)$. However, one calculates that the generating interval for $K_0^{\mathrm{gr}}(R)$ is $\{\sum_{i\in\mathbb{N}} x^i\}$, whereas for S the interval is $\{\delta + nx \mid \delta \in \{0, 1\}, n \in \mathbb{N}\}$. Thus there is no contractive order preserving isomorphism between $K_0^{\mathrm{gr}}(R)$ and $K_0^{\mathrm{gr}}(S)$. Therefore, by the nonunital version of Theorem 5.2.4, $\mathcal{L}(F)$ and $\mathcal{L}(E)$ are not graded isomorphic.

6

Graded versus nongraded (higher) K-theory

Recall that for a Γ-graded ring A, the category of finitely generated Γ-graded projective right A-modules is denoted by $\text{Pgr}^{\Gamma}\text{-}A$. This is an exact category with the usual notion of a (split) short exact sequence. Thus one can apply Quillen's Q-construction [81] to obtain K-groups

$$K_i(\text{Pgr}^{\Gamma}\text{-}A),$$

for $i \geq 0$, which we denote by $K_i^{\text{gr}}(A)$. If more than one grading is involved (which is the case in this chapter) we denote it by $K_i^{\Gamma}(A)$. The group Γ acts on the category $\text{Pgr}^{\Gamma}\text{-}A$ from the right via $(P, \alpha) \mapsto P(\alpha)$. By the functoriality of K-groups this gives $K_i^{\text{gr}}(A)$ the structure of a right $\mathbb{Z}[\Gamma]$-module.

The relation between graded K-groups and nongraded K-groups is not always apparent. For example, consider the $\mathbb{Z}$-graded matrix ring

$$A = \mathbb{M}_5(K)(0, 1, 2, 2, 3),$$

where K is a field. Using graded Morita theory one can show that $K_0^{\mathbb{Z}}(A) \cong \mathbb{Z}[x, x^{-1}]$ (see Example 3.7.3), whereas $K_0(A) \cong \mathbb{Z}$ and $K_0(A_0) \cong \mathbb{Z}\times\mathbb{Z}\times\mathbb{Z}\times\mathbb{Z}$ (see Example 1.3.6).

In this chapter we describe a theorem due to van den Bergh [20] relating the graded K-theory to nongraded K-theory. For this we need a generalisation of a result of Quillen on relating K_i^{gr}-groups of positively graded rings to their nongraded K_i-groups (§6.1). We also need to recall a proof of the fundamental theorem of K-theory, *i.e.*, for a regular Noetherian ring A, $K_i(A[x]) \cong K_i(A)$ (§6.2). We will see that in the proof of this theorem, graded K-theory appears in a crucial way. Putting all these together, we can relate the graded K-theory of regular Noetherian $\mathbb{Z}$-graded rings to their nongraded K-theory (§6.4).

As a conseqeunce, we will see that if A is a $\mathbb{Z}$-graded right regular Noetherian ring, then one has an exact sequence (Corollary 6.4.2),

$$K_0^{\mathrm{gr}}(A) \xrightarrow{[P]\mapsto[P(1)]-[P]} K_0^{\mathrm{gr}}(A) \xrightarrow{U} K_0(A) \longrightarrow 0,$$

where U is induced by the forgetful functor (§1.2.7). This shows that

$$K_0^{\mathrm{gr}}(A)/\langle [P]-[P(1)]\rangle \cong K_0(A),$$

where P is a graded finitely generated projective A-module. Further, the action of $\mathbb{Z}[x,x^{-1}]$ on the quotient group is trivial. This means, as soon as we discard the shift in K_0^{gr}, this group reduces to the usual K_0-group.

However, this is not always the case. For example, consider the group ring $\mathbb{Z}[G]$, where G is an abelian group. Since $\mathbb{Z}[G]$ has IBN, the canonical homomorphism $\theta : K_0(\mathbb{Z}) \to K_0(\mathbb{Z}[G])$ is injective (see §3.1.4). The augmentation map $\mathbb{Z}[G] \to \mathbb{Z}, g \mapsto 1$, induces $\vartheta : K_0(\mathbb{Z}[G]) \to K_0(\mathbb{Z})$, such that $\vartheta\theta = 1$. Thus

$$K_0(\mathbb{Z}[G]) \cong \mathbb{Z} \bigoplus \widetilde{K_0}(\mathbb{Z}[G]).$$

Now since $\mathbb{Z}[G]$ is a group ring (so a crossed product),

$$K_0^{\mathrm{gr}}(\mathbb{Z}[G]) \cong K_0(\mathbb{Z}) \cong \mathbb{Z}$$

and the action of $\mathbb{Z}[x,x^{-1}]$ on the $K_0^{\mathrm{gr}}(\mathbb{Z}[G])$ is trivial (see Example 3.1.9). So

$$K_0^{\mathrm{gr}}(\mathbb{Z}[G])/\langle [P]-[P(1)]\rangle \cong \mathbb{Z}.$$

This shows that $K_0^{\mathrm{gr}}(\mathbb{Z}[G])$ does not shed any light on the group $K_0(\mathbb{Z}[G])$. The Grothendieck group of group rings is not easy to compute (see for example [15] and [95, Example 2.4]).

In this chapter we compare the graded K-theory to its nongraded counterpart. We start with a generalisation of Quillen's theorem.

6.1 K_*^{gr} of positively graded rings

For a $\mathbb{Z}$-graded ring A with support in $\mathbb{N}$, in his seminal paper [81], Quillen calculated the graded K-theory of A in terms of K-theory of its zero homogeneous component ring.

Theorem 6.1.1 [81, Chapter 3, Proposition] *Let A be a $\mathbb{Z}$-graded ring with support in $\mathbb{N}$. Then for $i \geq 0$, there is a $\mathbb{Z}[t,t^{-1}]$-module isomorphism,*

$$K_i^{\mathrm{gr}}(A) \cong K_i(A_0) \otimes_{\mathbb{Z}} \mathbb{Z}[t,t^{-1}]. \tag{6.1}$$

Moreover, the elements of $K_i(A_0) \otimes t^r$ map to $K_i^{\mathrm{gr}}(A)$ by $\overline{f}_r$, where f_r is the exact functor

$$f_r : \text{Pr-}A_0 \longrightarrow \text{Pgr-}A, \qquad (6.2)$$
$$P \longmapsto (P \otimes_{A_0} A)(-r).$$

In Theorem 6.1.3, we prove a generalised version of this proposition. Theorem 6.1.1 was used in an essential way to prove the fundamental theorem of K-theory (see §6.2.4). In particular, Theorem 6.1.1 gives a powerful tool to calculate the graded Grothendieck group of positively graded rings. We demonstrate this by calculating the K_0^{gr} of path algebras.

Theorem 6.1.2 *Let E be a finite graph and $\mathcal{P}_K(E)$ be the path algebra with coefficients in the field K. Then*

$$K_0^{\mathrm{gr}}(\mathcal{P}_K(E)) \cong \bigoplus_{|E^0|} \mathbb{Z}[t, t^{-1}].$$

Proof Recall from §1.6.4, that $\mathcal{P}_K(E)$ is a graded ring with support $\mathbb{N}$. Moreover, $\mathcal{P}_K(E)_0 = \bigoplus_{|E^0|} K$. Then by (6.1)

$$\begin{aligned} K_0^{\mathrm{gr}}(\mathcal{P}_K(E)) &\cong K_0\Big(\bigoplus_{|E^0|} K\Big) \otimes_{\mathbb{Z}} \mathbb{Z}[t, t^{-1}] \\ &\cong \bigoplus_{|E^0|} \mathbb{Z} \otimes_{\mathbb{Z}} \mathbb{Z}[t, t^{-1}] \cong \bigoplus_{|E^0|} \mathbb{Z}[t, t^{-1}]. \end{aligned}$$

□

The proof of Proposition 6.1.2 shows that, in the case of $i = 0$, $[A] \in K_i^{\mathrm{gr}}(A)$ is sent to $[A_0] \otimes 1 \in K_i(A_0) \otimes_{\mathbb{Z}} \mathbb{Z}[t, t^{-1}]$ under the isomorphism (6.1). Thus for the graphs

$$E: \quad \bullet \longrightarrow \bullet \qquad\qquad F: \quad \bullet \rightleftarrows \bullet$$

we have an order isomorphism

$$\begin{aligned} (K_0^{\mathrm{gr}}(\mathcal{P}_K(E), [\mathcal{P}_K(E)]) &\cong (\mathbb{Z}[t, t^{-1}] \oplus \mathbb{Z}[t, t^{-1}], (1, 1)) \\ &\cong (K_0^{\mathrm{gr}}(\mathcal{P}_K(E), [\mathcal{P}_K(E)]). \end{aligned}$$

This shows in particular that the ordered group K_0^{gr} in itself does not classify the path algebras.

In contrast to other fundamental theorems in the subject, such as the fundamental theorem of K-theory (*i.e.*, $K_n(R[x, x^{-1}]) = K_n(R) \times K_{n-1}(R)$, for a regular ring R), one cannot use an easy induction on (6.1) to write a similar statement for multivariable rings. For example, it appears that there is no obvious inductive approach to generalise (6.1) to $\mathbb{Z}^m \times G$-graded rings. However, by generalising Quillen's argument to take account of gradings on both sides of the isomorphism (6.1), such a procedure becomes feasible. The details have been worked out in [49] and we present it here.

We will prove the following statement.

Theorem 6.1.3 *Let G be an arbitrary group, and let A be a $\mathbb{Z} \times G$-graded ring with support in $\mathbb{N} \times G$. Then there is a $\mathbb{Z}[\mathbb{Z} \times G]$-module isomorphism*

$$K_i^{\mathbb{Z}\times G}(A) \cong K_i^G(A_{(0,G)}) \otimes_{\mathbb{Z}[G]} \mathbb{Z}[\mathbb{Z} \times G],$$

where $A_{(0,G)} = \bigoplus_{g\in G} A_{(0,g)}$.

By a straightforward induction this now implies the following.

Corollary 6.1.4 *For a $\mathbb{Z}^m \times G$-graded ring A with support in $\mathbb{N}^m \times G$ there is a $\mathbb{Z}[t_1^{\pm 1}, \dots, t_m^{\pm 1}]$-module isomorphism*

$$K_i^{\mathbb{Z}^m\times G}(A) \cong K_i^G(A_{(0,G)}) \otimes_{\mathbb{Z}} \mathbb{Z}[t_1^{\pm 1}, \dots, t_m^{\pm 1}].$$

For the trivial group G this is a direct generalisation of Quillen's theorem to $\mathbb{Z}^m$-graded rings.

As in Quillen's calculation, the proof of Theorem 6.1.3 is based on a version of Swan's theorem, modified to the present situation: it provides a correspondence between isomorphism classes of $\mathbb{Z}\times G$-graded finitely generated projective A-modules and of G-graded finitely generated projective $A_{(0,G)}$-modules.

Proposition 6.1.5 *Let Γ be a (possibly nonabelian) group. Let A be a Γ-graded ring, A_0 a graded subring of A and $\pi\colon A \to A_0$ a graded ring homomorphism such that $\pi|_{A_0} = 1$. (In other words, A_0 is a retract of A in the category of Γ-graded rings.) We denote the kernel of π by A_+.*

Suppose that for any graded finitely generated right A-module M the condition $MA_+ = M$ implies $M = 0$. Then the natural functor

$$S = - \otimes_{A_0} A\colon \text{Pgr}^\Gamma\text{-}A_0 \longrightarrow \text{Pgr}^\Gamma\text{-}A$$

induces a bijective correspondence between the isomorphism classes of graded finitely generated projective A_0-modules and of graded finitely generated projective A-modules. An inverse of the bijection is given by the functor

$$T = - \otimes_A A_0\colon \text{Pgr}^\Gamma\text{-}A \longrightarrow \text{Pgr}^\Gamma\text{-}A_0.$$

There is a natural isomorphism $T \circ S \cong \text{id}$, and for each $P \in \text{Pgr}^\Gamma\text{-}A$ a noncanonical isomorphism $S \circ T(P) \cong P$. The latter is given by

$$T(P) \otimes_{A_0} A \longrightarrow P, \quad x \otimes a \longmapsto g(x) \cdot a, \tag{6.3}$$

where g is an A_0-linear section of the epimorphism $P \to T(P)$.

Proof For any graded finitely generated projective A_0-module Q we have a natural isomorphism $TS(Q) \cong Q$ given by

$$\nu_Q \colon TS(Q) = Q \otimes_{A_0} A \otimes_A A_0 \longrightarrow Q, \quad q \otimes a \otimes a_0 \mapsto q\pi(a)a_0. \tag{6.4}$$

We will show that for a graded projective A-module P there is a noncanonical graded isomorphism $ST(P) \cong_{\mathrm{gr}} P$. The lemma then follows.

Consider the natural graded A-module epimorphism

$$f \colon P \longrightarrow T(P) = P \otimes_A A_0, \quad p \longmapsto p \otimes 1.$$

Here $T(P)$ is considered as an A-module via the map π. Since $T(P)$ is a graded projective A_0-module, the map f has a graded A_0-linear section $g \colon T(P) \to P$. This section determines an A-linear graded map

$$\psi \colon ST(P) = P \otimes_A A_0 \otimes_{A_0} A \longrightarrow P, \quad p \otimes a_0 \otimes a \longmapsto g(p \otimes a_0) \cdot a,$$

and we will show that ψ is an isomorphism. First note that the map

$$T(f) \colon T(P) \longrightarrow TT(P), \quad p \otimes a_0 \longmapsto f(p) \otimes a_0 = p \otimes 1 \otimes a_0$$

is an isomorphism (here we consider $T(P)$ as an A-module via π). In fact the inverse is given by the isomorphism $TT(P) = P \otimes_A A_0 \otimes_A A_0 \to P \otimes_A A_0$ which maps $p \otimes a_0 \otimes b_0$ to $p \otimes (a_0 b_0)$. Tracing the definitions now shows that both composites

$$TST(P) \underset{\nu_{T(P)}}{\overset{T(\psi)}{\rightrightarrows}} T(P) \xrightarrow[\cong]{T(f)} TT(P)$$

map $p \otimes a_0 \otimes a \otimes b_0 \in P \otimes_A A_0 \otimes_{A_0} A \otimes_A A_0 = TST(P)$ to the element

$$f(g(p \otimes a_0) \cdot a) \otimes b_0 = p \otimes (a_0 \pi(a) b_0) \otimes 1 \in TT(P).$$

This implies that $T(\psi) = \nu_{T(P)}$, which is an isomorphism.

The exact sequence

$$0 \longrightarrow \ker \psi \longrightarrow ST(P) \xrightarrow{\psi} P \longrightarrow \operatorname{coker} \psi \longrightarrow 0 \tag{6.5}$$

gives rise, upon application of the right exact functor T, to an exact sequence

$$TST(P) \xrightarrow{T(\psi)} T(P) \longrightarrow T(\operatorname{coker} \psi) \longrightarrow 0.$$

Since $T(\psi)$ is an isomorphism we have $T(\operatorname{coker} \psi) = \operatorname{coker} T(\psi) = 0$. Since $\operatorname{coker} \psi$ is finitely generated by (6.5) this implies $\operatorname{coker} \psi = 0$ (note that $T(M) =$

MA_+ for every finitely generated module M). In other words, ψ is surjective and (6.5) becomes the short exact sequence

$$0 \to \ker\psi \to ST(P) \xrightarrow{\psi} P \to 0.$$

This latter sequence splits since P is projective; this immediately implies that $\ker\psi$ is finitely generated, and since $T(\psi)$ is injective we also have $T(\ker\psi) = \ker T(\psi) = 0$. The hypotheses guarantee $\ker\psi = 0$ now so that ψ is injective as well as surjective, and thus is an isomorphism as claimed. □

Lemma 6.1.6 *Let G and Γ be groups, and let A be a G-graded ring. Then, considering A as a $\Gamma \times G$-graded ring in a trivial way where necessary, the functorial assignment $(M, \gamma) \mapsto M(\gamma, 0)$ induces a $\mathbb{Z}[\Gamma \times G]$-module isomorphism*

$$K_i^{\Gamma\times G}(A) \cong K_i^G(A) \otimes_{\mathbb{Z}[G]} \mathbb{Z}[\Gamma \times G].$$

Proof Let $P = \bigoplus_{(\gamma,g)\in\Gamma\times G} P_{(\gamma,g)}$ be a $\Gamma \times G$-graded finitely generated projective A-module. Since the support of A is $G = 1 \times G$, there is a unique decomposition $P = \bigoplus_{\gamma\in\Gamma} P_\gamma$, where the $P_\gamma = \bigoplus_{g\in G} P_{(\gamma,g)}$ are finitely generated G-graded projective A-modules. This gives a natural isomorphism of categories (see Corollary 1.2.13 and Remark 1.2.16)

$$\Psi\colon \mathrm{Pgr}^{\Gamma\times G}\text{-}A \xrightarrow{\cong} \bigoplus_{\gamma\in\Gamma} \mathrm{Pgr}^{G}\text{-}A.$$

The natural right action of $\Gamma \times G$ on these categories is described as follows: for a given module $P \in \mathrm{Pgr}^{\Gamma\times G}\text{-}A$ as above and elements $(\gamma, g) \in \Gamma\times G$ we have

$$P(\gamma, g)_{(\delta,h)} = P_{(\gamma+\delta,g+h)} \text{ and } \Psi(P)_\delta = P_{\gamma+\delta}(g)$$

so that $\Psi(P(\gamma, g)) = \Psi(P)(\gamma, g)$. Since K-groups respect direct sums we thus have a chain of $\mathbb{Z}[\Gamma \times G]$-linear isomorphisms

$$\begin{aligned} K_i^{\Gamma\times G}(A) &= K_i(\mathrm{Pgr}^{\Gamma\times G}\text{-}A) \cong K_i\Big(\bigoplus_{\gamma\in\Gamma} \mathrm{Pgr}^{G}\text{-}A\Big) \\ &= \bigoplus_{\gamma\in\Gamma} K_i^G(A) \cong K_i^G(A) \otimes_{\mathbb{Z}} \mathbb{Z}[\Gamma] = K_i^G(A) \otimes_{\mathbb{Z}[G]} \mathbb{Z}[\Gamma \times G]. \end{aligned}$$

□

We are in a position to prove Theorem 6.1.3.

Proof of Theorem 6.1.3 Let A be a $\mathbb{Z} \times G$-graded ring with support in $\mathbb{N} \times G$. That is, A comes equipped with a decomposition

$$A = \bigoplus_{\omega\in\mathbb{N}} A_{(\omega,G)}, \quad \text{where} \quad A_{(\omega,G)} = \bigoplus_{g\in G} A_{(\omega,g)}.$$

The ring A has a $\mathbb{Z} \times G$ graded subring $A_{(0,G)}$ (with trivial grading in the $\mathbb{Z}$-direction). The projection map $A \to A_{(0,G)}$ is a $\mathbb{Z} \times G$-graded ring homomorphism; its kernel is denoted by A_+. Explicitly, A_+ is the two-sided ideal

$$A_+ = \bigoplus_{\omega > 0} A_{(\omega,G)}.$$

We identify the quotient ring A/A_+ with the subring $A_{(0,G)}$ via the projection.

If P is a graded finitely generated projective A-module, then $P \otimes_A A_{(0,G)}$ is a finitely generated $\mathbb{Z} \times G$-graded projective $A_{(0,G)}$-module. Similarly, if Q is a graded finitely generated projective $A_{(0,G)}$-module then $Q \otimes_{A_{(0,G)}} A$ is a $\mathbb{Z} \times G$-graded finitely generated projective A-module. We can thus define functors

$$T = - \otimes_A A_{(0,G)} \colon \mathrm{Pgr}^{\mathbb{Z}\times G}\text{-}A \longrightarrow \mathrm{Pgr}^{\mathbb{Z}\times G}\text{-}A_{(0,G)}$$
$$\text{and} \quad S = - \otimes_{A_{(0,G)}} A \colon \mathrm{Pgr}^{\mathbb{Z}\times G}\text{-}A_{(0,G)} \longrightarrow \mathrm{Pgr}^{\mathbb{Z}\times G}\text{-}A.$$

Since $T(P) = P/PA_+$ we see that *the support of $T(P)$ is contained in the support of P.*

Observe now that *if M is a finitely generated $\mathbb{Z} \times G$-graded A-module and $MA_+ = M$ then $M = 0$*; for if $M \neq 0$ there is a minimal $\omega \in \mathbb{Z}$ such that $M_{(\omega,G)} \neq 0$, but $(MA_+)_{(\omega,G)} = 0$. It follows from Proposition 6.1.5 that for each graded finitely generated projective A-module P there is a noncanonical isomorphism $P \cong T(P) \otimes_{A_{(0,G)}} A$ as in (6.3) which respects the $\mathbb{Z} \times G$-grading. Explicitly, for a given $(\omega, g) \in \mathbb{Z}\times G$ we have an isomorphism of abelian groups

$$P_{(\omega,g)} \cong \bigoplus_{(\kappa,h)} T(P)_{(\kappa,h)} \otimes A_{(-\kappa+\omega,-h+g)}; \tag{6.6}$$

the tensor product $T(P)_{(\kappa,h)} \otimes A_{(-\kappa+\omega,-h+g)}$ denotes, by convention, the abelian subgroup of $T(P) \otimes_{A_{(0,G)}} A$ generated by primitive tensors of the form $x \otimes y$ with homogeneous elements $x \in T(P)$ of degree (κ, h) and $y \in A$ of degree $(-\kappa + \omega, -h + g)$.

For a $\mathbb{Z} \times G$-graded A-module P write $P = \bigoplus_{\omega\in\mathbb{Z}} P_{(\omega,G)}$, where $P_{(\omega,G)} = \bigoplus_{g\in G} P_{(\omega,g)}$. For $\lambda \in \mathbb{Z}$ let $F^\lambda P$ denote the A-submodule of P generated by the elements of $\bigcup_{\omega\leq\lambda} P_{(\omega,G)}$; this is $\mathbb{Z} \times G$-graded again. As an explicit example, we have

$$F^\lambda A(\omega, g) = \begin{cases} A(\omega, g) & \text{if } \lambda \geq -\omega, \\ 0 & \text{otherwise.} \end{cases}$$

Suppose that P is a graded finitely generated projective A-module. Since the support of A is contained in $\mathbb{N} \times G$ there exists $n \in \mathbb{Z}$ such that $F^{-n}P = 0$ and $F^n P = P$. Write $\mathrm{Pgr}_n^{\mathbb{Z}\times G}\text{-}A$ for the full subcategory of $\mathrm{Pgr}^{\mathbb{Z}\times G}\text{-}A$ spanned by those modules P which satisfy $F^{-n}P = 0$ and $F^n P = P$. Then $\mathrm{Pgr}^{\mathbb{Z}\times G}\text{-}A$ is the filtered union of the $\mathrm{Pgr}_n^{\mathbb{Z}\times G}\text{-}A$.

Let $P \in \mathrm{Pgr}_{\mathrm{n}}^{\mathbb{Z}\times\mathrm{G}}\text{-}A$; we want to identify $F^{\lambda}P$. By definition, the A-module $F^{\lambda}P$ is generated by the elements of $P_{(\omega,g)}$ for $\omega \leq \lambda$, with $P_{(\omega,g)}$ having been identified in (6.6). We remark that the direct summands in (6.6) indexed by $\kappa > \omega$ are trivial as A has support in $\mathbb{N} \times G$. On the other hand, for $\omega \geq \kappa$ a given primitive tensor, $x \otimes y \in P_{(\omega,g)}$ with $x \in T(P)_{(\kappa,h)}$ and $y \in A_{(-\kappa+\omega,-h+g)}$ can always be rewritten, using the right A-module structure of $T(P) \otimes_{A_{(0,G)}} A$, as

$$x \otimes y = (x \otimes 1) \cdot y, \quad \text{where} \quad x \otimes 1 \in T(P)_{(\kappa,h)} \otimes A_{(0,0)} \subseteq P_{(\kappa,h)}.$$

That is, the A-module $F^{\lambda}P$ is generated by those summands of (6.6) with $\kappa = \omega \leq \lambda$. We claim now that $F^{\lambda}P$ is isomorphic to

$$M^{(\lambda)} = \bigoplus_{\kappa \leq \lambda} T(P)_{(\kappa,-)} \otimes_{A_{(0,G)}} A(-\kappa, 0), \tag{6.7}$$

considering $T(P)_{(\kappa,-)}$ as a $\mathbb{Z} \times G$-graded $A_{(0,G)}$-module with support in $\{0\} \times G$. The homogeneous components of $M^{(\lambda)}$ are given by

$$M^{(\lambda)}_{(\omega,g)} = \bigoplus_{\kappa \leq \lambda} \bigoplus_{h \in G} T(P)_{(\kappa,h)} \otimes A(-\kappa, 0)_{(\omega,-h+g)}.$$

Now elements of the form

$$x \otimes 1 \in T(P)_{(\kappa,h)} \otimes A(-\kappa, 0)_{(\kappa,-h+h)} \subseteq M^{(\lambda)}_{(\kappa,h)}$$

clearly form a set of A-module generators for $M^{(\lambda)}$ so that, by the argument given above, $F^{\lambda}P$ and $M^{(\lambda)}$ have the same generators in the same degrees. The claim follows. The module $F^{\lambda}P$ is finitely generated (by those generators of P that have $\mathbb{Z}$-degree at most λ). Since $T(P)$ is a finitely generated projective $A_{(0,G)}$-module so is its summand $T(P)_{(\lambda_k,G)}$; consequently, $P \mapsto F^k P$ is an endofunctor of $\mathrm{Pgr}_{\mathrm{n}}^{\mathbb{Z}\times\mathrm{G}}\text{-}A$. It is exact, as can be deduced from the (noncanonical) isomorphism in (6.7), using exactness of tensor products.

From the isomorphism $F^k P \cong M^{(\lambda_k)}$, see (6.7), we obtain an isomorphism

$$F^{k+1}P/F^k P \cong T(P)_{(\lambda_k,G)} \otimes_{A_{(0,G)}} A(-\lambda_k, G); \tag{6.8}$$

in particular, $F^{k+1}P/F^k P \in \mathrm{Pgr}_{\mathrm{n}}^{\mathbb{Z}\times\mathrm{G}}\text{-}A$.

The isomorphism (6.8) depends on the isomorphism (6.7), and thus ultimately on (6.3). The latter depends on a choice of a section g of $P \to T(P)$. Given another section g_0, the difference $g - g_0$ has image in $\ker(P \to T(P)) = PA_+$. Since A_+ consists of elements of positive $\mathbb{Z}$-degree only, this implies that the isomorphism $F^{k+1}P \cong M^{(\lambda_{k+1})}$ does not depend on g up to elements in $F^k P$; in other words, the quotient $F^{k+1}P/F^k P$ is independent of the choice of g. Thus the isomorphism (6.8) is, in fact, a natural isomorphism of functors.

We are now in a position to perform the K-theoretical calculations. First define the exact functor

$$\Theta_q : \mathrm{Pgr}_{\mathrm{q}}^{\mathbb{Z}\times \mathrm{G}}\text{-}A_{(0,G)} \longrightarrow \mathrm{Pgr}_{\mathrm{q}}^{\mathbb{Z}\times \mathrm{G}}\text{-}A,$$
$$P = \bigoplus_{\omega} P_{(\omega,G)} \longmapsto \bigoplus_{\omega} P_{(\omega,G)} \otimes_{A_{(0,G)}} A(-\omega, G);$$

here $\mathrm{Pgr}_{\mathrm{q}}^{\mathbb{Z}\times \mathrm{G}}\text{-}A_{(0,G)}$ denotes the full subcategory of $\mathrm{Pgr}^{\mathbb{Z}\times \mathrm{G}}\text{-}A_{(0,G)}$ spanned by modules with support in $[-q, q] \times G$, and $P_{(\omega,G)}$ on the right is considered as a $\mathbb{Z} \times G$-graded $A_{(0,G)}$-module with support in $\{0\} \times G$.

Next define the exact functor

$$\Psi_q : \mathrm{Pgr}_{\mathrm{q}}^{\mathbb{Z}\times \mathrm{G}}\text{-}A \longrightarrow \mathrm{Pgr}_{\mathrm{q}}^{\mathbb{Z}\times \mathrm{G}}\text{-}A_{(0,G)},$$
$$P \longmapsto \bigoplus_{\omega\in\mathbb{Z}} T(P)_{(\omega,G)};$$

here $T(P)_{(\omega,G)}$ is considered as an $A_{(0,G)}$-module with support in $\{\omega\} \times G$.

Now $\Psi_q \circ \Theta_q \cong \mathrm{id}$; indeed, the composition sends the summand $P_{(\omega,G)}$ of P to the κ-indexed direct sum of

$$T(P_{(\omega,G)} \otimes_{A_{(0,G)}} A(-\omega, G))_{(\kappa,G)} \cong \begin{cases} P_{(\omega,G)} & \text{if } \kappa = \omega, \\ 0 & \text{otherwise.} \end{cases}$$

In particular, $\Psi_q \circ \Theta_q$ induces the identity on K-groups. As for the other composition, we have

$$\Theta_q \circ \Psi_q(P) = \bigoplus_{\omega} T(P)_{(\omega,G)} \otimes_{A_{(0,G)}} A(-\omega, G) \underset{(6.8)}{=} \bigoplus_{j=-q}^{q-1} F^{j+1}P/F^jP.$$

Since $F^q = \mathrm{id}$, additivity for characteristic filtrations [81, p. 107, Corollary 2] implies that $\Theta_q \circ \Psi_q$ induces the identity on K-groups.

For any $P \in \mathrm{Pgr}^{\mathbb{Z}\times \mathrm{G}}\text{-}A_{(0,G)}$ we have

$$(P \otimes_{A_{(0,G)}} A(-\omega, G))(0, g) = P(0, g) \otimes_{A_{(0,G)}} A(-\omega, G),$$

by direct calculation. Hence the functor Θ_q induces a $\mathbb{Z}[G]$-linear isomorphism on K-groups. Since K-groups are compatible with direct limits, letting $q \to \infty$ yields a $\mathbb{Z}[G]$-linear isomorphism

$$K_i^{\mathbb{Z}\times G}(A) \cong K_i^{\mathbb{Z}\times G}(A_{(0,G)})$$

and thus, by Lemma 6.1.6, a $\mathbb{Z}[\mathbb{Z} \times G]$-module isomorphism

$$K_i^{\mathbb{Z}\times G}(A) \cong K_i^{G}(A_{(0,G)}) \otimes_{\mathbb{Z}[G]} \mathbb{Z}[\mathbb{Z} \times G]. \qquad \square$$

6.2 The fundamental theorem of K-theory

6.2.1 Quillen's K-theory of exact categories

Recall that an exact category $\mathcal{P}$ is a full additive subcategory of an abelian category $\mathcal{A}$ which is closed under extension (see Definition 3.12.1). An *exact functor* $\mathcal{F} : \mathcal{P} \to \mathcal{P}'$, between exact categories $\mathcal{P}$ and $\mathcal{P}'$, is an additive functor which preserves the short exact sequences.

From an exact category $\mathcal{P}$, Quillen constructed abelian groups $K_n(\mathcal{P})$, $n \geq 0$ (see [81]). In fact, for a fixed n, K_n is a functor from the category of exact categories, with exact functors as morphisms to the category of abelian groups. If $f : \mathcal{A} \to \mathcal{B}$ is an exact functor, then we denote the corresponding group homomorphism by $\overline{f} : K_n(\mathcal{A}) \longrightarrow K_n(\mathcal{B})$.

When $\mathcal{P} = \text{Pr-}A$, the category of finitely generated projective A-modules, then $K_0(\mathcal{P}) = K_0(A)$. Quillen proved that basic theorems established for K_0-group, such as Dévissage, resolution and localisation theorems ([89, 95]), can be established for these higher K-groups. We briefly mention the main theorems of K-theory which will be used in §6.2.4. The following statements were established by Quillen in [81] (see also [88, 95]).

1 **The Grothendieck group** For an exact category $\mathcal{P}$, the abelian group $K_0(\mathcal{P})$ coincides with the construction given in Definition 3.12.1(1). In particular, Quillen's construction gives the Grothendieck group $K_0(A) = K_0(\text{Pr-}A)$ and the graded Grothendieck group $K_0^{\text{gr}}(A) = K_0(\text{Pgr-}A)$, for nongraded and graded rings, respectively.

2 **Dévissage** Let $\mathcal{A}$ be an abelian category and $\mathcal{B}$ be a nonempty full subcategory closed under subobjects, quotient objects and finite products in $\mathcal{A}$. Thus $\mathcal{B}$ is an abelian category and the inclusion $\mathcal{B} \hookrightarrow \mathcal{A}$ is exact. If any element A in $\mathcal{A}$ has a finite filtration

$$0 = A_0 \subseteq A_1 \subseteq \cdots \subseteq A_n = A,$$

with A_i/A_{i-1} in $\mathcal{B}$, $1 \leq i \leq n$, then the inclusion $\mathcal{B} \hookrightarrow \mathcal{A}$ induces isomorphisms $K_n(\mathcal{B}) \cong K_n(\mathcal{A})$, $n \geq 0$.

3 **Resolution** Let $\mathcal{M}$ be an exact category and $\mathcal{P}$ be a full subcategory closed under extensions in $\mathcal{M}$. Suppose that if $0 \to M \to P \to P' \to 0$ is exact in $\mathcal{M}$ with P and P' in $\mathcal{P}$, then M is in $\mathcal{P}$. Moreover, for every M in $\mathcal{M}$, there is a finite $\mathcal{P}$ resolution of M,

$$0 \longrightarrow P_n \longrightarrow \cdots \longrightarrow P_1 \longrightarrow P_0 \longrightarrow M \longrightarrow 0.$$

Then the inclusion $\mathcal{P} \hookrightarrow \mathcal{M}$ induces isomorphisms $K_n(\mathcal{P}) \cong K_n(\mathcal{M})$, $n \geq 0$.

4 **Localisation** Let $\mathcal{S}$ be a *Serre subcategory* of an abelian category $\mathcal{A}$ (*i.e.*, $\mathcal{S}$ is an abelian subcategory of $\mathcal{A}$, closed under subobjects, quotient objects and extensions in $\mathcal{A}$), and let $\mathcal{A}/\mathcal{S}$ be the quotient abelian category ([94, Exercise 10.3.2]). Then there is a natural long exact sequence

$$\cdots \longrightarrow K_{n+1}(\mathcal{A}/\mathcal{S}) \xrightarrow{\delta} K_n(\mathcal{S}) \longrightarrow K_n(\mathcal{A}) \longrightarrow K_n(\mathcal{A}/\mathcal{S}) \longrightarrow \cdots.$$

5 **Exact sequence of functors** Let $\mathcal{P}$ and $\mathcal{P}'$ be exact categories and

$$0 \longrightarrow f' \longrightarrow f \longrightarrow f'' \longrightarrow 0$$

be an exact sequence of exact functors from $\mathcal{P}$ to $\mathcal{P}'$. Then

$$\overline{f} = \overline{f}' + \overline{f}'' : K_n(\mathcal{P}) \longrightarrow K_n(\mathcal{P}').$$

6.2.2 Base change and transfer functors

Let A be a ring with identity. If A is a right Noetherian ring then the category of finitely generated right A-modules, mod-A, form an abelian category. Thus one can define the K-groups of this category, $G_n(A) := K_n(\text{mod-}A)$, $n \in \mathbb{N}$. If, further, A is regular, then by the resolution theorem (§6.2.1), $G_n(A) = K_n(A)$.

Let $A \to B$ be a ring homomorphism. This makes B an A-module. If B is a finitely generated A-module, then any finitely generated B-module can be considered as a finitely generated A-module. This induces an exact functor, called a *transfer functor*, mod-$B \to$ mod-A. Since K-theory is a functor from the category of exact categories with exact functors as morphisms, we get a transfer map $G_n(B) \to G_n(A)$, $n \in \mathbb{N}$. A similar argument shows that, if $A \to B$ is a graded homomorphism, then we have a transfer map $G_n^{\mathrm{gr}}(B) \to G_n^{\mathrm{gr}}(A)$, $n \in \mathbb{N}$.

On the other hand, if under the ring homomorphism $A \to B$, B is a flat A-module, then the *base change functor* $- \otimes_A B$: Mod-$A \to$ Mod-B is exact, which restricts to $- \otimes_A B$: mod-$A \to$ mod-B. This in turn induces the group homomorphism $G_n(A) \to G_n(B)$, $n \in \mathbb{N}$. Note that any ring homomorphism $A \to B$ induces an exact functor $- \otimes_A B$: Pr-$A \to$ Pr-B, which induces the homomorphism $K_n(A) \to K_n(B)$. Similar statements can be written for graded rings A and B and the graded homomorphism $A \to B$. However, to prove the fundamental theorem of K-theory (see (6.13)), we need to use the Dévissage and localisation theorems, which are only valid for abelian categories. This forces us to work in the abelian category gr-A (assuming A is right Noetherian), rather than the exact category Pgr-A.

6.2.3 A localisation exact sequence for graded rings

Let A be a right regular Noetherian Γ-graded ring, *i.e.*, A is graded right regular and graded right Noetherian. We assume the ring is graded regular so that we can work with K-theory instead of G-theory. We recall the concept of graded regular rings. For a graded right module M, the minimum length of all graded projective resolutions of M is defined as the *graded projective dimension* of M and denoted by $\operatorname{pdim}^{\mathrm{gr}}(M)$. If M does not admit a finite graded projective resolution, then we set $\operatorname{pdim}^{\mathrm{gr}}(M) = \infty$. The *graded right global dimension* of the graded ring A is the supremum of the graded projective dimension of all the graded right modules over A and denoted by $\operatorname{grdim}^{\mathrm{gr}}(A)$. We say A is *graded right regular* if the graded right global dimension is finite. As soon as Γ is a trivial group, the above definitions become the standard definitions of projective dimension, denoted by pdim, and the global dimension, denoted by grdim, in ring theory. In the case of a $\mathbb{Z}$-graded ring, we have the following relation between the dimensions (see [74, Theorem II.8.2]):

$$\operatorname{grdim}^{\mathrm{gr}}(A) \leq \operatorname{grdim}(A) \leq 1 + \operatorname{grdim}^{\mathrm{gr}}(A). \tag{6.9}$$

This shows that if A is graded regular, then A is regular. This will be used in §6.4.

Let S be a central multiplicative closed subset of A, consisting of homogeneous elements. Then $S^{-1}A$ is a Γ-graded ring (see Example 1.1.11). Let gr_S-A be the category of S-torsion graded finitely generated right A-modules, *i.e.*, a graded finitely generated A-module M such that $Ms = 0$ for some $s \in S$. Clearly gr_S-A is a Serre subcategory of the abelian category gr-A and

$$\text{gr-}A/\,\mathrm{gr}_S\text{-}A \approx \text{gr-}(S^{-1}A) \tag{6.10}$$

(see [94, Exercise 10.3.2]).

Now let $S = \{s^n \mid n \in \mathbb{N}\}$, where s is a homogeneous element in the centre of A. Then any module M in gr_S-A has a filtration

$$0 = Ms^k \subseteq \cdots \subseteq Ms \subseteq M,$$

and clearly the consecutive quotients in the filtration are Γ-graded A/sA-modules. Considering gr-A/sA as a full subcategory of gr_S-A via the transfer map $A \to A/sA$ (see §6.2.2), and using the Dévissage theorem, we get, for any $n \in \mathbb{N}$,

$$K_n(\mathrm{gr}_S\text{-}A) \cong K_n^{\mathrm{gr}}(A/sA).$$

Now using the localisation theorem for (6.10), we obtain a long localisation exact sequence

$$\cdots \longrightarrow K_{n+1}^{\mathrm{gr}}(A_s) \longrightarrow K_n^{\mathrm{gr}}(A/sA) \longrightarrow K_n^{\mathrm{gr}}(A) \longrightarrow K_n^{\mathrm{gr}}(A_s) \longrightarrow \cdots, \tag{6.11}$$

where $A_s := S^{-1}A$.

In particular, consider the Γ-graded rings $A[y]$ and $A[y, y^{-1}]$, where $\deg(y) = \alpha$, $\deg(y^{-1}) = -\alpha$, $\alpha \in \Gamma$. For $A[y]$ and $S = \{y^n \mid n \in \mathbb{N}\}$, the sequence 6.11 reduces to

$$\cdots \longrightarrow K^{\mathrm{gr}}_{n+1}(A[y, y^{-1}]) \longrightarrow K^{\mathrm{gr}}_n(A) \longrightarrow K^{\mathrm{gr}}_n(A[y]) \longrightarrow K^{\mathrm{gr}}_n(A[y, y^{-1}]) \longrightarrow \cdots. \tag{6.12}$$

This long exact sequence will be used in §6.4 in two occasions (with $\deg(y) = 1 \in \mathbb{Z}$ and $\deg(y) = (1, -1) \in \mathbb{Z} \times \mathbb{Z}$) to relate K^{gr}_n-groups to K_n-groups.

6.2.4 The fundamental theorem

Let A be a right Noetherian and regular ring. Consider the polynomial ring $A[x]$. The fundamental theorem of K-theory gives that, for any $n \in \mathbb{N}$,

$$K_n(A[x]) \cong K_n(A). \tag{6.13}$$

In this section we prove that if A is a positively graded $\mathbb{Z}$-graded ring, then

$$\boxed{K_n(A) \cong K_n(A_0).}$$

This, in particular, gives (6.13). We will follow Gersten's treatment in [39]. The proof shows how the graded K-theory is effectively used to prove a theorem in (nongraded) K-theory.

Let A be a positively $\mathbb{Z}$-graded ring, *i.e.*, $A = \bigoplus_{i \in \mathbb{N}} A_i$. Define a $\mathbb{Z}$-grading on the polynomial ring $A[x]$ as follows (see Example 1.1.7):

$$A[x]_n = \bigoplus_{i+j=n} A_i x^j. \tag{6.14}$$

Clearly this is also a positively graded ring and $A[x]_0 = A_0$.

Throughout the proof of the fundamental theorem we consider two evaluation maps on $A[x]$ as follows.

1 Consider the evaluation homomorphism at 0, *i.e.*,

$$\begin{aligned} e_0 : A[x] &\longrightarrow A, \\ f(x) &\longmapsto f(0). \end{aligned} \tag{6.15}$$

This is a graded homomorphism (with the grading defined in (6.14)), and A becomes a graded finitely generated $A[x]$-module. This induces an exact transfer functor (see §6.2.2)

$$\begin{aligned} i : \text{gr-}A &\longrightarrow \text{gr-}A[x], \\ M &\longmapsto M \end{aligned} \tag{6.16}$$

and in turn a homomorphism

$$\bar{i} : G_n^{\mathrm{gr}}(A) \to G_n^{\mathrm{gr}}(A[x]).$$

Let M be a graded A-module. Under the homomorphism 6.16, M is also a grade $A[x]$-module. On the other hand, the canonical graded homomorphism $A \to A[x]$, gives the graded $A[x]$-module $M \otimes_A A[x] \cong_{\mathrm{gr}} M[x]$. Then we have the following short exact sequence of graded $A[x]$-modules:

$$0 \longrightarrow M[x](-1) \xrightarrow{x} M[x] \xrightarrow{e_0} M \longrightarrow 0. \tag{6.17}$$

Note that this short exact sequence is not a split sequence.

2 Consider the evaluation homomorphism at 1, *i.e.*,

$$\begin{aligned} e_1 : A[x] &\longrightarrow A, \\ f(x) &\longmapsto f(1). \end{aligned} \tag{6.18}$$

Note that this homomorphism is not graded. Clearly $\ker(e_1) = (1 - x)$ (*i.e.*, the ideal generated by $1 - x$). Thus $A[x]/(1 - x) \cong A$. If M is a (graded) $A[x]$-module, then

$$M \otimes_{A[x]} A \cong M \otimes_{A[x]} A[x]/(1 - x) \cong M/M(1 - x)$$

is an A-module. Thus the homomorphism e_1 induces the functor

$$\begin{aligned} \mathcal{F} := - \otimes_{A[x]} A : \text{gr-}A[x] &\longrightarrow \text{mod-}A, \\ M &\longmapsto M/M(1 - x). \end{aligned} \tag{6.19}$$

We need the following lemma (see [74, Lemma II.8.1]).

Lemma 6.2.1 *Consider the functor $\mathcal{F}$ defined in* (6.19). *Then we have the following.*

(1) *$\mathcal{F}$ is an exact functor.*
(2) *For a graded finitely generated $A[x]$-module M, $\mathcal{F}(M) = 0$ if and only if $Mx^n = 0$ for some $n \in \mathbb{N}$.*

Proof (1) Since the tensor product is a right exact functor, we are left to show that if $0 \to M' \to M$ is exact, then by (6.19),

$$0 \longrightarrow M'/M'(1 - x) \longrightarrow M/M(1 - x)$$

is exact, *i.e.*, $M' \cap M(1 - x) = M'(1 - x)$.

Let $m' \in M' \cap M(1 - x)$. Then

$$m' = m(1 - x), \tag{6.20}$$

for some $m \in M$. Since M' and M are graded modules, $m' = \sum m'_i$ and $m =$

$\sum m_i$, with $\deg(m'_i) = \deg(m_i) = i$. Comparing the degrees in Equation (6.20), we have $m_i - m_{i-1}x = m'_i \in M'_i$. Let j be the smallest i in the support of m such that $m_j \notin M'_j$. Then $m_{j-1}x \in M'_j$ which implies that $m_j = m'_j - m_{j-1}x \in M'_j$, which is a contradiction. Thus $m \in M'$, and so $m' \in M'(1-x)$.

(2) The proof is elementary and left to the reader. □

By Lemma 6.2.1(2), $\ker(\mathcal{F})$ is an (abelian) full subcategory of gr-$A[x]$, with objects the graded $A[x]$-modules M such that $Mx^n = 0$, for some $n \in \mathbb{N}$. Define $\mathcal{C}_1$ as the (abelian) full subcategory of gr-$A[x]$ with objects the graded $A[x]$-modules M such that $Mx = 0$. Thus $\mathcal{C}_1 \hookrightarrow \ker(\mathcal{F})$ is an exact functor. Moreover, for any object M in $\ker(\mathcal{F})$, we have a finite filtration of graded $A[x]$-modules

$$0 = Mx^n \subseteq Mx^{n-1} \subseteq \cdots \subseteq Mx \subseteq M,$$

such that $Mx^i/Mx^{i+1} \in \mathcal{C}_1$. Thus, by the Dévissage theorem (§6.2.1), we have

$$K_n(\mathcal{C}_1) \cong K_n(\ker(\mathcal{F})). \tag{6.21}$$

Next recall the exact functor $i : \text{gr-}A \longrightarrow \text{gr-}A[x]$ from (6.16). It is easy to see that this functor induces an equivalence between the categories

$$i : \text{gr-}A \longrightarrow \mathcal{C}_1, \tag{6.22}$$

and thus on the level of K-groups. Combining this with (6.21) we get

$$K_n^{\text{gr}}(A) = K_n(\mathcal{C}_1) \cong K_n(\ker(\mathcal{F})). \tag{6.23}$$

Since by Lemma 6.2.1, $\mathcal{F}$ is exact, $\ker(\mathcal{F})$ is closed under subobjects, quotients and extensions, i.e, it is a Serre subcategory of gr-$A[x]$. We next prove that

$$\text{gr-}A[x]/\ker(\mathcal{F}) \cong \text{mod-}A. \tag{6.24}$$

Any finitely generated module over a Noetherian ring is finitely presented. We need the following general lemma to invoke the localisation.

Lemma 6.2.2 *Let M be a finitely presented A-module. Then there is a graded finitely presented $A[x]$-module N such that $\mathcal{F}(N) \cong M$, where $\mathcal{F}$ is the functor defined in* (6.19).

Proof Since $\mathcal{F} = - \otimes_{A[x]} A$, for any free A-module M of rank k we have $\mathcal{F}(\bigoplus_k A[x]) \cong \bigoplus_k A \cong M$. Let M be a finitely presented A-module. Then there is an exact sequence

$$\bigoplus_m A \xrightarrow{f} \bigoplus_n A \longrightarrow M \longrightarrow 0.$$

If we show that there are graded free $A[x]$-modules F_1 and F_2 and a graded homomorphism $g : F_1 \to F_2$ such that the following diagram is commutative,

$$\begin{array}{ccc} \mathcal{F}(F_1) & \xrightarrow{\mathcal{F}(g)} & \mathcal{F}(F_2) \\ \downarrow \cong & & \downarrow \cong \\ \bigoplus_m A & \xrightarrow{f} & \bigoplus_n A, \end{array}$$

then since $\mathcal{F}$ is right exact, one can complete the diagram

$$\begin{array}{ccccccc} \mathcal{F}(F_1) & \xrightarrow{\mathcal{F}(g)} & \mathcal{F}(F_2) & \longrightarrow & \mathcal{F}(\operatorname{coker}(g)) & \longrightarrow & 0 \\ \downarrow \cong & & \downarrow \cong & & \vdots \cong & & \\ \bigoplus_m A & \xrightarrow{f} & \bigoplus_n A & \longrightarrow & M & \longrightarrow & 0. \end{array}$$

This shows that $\mathcal{F}(\operatorname{coker}(g)) \cong M$ and we will be done.

Thus, suppose $f : \bigoplus_m A \to \bigoplus_n A$ is an A-module homomorphism. With a misuse of notation, denote the standard basis for both A-modules A^m and A^n by e_i. Then

$$f(e_i) = \sum_{j=1}^{n} a_{ij} e_j,$$

where $1 \leq i \leq m$ and $a_{ij} \in A$. Decomposing a_{ij} into its homogeneous components, $a_{ij} = \sum_k a_{ij_k}$, since the number of a_{ij}, $1 \leq i \leq m$, $1 \leq j \leq n$, is finite, there is an $N \in \mathbb{N}$ such that $a_{ij_N} = 0$ for all i, j. Now consider the graded free $A[x]$-modules $\bigoplus_m A[x](-N)$ and $\bigoplus_n A[x]$, with the standard bases (which are of degrees N and 0 in $\bigoplus_m A[x](-N)$ and $\bigoplus_n A[x]$, respectively). Define

$$\begin{aligned} g : \bigoplus_m A[x](-N) &\longrightarrow \bigoplus_n A[x], \\ e_i &\longmapsto \sum_j \Big(\sum_k x^{N-k} a_{ij_k} \Big) e_j. \end{aligned}$$

This is a graded $A[x]$-module homomorphism. Further, $\mathcal{F}(g)$ amounts to evaluation of g at $x = 1$ which coincides with the map f. This finishes the proof. □

Remark 6.2.3 The proof of Lemma 6.2.2, in particular, shows that if $f \in \mathbb{M}_n(A)$, then there is a $g \in \mathbb{M}_n(A[x])(\overline{\delta})$ such that evaluation of g at 1 gives f.

Lemma 6.2.2, along with standard results of localisation theory (see [89, p. 114, Theorem 5.11]) implies that

$$\text{gr-}A[x]/\ker(\mathcal{F}) \cong \text{mod-}A.$$

Now we are ready to apply the localisation theorem (§6.2.1) to the sequence

$$\ker(\mathcal{F}) \hookrightarrow \text{gr-}A[x] \longrightarrow \text{gr-}A[x]/\ker(\mathcal{F}) \cong \text{mod-}A,$$

to obtain a long exact sequence

$$\cdots \longrightarrow K_{n+1}(A) \longrightarrow K_n(\ker(\mathcal{F})) \longrightarrow K_n^{\text{gr}}(A[x]) \longrightarrow K_n(A) \longrightarrow \cdots.$$

Now, using (6.23) to replace $K_n(\ker(\mathcal{F}))$, we get

$$\cdots \longrightarrow K_{n+1}(A) \longrightarrow K_n^{\text{gr}}(A) \xrightarrow{\bar{i}} K_n^{\text{gr}}(A[x]) \longrightarrow K_n(A) \longrightarrow \cdots. \tag{6.25}$$

Next, we show that the homomorphism $\bar{i}$ is injective. Note that the homomorphism $\bar{i}$ on the level of K-groups induced by the composition of exact functors (see (6.16) and (6.22)),

$$\text{Pgr-}A \hookrightarrow \text{gr-}A \longrightarrow \mathcal{C}_1 \hookrightarrow \ker(\mathcal{F}) \hookrightarrow \text{gr-}A[x],$$

which sends a graded finitely generated projective A-module M, to M considered as graded $A[x]$-module. Define three exact functors $\phi_i : \text{Pgr-}A \to \text{gr-}A[x]$, $1 \leq i \leq 3$, as follows:

$$\begin{aligned} \phi_1(M) &= M[x](-1), \\ \phi_2(M) &= M[x], \\ \phi_3(M) &= M. \end{aligned} \tag{6.26}$$

Now the exact sequence (6.17) and the exact sequence of functors theorem (§6.2.1) immediately give

$$\overline{\phi}_3 = \overline{\phi}_2 - \overline{\phi}_1. \tag{6.27}$$

Note that $\overline{\phi}_3 = \bar{i}$.

Now invoking Quillen's Theorem 6.1.1, from the exact sequence (6.25), we get

$$\cdots \longrightarrow K_{n+1}(A) \longrightarrow \mathbb{Z}[t,t^{-1}]\otimes K_n(A_0) \xrightarrow{\bar{i}} \mathbb{Z}[t,t^{-1}]\otimes K_n(A_0) \longrightarrow K_n(A) \longrightarrow \cdots. \tag{6.28}$$

Further, the maps $\overline{\phi}_2$ and $\overline{\phi}_1$ becomes $1 \otimes 1$ and $t \otimes 1$, respectively (see (6.2)). Thus from (6.27) we get

$$\bar{i} = (1-t)\otimes 1 : \mathbb{Z}[t,t^{-1}] \otimes_{\mathbb{Z}} K_n(A_0) \longrightarrow \mathbb{Z}[t,t^{-1}] \otimes_{\mathbb{Z}} K_n(A_0).$$

This shows that $\bar{i}$ is an injective map. Thus the exact sequence (6.28) reduces to

$$0 \longrightarrow \mathbb{Z}[t,t^{-1}] \otimes_{\mathbb{Z}} K_n(A_0) \xrightarrow{\bar{i}} \mathbb{Z}[t,t^{-1}] \otimes_{\mathbb{Z}} K_n(A_0) \longrightarrow K_n(A) \longrightarrow 0. \tag{6.29}$$

Again, from the description of $\bar{i}$ it follows that

$$K_n(A) \cong K_n(A_0).$$

6.3 Relating $K_*^{\mathrm{gr}}(A)$ to $K_*(A_0)$

If A is a strongly Γ-graded ring, then Gr-A is equivalent to Mod-A_0. Thus we have $K_n^{\mathrm{gr}}(A) \cong K_n(A_0)$, for any $n \in \mathbb{N}$ (see §3.1.3). In general (for graded Noetherian regular rings) these two groups are related via a long exact sequence which we will establish in this section (see (6.31)).

Let A be a Γ-graded ring. For a subset $\Omega \subseteq \Gamma$, consider the full subcategory Gr_Ω-A of Gr-A, of all graded right A-modules M as objects such that $M_\omega = 0$, for all $\omega \in \Omega$. This is a Serre subcategory of Gr-A. We need the following lemma.

Lemma 6.3.1 *Let A be a Γ-graded ring and Ω a subgroup of Γ. Then we have*

$$\mathrm{Gr}^\Gamma\text{-}A/\,\mathrm{Gr}_\Omega\text{-}A \cong \mathrm{Gr}^\Omega\text{-}A_\Omega. \tag{6.30}$$

In particular, $\mathrm{Gr}\text{-}A/\,\mathrm{Gr}_0\text{-}A \cong \mathrm{Mod}\text{-}A_0$.

Proof Consider a family Σ of morphisms f in Gr^Γ-A such that $\ker(f)$ and $\mathrm{coker}(f)$ are in Gr_Ω-A. By definition

$$\mathrm{Gr}^\Gamma\text{-}A/\,\mathrm{Gr}_\Omega\text{-}A = \Sigma^{-1}A.$$

Note that $f : M \to N$ is in Σ if and only if $f|_{M_\Omega} : M_\Omega \to N_\Omega$ is an A_Ω-module isomorphism. Consider the (forgetful) functor

$$\begin{aligned}(-)_\Omega : \mathrm{Gr}^\Gamma\text{-}A &\longrightarrow \mathrm{Gr}^\Omega\text{-}A_\Omega,\\ M &\longmapsto M_\Omega\end{aligned}$$

(see §1.2.8). Under $(-)_\Omega$ the morphisms of Σ are sent to invertible morphisms in Gr^Ω-A_Ω. Thus by the property of quotient categories, there is an induced functor ψ which makes the following diagram commutative:

$$\begin{array}{ccc}\mathrm{Gr}^\Gamma\text{-}A & \xrightarrow{\ \phi\ } & \mathrm{Gr}^\Gamma\text{-}A/\,\mathrm{Gr}_\Omega\text{-}A\\ {\scriptstyle (-)_\Omega}\downarrow & \swarrow{\scriptstyle \psi} & \\ \mathrm{Gr}^\Omega\text{-}A_\Omega & & \end{array}$$

The functor ϕ is defined as follows:

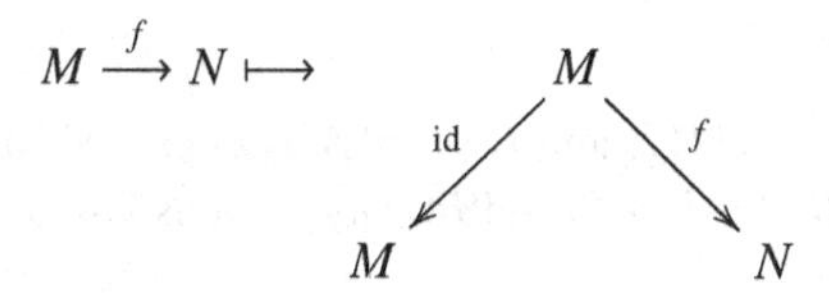

and the functor ψ is defined as follows:

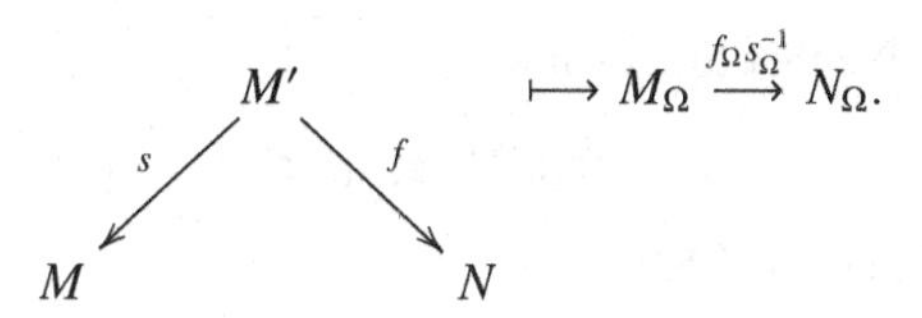

One can now check that ψ is an equivalence of categories. This finishes the proof. □

When A is a (right) regular Noetherian graded ring, the quotient category identity (6.30) holds for the corresponding graded finitely generated modules, *i.e.*, $\text{gr-}A/\,\text{gr}_0\text{-}A \cong \text{mod-}A_0$ and the localisation theorem gives a long exact sequence,

$$\cdots \longrightarrow K_{n+1}(A_0) \overset{\delta}{\longrightarrow} K_n(\text{gr}_0\text{-}A) \longrightarrow K_n^{\mathrm{gr}}(A) \longrightarrow K_n(A_0) \longrightarrow \cdots . \qquad (6.31)$$

Note that if A is a strongly graded ring, then $\text{Gr}_0\text{-}A = 0$, for any $\alpha \in \Gamma$. In particular, the long exact sequence 6.31 gives $K_n^{\mathrm{gr}}(A) \cong K_n(A_0)$, for any $n \in \mathbb{N}$.

6.4 Relating $K_*^{\mathrm{gr}}(A)$ to $K_*(A)$

In [20], van den Bergh, following the methods used to prove the fundamental theorem of K-theory (§6.2.4), established a long exact sequence relating the graded K-theory of a ring to the nongraded K-theory in the case of right regular Noetherian $\mathbb{Z}$-graded rings. In this section we will present van den Bergh's observation. As a consequence, we will see that in the case of graded Grothendieck group, the shift of modules is the only difference between the K_0^{gr} and K_0-groups (Corollary 6.4.2).

Let A be a right regular Noetherian $\mathbb{Z}$-graded ring. Thus G^{gr}-theory and G-theory become K^{gr}-theory and K-theory, respectively (see §6.2.3). Consider the $\mathbb{Z}$-graded ring $A[y]$, where $\deg(y) = 1$ and $\mathbb{Z} \times \mathbb{Z}$-graded ring $B = A[s]$, where the homogeneous components are defined by $B_{(n,m)} = A_m s^n$. Moreover, consider the $\mathbb{Z} \times \mathbb{Z}$-graded rings $B[z]$ and $B[z, z^{-1}]$, where $\deg(z) = (1, -1)$. It is easy to check that

(1) the support of B is $\mathbb{N} \times \mathbb{Z}$ and $B_{(0,\mathbb{Z})} = A$ as $\mathbb{Z}$-graded rings;
(2) the support of $B[z]$ is $\mathbb{N} \times \mathbb{Z}$ and $B[z]_{(0,\mathbb{Z})} = A$ as $\mathbb{Z}$-graded rings.

Moreover,

(3) $B[z, z^{-1}]$ is a $\mathbb{Z} \times \mathbb{Z}$-graded ring such that $\mathrm{Gr}^{\mathbb{Z}\times\mathbb{Z}}\text{-}B[z, z^{-1}] \approx \mathrm{Gr}^{\mathbb{Z}}\text{-}A[y]$.

Proof Observe that for any $n \in \mathbb{Z}$,

$$1 \in B[z, z^{-1}]_{(n,\mathbb{Z})} B[z, z^{-1}]_{(-n,\mathbb{Z})}.$$

Now, by Example 1.5.9,

$$\mathrm{Gr}^{\mathbb{Z}\times\mathbb{Z}}\text{-}B[z, z^{-1}] \approx \mathrm{Gr}^{\mathbb{Z}}\text{-}B[z, z^{-1}]_{(0,\mathbb{Z})}.$$

But it is easy to see that $B[z, z^{-1}]_{(0,\mathbb{Z})} \cong_{\mathrm{gr}} A[y]$ as $\mathbb{Z}$-graded rings. □

By §6.2.3 (in particular (6.12) for $\deg(z) = (1, -1)$), we have a long exact sequence

$$\cdots \longrightarrow K_{n+1}^{\mathrm{gr}}(B[z, z^{-1}]) \longrightarrow K_n^{\mathrm{gr}}(B) \xrightarrow{\bar{i}} K_n^{\mathrm{gr}}(B[z]) \longrightarrow K_n^{\mathrm{gr}}(B[z, z^{-1}]) \longrightarrow \cdots. \tag{6.32}$$

By Corollary 6.1.4 and (1) and (2) above,

$$K_n^{\mathrm{gr}}(B) = K_n^{\mathbb{Z}\times\mathbb{Z}}(B) \cong \mathbb{Z}[t, t^{-1}] \otimes K_n^{\mathbb{Z}}(B_{(0,\mathbb{Z})}) \cong \mathbb{Z}[t, t^{-1}] \otimes K_n^{\mathrm{gr}}(A)$$

and

$$K_n^{\mathrm{gr}}(B[z]) = K_n^{\mathbb{Z}\times\mathbb{Z}}(B[z]) \cong \mathbb{Z}[t, t^{-1}] \otimes K_n^{\mathbb{Z}}(B[z]_{(0,\mathbb{Z})}) \cong \mathbb{Z}[t, t^{-1}] \otimes K_n^{\mathrm{gr}}(A).$$

Moreover, from (3) we have

$$K_n^{\mathrm{gr}}(B[z, z^{-1}]) = K_n^{\mathbb{Z}\times\mathbb{Z}}(B[z, z^{-1}]) \cong K_n^{\mathbb{Z}}(A[y]) = K_n^{\mathrm{gr}}(A[y]).$$

So, from the sequence (6.32), we get

$$\cdots \longrightarrow K_{n+1}^{\mathrm{gr}}(A[y]) \longrightarrow \mathbb{Z}[t, t^{-1}] \otimes K_n^{\mathrm{gr}}(A) \xrightarrow{\bar{i}} \mathbb{Z}[t, t^{-1}] \otimes K_n^{\mathrm{gr}}(A) \longrightarrow K_n^{\mathrm{gr}}(A[y]) \longrightarrow \cdots. \tag{6.33}$$

The rest is similar to the method used to prove the fundamental theorem in §6.2.4. First note that we have a short exact sequence of $\mathbb{Z} \times \mathbb{Z}$-graded $B[z]$-modules (compare this with (6.17))

$$0 \longrightarrow M[z](-1, 1) \xrightarrow{z} M[z] \xrightarrow{e_0} M \longrightarrow 0. \tag{6.34}$$

Here M is a graded B-module which becomes a graded $B[z]$-module under the evaluation map $e_0 : B[z] \to S, z \mapsto 0$. Again, similarly to (6.26), for $1 \le i \le 3$, defining

$$\phi_i : \mathrm{gr}\, B \longrightarrow \mathrm{gr}\, B[z]$$

by

$$\begin{aligned}\phi_1(M) &= M[z](-1,1),\\ \phi_2(M) &= M[z],\\ \phi_3(M) &= M,\end{aligned}$$

the exact sequence (6.34) and the exact sequence of functors theorem (§6.2.1) immediately give

$$\overline{\phi}_3 = \overline{\phi}_2 - \overline{\phi}_1, \tag{6.35}$$

and $\overline{\phi}_3 = \overline{i}$. One can then observe that $\overline{i}$ is injective, so the exact sequence (6.33) reduces to

$$0 \longrightarrow \mathbb{Z}[t,t^{-1}] \otimes K_n^{\mathrm{gr}}(A) \xrightarrow{\overline{i}} \mathbb{Z}[t,t^{-1}] \otimes K_n^{\mathrm{gr}}(A) \longrightarrow K_n^{\mathrm{gr}}(A[y]) \longrightarrow 0.$$

Finally, from this exact sequence it follows that

$$K_n^{\mathrm{gr}}(A[y]) \cong K_n^{\mathrm{gr}}(A). \tag{6.36}$$

We are in a position to relate graded K-theory to nongraded K-theory.

Theorem 6.4.1 *Let A be a right regular Noetherian $\mathbb{Z}$-graded ring. Then there is a long exact sequence*

$$\cdots \longrightarrow K_{n+1}(A) \longrightarrow K_n^{\mathrm{gr}}(A) \xrightarrow{\overline{i}} K_n^{\mathrm{gr}}(A) \xrightarrow{U} K_n(A) \longrightarrow \cdots. \tag{6.37}$$

Here $\overline{i} = \overline{\mathcal{T}_1} - \overline{\mathcal{T}_0} = \overline{\mathcal{T}_1} - 1$, where $\mathcal{T}$ is the shift functor (1.16) *and U is the forgetful functor.*

Proof By §6.2.3 (in particular (6.12) for $\deg(y) = 1$), we have a long exact sequence

$$\cdots \longrightarrow K_{n+1}^{\mathrm{gr}}(A[y,y^{-1}]) \longrightarrow K_n^{\mathrm{gr}}(A) \xrightarrow{i} K_n^{\mathrm{gr}}(A[y]) \longrightarrow K_n^{\mathrm{gr}}(A[y,y^{-1}]) \longrightarrow \cdots.$$

By (6.36),

$$K_n^{\mathrm{gr}}(A[y]) \cong K_n^{\mathrm{gr}}(A).$$

Since $A[y,y^{-1}]$ is strongly graded, by Dade's Theorem 1.5.1,

$$K_n^{\mathrm{gr}}(A[y,y^{-1}]) \cong K_n(A).$$

Replacing these into the above long exact sequence, the theorem follows. □

The following is an immediate corollary of Theorem 6.4.1. It shows that for the graded regular Noetherian rings, the shift of modules is the only difference between the graded Grothendieck group and the usual Grothendieck group.

Corollary 6.4.2 *Let A be a right regular Noetherian $\mathbb{Z}$-graded ring. Then*

$$K_0^{\mathrm{gr}}(A)/\langle [P(1)] - [P] \rangle \cong K_0(A),$$

where P is a graded finitely generated projective A-module.

Proof The long exact sequence (6.37), for $n = 0$, reduces to

$$K_0^{\mathrm{gr}}(A) \xrightarrow{[P] \mapsto [P(1)] - [P]} K_0^{\mathrm{gr}}(A) \xrightarrow{U} K_0(A) \longrightarrow 0.$$

The corollary is now immediate. □

Remark 6.4.3 Both the fundamental theorem (§6.2.4) and Theorem 6.4.1 can be written for the case of graded coherent regular rings, as was demonstrated by Gersten in [39] for the nongraded case.

References

[1] G. Abrams, C. Menini, *Categorical equivalences and realization theorems*, J. Pure Appl. Algebra **113** (1996), no. 2, 107–120.

[2] G. Abrams, G. Aranda Pino, *The Leavitt path algebra of a graph,* J. Algebra **293** (2005), no. 2, 319–334.

[3] G. Abrams, P.N. Ánh, A. Louly, E. Pardo, *The classification question for Leavitt path algebras*, J. Algebra **320** (2008), no. 5, 1983–2026.

[4] F.W. Anderson, K.R. Fuller, *Rings and categories of modules*, Second Edition, Graduate Texts in Mathematics, Springer-Verlag 1992.

[5] P. Ara, M.A. Moreno, E. Pardo, *Nonstable K-theory for graph algebras,* Algebr. Represent. Theory **10** (2007), no. 2, 157–178.

[6] P. Ara, M. Brustenga, *K_1 of corner skew Laurent polynomial rings and applications*, Comm. Algebra **33** (2005), no. 7, 2231–2252.

[7] P. Ara, M.A. González-Barroso, K.R. Goodearl, E. Pardo, *Fractional skew monoid rings*, J. Algebra **278** (2004), no. 1, 104–126.

[8] P. Ara, A. Facchini, *Direct sum decompositions of modules, almost trace ideals, and pullbacks of monoids*, Forum Math. **18** (2006), 365–389.

[9] M. Artin, J.J. Zhang, *Noncommutative projective schemes*, Adv. Math. **109** (1994), no. 2, 228–287.

[10] Yu.A. Bahturin, S.K. Sehgal, M.V. Zaicev, *Group gradings on associative algebras*, J. Algebra **241** (2001), 677–698.

[11] I.N. Balaba, A.L. Kanunnikov, A.V. Mikhalev, *Quotient rings of graded associative rings I*, Journal of Mathematical Sciences **186** No. 4, (2012), 531–577.

[12] C. Barnett, V. Camillo, *Idempotents in matrix rings*, Proc. Amer. Math. Soc. **122.4** (1994), 965–969.

[13] H. Bass, *Algebraic K-theory*, W. A. Benjamin, Inc., New York, Amsterdam 1968.

[14] H. Bass, *K_1-theory and stable algebra,* Publ. Math. IHES **22** (1964), 5–60.

[15] H. Bass, M. Pavaman Murthy, *Grothendieck groups and Picard groups of abelian group rings*, Ann. of Math. (2) **86** (1967) 16–73.

[16] H. Bass, *Lectures on topics in algebraic K-theory*, Notes by Amit Roy, Tata Inst. Research Lectures on Math., No. 41, Tata Institute of Fundamental Research, Bombay, 1967.

[17] M. Beattie, A. del Río, *The Picard group of a category of graded modules,* Comm. Algebra **24** (1996), no. 14, 4397–4414.

[18] M. Beattie, A. del Río, *Graded equivalences and Picard groups*, J. Pure Appl. Algebra **141** (1999), no. 2, 131–152.

[19] G. Bergman, *On Jacobson radicals of graded rings*, unpublished notes. Available from `math.berkeley.edu/~gbergman/papers/unpub/`

[20] M. van den Bergh, *A note on graded K-theory*, Comm. Algebra **14** (1986), no. 8, 1561–1564.

[21] A.J. Berrick, M.E. Keating, *Rectangular invertible matrices*, Amer. Math. Monthly **104** (1997), no. 4, 297–302.

[22] P. Boisen, *Graded Morita theory*, J. Algebra **164** (1994), no. 1, 1–25.

[23] A. Borovik, *Mathematics under the microscope*, American Mathematical Society Publication, Providence, 2010.

[24] N. Bourbaki, *Algebra I. Chapters 1–3*. Translated from the French. Elements of Mathematics, Springer-Verlag, Berlin, 1998.

[25] M. Boyle *Symbolic dynamics and matrices*, Combinatorial and Graph-Theoretic Problems in Linear Algebra (IMA Volumes in Math and Appl., 50). Eds. R. Brualdi, S. Friedland and V. Klee. Springer, 1993, 1–38.

[26] S. Caenepeel, F. Van Oystaeyen, *Brauer groups and the cohomology of graded rings*, Monographs and Textbooks in Pure and Appl. Math., 121, Marcel Dekker, Inc., New York, 1988.

[27] S. Caenepeel, S. Dăscălescu, C. Năstăsescu, *On gradings of matrix algebras and descent theory*, Comm. Algebra **30** (2002), no. 12, 5901–5920.

[28] P.M. Cohn, *Some remarks on the invariant basis property*, Topology **5** (1966), 215–228.

[29] M. Cohen, S. Montgomery, *Group-graded rings, smash products, and group actions*, Trans. Amer. Math. Soc. **282** (1984), no. 1, 237–258. Addendum: Trans. Amer. Math. Soc. **300** (1987), no. 2, 810–811.

[30] J. Cuntz, R. Meyer, J. Rosenberg, *Topological and bivariant K-theory*, Oberwolfach Seminars, 36. Birkhäuser Verlag, Basel, 2007.

[31] C.W. Curtis, I. Reiner, *Methods of representation theory with applications to finite groups and orders,* Vol. II, Pure Appl. Math., Wiley-Interscience, 1987.

[32] E. Dade, *Group-graded rings and modules*, Math. Z. **174** (1980), no. 3, 241–262.

[33] E. Dade, *The equivalence of various generalizations of group rings and modules*, Math. Z. **181** (1982), no. 3, 335–344.

[34] S. Dăscălescu, B. Ion, C. Năstăsescu, J. Rios Montes, *Group gradings on full matrix rings*, J. Algebra **220** (1999), no. 2, 709–728.

[35] P.K. Draxl, *Skew fields*, London Mathematical Society Lecture Note Series, 81, Cambridge University Press, Cambridge, 1983.

[36] G.A. Elliott, *On the classification of inductive limits of sequences of semisimple finite-dimensional algebras*, J. Algebra **38** (1976), no. 1, 29–44.

[37] E.G. Effros, *Dimensions and C^*-algebras*, CBMS Regional Conference Series in Mathematics, **46** AMS, 1981.

[38] A. Fröhlich, *The Picard group of nonzero rings, in particular of orders*, Trans. Amer. Math. Soc. **180**, (1973) 1–46.

[39] S.M. Gersten, *K-theory of free rings*, Comm. Algebra **1** (1974), 39–64.

[40] K.R. Goodearl, *Von Neumann regular rings*, Second Edition, Krieger Publishing Co., Malabar, FL, 1991.

[41] K.R. Goodearl, D.E. Handelman, *Classification of ring and C^*-algebra direct limits of finite-dimensional semisimple real algebras*, Mem. Amer. Math. Soc. **69** (1987), no. 372.

[42] R. Gordon, E.L. Green, *Graded Artin algebras*, J. Algebra **76** (1982), 111–137.

[43] J. Haefner, A. del Río, *Actions of Picard groups on graded rings*, J. Algebra **218** (1999), no. 2, 573–607.

[44] J. Haefner, *Graded equivalence theory with applications*, J. Algebra **172** (1995), no. 2, 385–424.

[45] D. Handelman, W. Rossmann, *Actions of compact groups on AF C^*-algebras*, Illinois J. Math. **29** (1985), no. 1, 51–95.

[46] D. Happel, *Triangulated categories in representation theory of finite dimensional algebras*, London Math. Soc., Lecture Notes Ser. **119** (1988) Cambridge University Press.

[47] R. Hazrat, *The graded structure of Leavitt path algebras*, Israel J. Math. **195** (2013), 833–895.

[48] R. Hazrat, *The graded Grothendieck group and the classification of Leavitt path algebras*, Math. Annalen **355** (2013), no. 1, 273–325.

[49] R. Hazrat, T. Hüttemann, *On Quillen's calculation of graded K-theory*, J. Homotopy Relat. Struct. **8** (2013) 231–238.

[50] T.W. Hungerford, *Algebra*, Graduate Texts in Mathematics 73, Springer-Verlag, Berlin, 1974.

[51] D. Lind, B. Marcus, *An introduction to symbolic dynamics and coding*, Cambridge University Press, 1995.

[52] E. Jespers, *Simple graded rings*, Comm. Algebra **21:7** (1993), 2437–2444.

[53] A.V. Kelarev, *Ring constructions and applications*, World Scientific, 2001.

[54] A. Kleshchev, *Representation theory of symmetric groups and related Hecke algebras*, Bull. Amer. Math. Soc. (N.S.) **47** (2010), no. 3, 419–481.

[55] M.-A. Knus, *Quadratic and Hermitian forms over rings*, Grundlehren der Mathematischen Wissenschaften [Fundamental Principles of Mathematical Sciences], 294, Springer-Verlag, Berlin, 1991.

[56] W. Krieger, *On dimension functions and topological Markov chains*, Invent. Math. **56** (1980), no. 3, 239–250.

[57] P.A. Krylov, A.A. Tuganbaev, *Modules over formal matrix rings*, J. Math. Sci. (N.Y.) **171** (2010), no. 2, 248–295.

[58] T.-Y. Lam, M.K. Siu, *K_0 and K_1 – an introduction to algebraic K-theory*, Amer. Math. Monthly **82** (1975), 329–364.

[59] T.-Y. Lam, *A crash course on stable range, cancellation, substitution and exchange*, J. Algebra Appl. **03** (2004), 301–343.

[60] T.-Y. Lam, *A first course in noncommutative rings*, Graduate Texts in Mathematics, 131. Springer-Verlag, New York, 1991.

[61] T.-Y. Lam, *Lectures on modules and rings*, Graduate Texts in Mathematics, 189. Springer-Verlag, New York, 1999.

[62] M. Lawson, *Inverse semigroups. The theory of partial symmetries*, World Scientific Publishing Co., Inc., River Edge, NJ, 1998

[63] W.G. Leavitt, *The module type of a ring,* Trans. Amer. Math. Soc. **103** (1962) 113–130.

[64] H. Li, *On monoid graded local rings*, J. Pure Appl. Algebra **216** (2012), no. 12, 2697–2708.

[65] G. Liu, F. Li, *Strongly groupoid graded rings and the corresponding Clifford theorem*, Algebra Colloq. **13** (2006), no. 2, 181–196.

[66] P. Lundström, *The category of groupoid graded modules*, Colloq. Math. **100** (2004), 195–211.

[67] B. Magurn, *An algebraic introduction to K-theory*, Cambridge University Press, 2002.

[68] A. Marcus, *On Picard groups and graded rings*, Comm. Algebra **26** (1998), no. 7, 2211–2219.

[69] P. Menal, J. Moncasi, *Lifting units in self-injective rings and an index theory for Rickart C^*-algebras*, Pacific J. Math. **126** (1987), 295–329.

[70] C. Menini, C. Năstăsescu, *When is R-gr equivalent to the category of modules?* J. Pure Appl. Algebra **51** (1988), no. 3, 277–291.

[71] J.R. Millar, *K-theory of Azumaya algebras*, Ph.D. thesis, Queen's University Belfast, United Kingdom 2010, `arXiv:1101.1468`.

[72] C. Năstăsescu, *Group rings of graded rings. Applications*, J. Pure Appl. Algebra **33** (1984), no. 3, 313–335.

[73] C. Năstăsescu, F. van Oystaeyen, *Graded and filtered rings and modules*, Lecture Notes in Mathematics, **758**, Springer, Berlin, 1979.

[74] C. Năstăsescu, F. van Oystaeyen, *Graded ring theory*, North-Holland, Amsterdam, 1982.

[75] C. Năstăsescu, F. van Oystaeyen, *Methods of graded rings*, Lecture Notes in Mathematics, 1836, Springer-Verlag, Berlin, 2004.

[76] C. Năstăsescu, N. Rodinó, *Group graded rings and smash products*, Rend. Sem. Mat. Univ. Padova **74** (1985), 129–137.

[77] C. Năstăsescu, B. Torrecillas, F. Van Oystaeyen, *IBN for graded rings*, Comm. Algebra **28** (2000) 1351–1360.

[78] F. Van Oystaeyen, *On Clifford systems and generalized crossed products*, J. Algebra **87** (1984), 396–415.

[79] S. Paul Smith, *Category equivalences involving graded modules over path algebras of quivers*, Adv. in Math. **230** (2012) 1780–1810.

[80] N.C. Phillips, *A classification theorem for nuclear purely infinite simple C^*-algebras*, Doc. Math. **5** (2000), 49–114.

[81] D. Quillen, *Higher algebraic K-theory. I.* Algebraic K-theory, I: Higher K-theories (Proc. Conf., Battelle Memorial Inst., Seattle, WA, 1972), pp. 85–147. Lecture Notes in Math., Vol. 341, Springer, Berlin 1973.

[82] A. del Río, *Graded rings and equivalences of categories*, Comm. Algebra **19** (1991), no. 3, 997–1012. Correction: Comm. Algebra **23** (1995), no. 10, 3943–3946.

[83] A. del Río, *Categorical methods in graded ring theory*, Publ. Mat. **36** (1992), no. 2A, 489–531.

[84] J. Rosenberg, *Algebraic K-theory and its applications*, Springer-Verlag, New York, 1994.

[85] J.-P. Serre, *Faisceaux alǵebrique cohérents,* Ann. of Math. **61** (1955), 197–278.

[86] S. Sierra, *G-algebras, twistings, and equivalences of graded categories*, Algebr. Represent. Theory **14** (2011), no. 2, 377–390.

[87] S. Sierra, *Rings graded equivalent to the Weyl algebra*, J. Algebra **321** (2009) 495–531.

[88] V. Srinivas, *Algebraic K-theory,* Second Edition. Progress in Mathematics, 90. Birkhauser Boston, Inc., Boston, MA, 1996.

[89] R. Swan, *Algebraic K-theory*, Lecture Notes in Mathematics, No. 76, Springer-Verlag, Berlin, New York 1968.

[90] J.-P. Tignol, A.R. Wadsworth, *Value functions on simple algebras, and associated graded rings*, Springer's Monographs in Mathematics, 2015.

[91] A.M. Vershik, S.V. Kerov, *Locally semi-simple algebras: Combinatorial theory and K_0-functor*, J. Sov. Math. **38** (1987), 1701–1734.

[92] J.B. Wagoner, *Markov partitions and K_2*, Publ. Math. IHES **65** (1987), 91–129.

[93] J.B. Wagoner, *Topological Markov chains, C^*-algebras, and K_2*, Adv. in Math. **71** (1988), no. 2, 133–185.

[94] C.A. Weibel, *An introduction to homological algebra*, Cambridge studies in advanced mathematics, **38**, Cambridge University Press, 1994.

[95] C.A. Weibel, *The K-book: an introduction to algebraic K-theory*, Graduate Studies in Math. **145** AMS, 2013.

[96] H. Yahya, *A note on graded regular rings*, Comm. Algebra **25:1** (1997), 223–228.

[97] R.F. Williams *Classification of subshifts of finite type*, Ann. Math. **98** (2) (1973), 120–153.

[98] A. Zalesskii, *Direct limits of finite dimensional algebras and finite groups*, Proc. Miskolc Ring Theory Conf. 1996, Canadian Math. Soc. Conference Proceedings **22** (1998), 221–239.

[99] J.J. Zhang, *Twisted graded algebras and equivalences of graded categories*, Proc. London Math. Soc. **72** (1996) 281–311.

[100] Z. Zhang, *A matrix description of K_1 of graded rings*, Israel J. Math. **211** (2016), 45–66.

Index

Printed in the United States
by Baker & Taylor Publisher Services

Printed in the United States
by Baker & Taylor Publisher Services